Electron Microscopy
Methods and Protocols

METHODS IN MOLECULAR BIOLOGY™

John M. Walker, SERIES EDITOR

METHODS IN MOLECULAR BIOLOGY™

Electron Microscopy Methods and Protocols

Edited by

M. A. Nasser Hajibagheri

Imperial Cancer Research Fund, London, UK

Humana Press ✳ **Totowa, New Jersey**

This publication is printed on acid-free paper. ∞
ANSI Z39.48-1984 (American Standards Institute) Permanence of Paper for Printed Library Materials.

Cover illustration: Color Plate 1, following p. 180. Scanning electron micrograph showing a population of fission yeast *S. pombe*. *See* discussion in Chapter 12 and full caption on p. 184.

Cover design by Patricia F. Cleary.

For additional copies, pricing for bulk purchases, and/or information about other Humana titles, contact Humana at the above address or at any of the following numbers: Tel: 973-256-1699; Fax: 973-256-8341; E-mail: humana@humanapr.com, or visit our Website at www.humanapress.com

Library of Congress Cataloging-in-Publication Data

Electron microscopy methods and protocols / edited by M.A. Nasser Hajibagheri.
     p.   cm. -- (Methods in molecular biology ; v. 117)
     Includes index.
     ISBN 0-89603-640-5 (alk. paper)
     1. Electron microscopy--Laboratory manuals. 2. Histology--Technique I. Hajibagheri, M. A. Nasser. II. Series: Methods in molecular biology (Totowa, N.J.) ; 117.
     [DNLM: 1. Microscopy, electron laboratory manuals. 2. Histocytological Preparation Techniques. QS 525 E38 1999 / W1 ME9616J v. 177 1999]
QH212.E4E39824   1999
570'.28'25--dc21
DNLM/DL   98-37685
for Library of Congress   CIP

Preface

Electron Microscopy Methods and Protocols is designed for the established researcher as a manual for extending knowledge of the field. It is also for the newcomer who wishes to move into the field. A wide range of applications for the examination of cells, tissues, biological macromolecules, molecular structures, and their interactions are discussed. We have tried to gather together methods that we consider to be those most generally applicable to current research in both cell and molecular biology. Each chapter contains a set of related practical protocols with examples provided by experts who have first-hand knowledge of the techniques they describe. The individual chapters are grouped according to similarities in their specimen preparation and methodology. Methods are presented in detail, in a step-by-step fashion, using reproducible protocols the authors have personally checked.

During the last decade, the scientific literature describing the use of colloidal gold as an immunocytochemical marker has increased at an exponential rate, and this trend is expected to continue. We have included a large number of variations on the immunogold labeling technique. In both the negative staining and cryo chapters, authors emphasize the "immunological applications" in order to correlate as fully as possible with the emphasis on immunogold labeling in the other chapters.

Electron Microscopy Methods and Protocols commences with the routine preparation of biological material for classical transmission electron microscopy involving tissue fixation, embedding, and sectioning (Chap. 1). Chapters from Robin Harris and Marc Adrian deal with negative staining of thinly spread biological particulates and the preparation of thin-film frozen-hydrated/vitrified specimens (Chaps. 2 and 3). The production of cryosections from fixed, cryoprotected biological material and their use in immunocytochemistry is covered by Paul Webster in Chapter 4. Ken McDonald in Chapter 5 describes high-pressure freezing and freeze-substitution with respect to morphology and antigenicity in immunolabeling (Chap. 5). Since the method of specimen embedding influences labeling with colloidal gold, a detailed protocol for the use of LR Gold and Lowicryl resins is presented in Chapters 6 and 7, respectively. Immunogold labeling following the progressive-lowering-of-temperature method is also detailed. Catherine Rabouille deals with quantitative aspects of immunogold labeling in embedded and nonembedded

v

sections in Chapter 8. Microwave processing techniques are covered in Chapter 9, followed by enzyme cytochemistry (Chap. 10) and *in situ* molecular hybridization (Chap. 11).

Our knowledge of the genetics, cell biology, and molecular biology of yeast continues to advance at an encouraging rate and with considerable excitement. In Chapter 12 we have emphasized the preparation of fission yeast for ultrastructure and immunocytochemistry in order to demonstrate the value of electron microscopy in yeast biomedical research. The preparation and analysis of structures, such as nucleic acids and proteins, as well as the binding of proteins to nucleic acids and protein-to-protein interaction are discussed in detail in Chapters 13 and 14. The emphasis on cryotechniques continues in this volume, in Chapter 15, with a detailed account of the procedures for quantitative biological X-ray microanalysis by A. John Morgan and his colleagues.

Electron Microscopy Methods and Protocols has been developed through the efforts of twenty-three scientists, representing seven countries. All the contributors are eminent authorities in their respective fields. I hope this volume will prove useful both for the novice and experienced worker carrying out high resolution microscopy in their research. My thanks are owed the authors for their excellent contributions and professional cooperation. Thanks also to my staff—Steve Gschmeissner, Carol Upton, and Ken Blight—for their valuable comments on some of the chapters.

M. A. Nasser Hajibagheri

Contents

Contributors

MARC ADRIAN • *Institute of Zoology, University of Mainz, Mainz, Germany*

NOBUKAZU ARAKI • *Department of Anatomy, Kagawa Medical University, Japan*

KEN BLIGHT • *Electron Microscopy Unit, Imperial Cancer Research Fund, London*

HEATHER A. DAVIES • *Biology Department, Open University, Milton Keynes, UK*

RICHARD S. DEMAREE, JR.• *Department of Biological Sciences, California State University, Chico, CA*

JEAN-GUY FOURNIER • *Institut de Mycologie, Hôpital Pitié-Salpêtrière, Paris, France*

STEVE GSCHMEISSNER • *Electron Microscopy Unit, Imperial Cancer Research Fund, London*

RICK T. GIBERSON • *Ted Pella, Inc., Redding, CA*

PIERRE GOUNON • *Station Centrale de Microscopie Electronique, Institut Pasteur, Paris, France*

M. A. NASSER HAJIBAGHERI • *Electron Microscopy Unit, Imperial Cancer Research Fund, London*

J. ROBIN HARRIS • *Institute of Zoology, University of Mainz, Mainz, Germany*

TANENORI HATAE • *Department of Anatomy, Kagawa Medical University, Kagawa, Japan*

ROSS B. INMAN • *Institute for Molecular Virology, University of Wisconsin-Madison, Madison, WI*

KENT MCDONALD • *Electron Microscope Lab, University of California, Berkeley, CA*

A. JOHN MORGAN • *Cardiff School of Biosciences, Cardiff University, Cardiff, Wales*

CATHERINE RABOUILLE • *Cell Biology, Imperial Cancer Research Fund, London*

KENNETH SAWIN • *Cell Cycle Laboratory, Imperial Cancer Research Fund, London*

MARIA SCHNOS • *Institute for Molecular Virology, University of Wisconsin-Madison, Madison, WI*

STEPHEN STÜRZENBAUM • *Cardiff School of Biosciences, Cardiff University, Cardiff, Wales*

J. R. THORPE • *EM and FACS Laboratory, Biological Sciences, University of Sussex, Falmer • Brighton, UK*

CAROL UPTON • *Electron Microscopy Unit, Imperial Cancer Research Fund, London*

PAUL WEBSTER • *House Ear Institute, Los Angeles, CA*

CAROLE WINTERS • *Cardiff School of Biosciences, Cardiff University, Cardiff, Wales*

Color Plates

The Color Plates appear after p. 180. The images were edited with Adobe Photoshop®.

PLATE 1, Fig. 1, p. 184: Scanning electron micrograph showing a population of fission yeast *S. pombe*. *See* discussion in Chapter 12.

PLATE 2, Fig. 9, p. 199: Immunolocalization of Tubulin on a mitotic spindle of fission yeast. *See* full caption on p. 199 and discussion in Chapter 12.

PLATE 3, Fig. 2, p. 192: Labeling of the Golgi stack and cell wall of Lowicryl embedded fission yeast. *See* full caption on p. 192 and discussion in Chapter 12.

PLATE 4, Fig. 2, p. 130: The boundaries of seven cell compartments distinguished by different colors for quantification. *See* full caption on p. 130 and discussion in Chapter 8.

PLATE 5, Fig. 4, p. 135: Point-hit method on intracellular compartments distinguished by different colors for counting gold particles. *See* full caption on p. 135 and discussion in Chapter 8.

1

General Preparation of Material and Staining of Sections

Heather A. Davies

1. Introduction

This chapter is aimed at those who have not previously done any processing for electron microscopy (EM). It deals with basic preparation of many different types of mammalian material for ultrastructural examination; for processing of plant material (*see* Hall and Hawes, **ref. *1***). The material to be processed may be cell suspensions, particulates, monolayer cultures, or tissue derived from organs. The former three must initially be processed differently from the latter.

For EM, the ultrastructure must be preserved as close to the in vivo situation as possible. This is done by either chemical or cryofixation; the latter will be dealt with in later chapters. Aldehydes that crosslink proteins are used for chemical fixation. Glutaraldehyde, a dialdehyde preserves ultrastucture well but penetrates slower than the monoaldehyde, paraformaldehyde. Glutaraldehyde is used alone for small pieces of material, but a mixture of the two aldehydes may be used for perfusion fixation or fixation of larger items.

All reagents used for EM processing must be of high purity. Analytical grade reagents must be used for all solutions, e.g., buffers and stains. Glutaraldehyde must be EM grade. For higher purity, distilled or vacuum distilled qualities are available. Secondary fixation is by osmium tetroxide which reacts with unsaturated lipids, is electron-dense and thus stains phospholipids of the cell membrane. This step is followed by dehydration through an ascending concentration series of solvent before embedding in resin. For simplicity, epoxy resin (Epon) embedding is described in this chapter; other resins are detailed in Glauert *(2)* and Chapters 6 and 7.

Ultramicrotomy and staining ultrathin sections are dealt with briefly; for a detailed account of the procedure and trouble-shooting *(3)*. The ultrathin sec-

From: *Methods in Molecular Biology*, vol. 117: *Electron Microscopy Methods and Protocols*
Edited by: N. Hajibagheri © Humana Press Inc., Totowa, NJ

tions are collected on grids for examination. Grids are manufactured of various metals, e.g., copper, nickel, and gold and are available in different designs including square mesh, hexagonal mesh, and parallel bars. Copper is the most common choice for grids and may be used with or without a support film. If there are holes in the ultrathin sections, a film on the grid provides extra support to prevent movement of the section in the electron beam. The size of mesh is a compromise between support of the section and the viewing area between the bars; hexagonal mesh gives a larger viewing area than square mesh. Slot grids covered with a support film are more difficult to prepare, but are ideal for viewing the whole section.

It may be necessary to section several blocks to be representative of the material or with pelleted material, differential layering can be assessed by sectioning transversely through the thickness of the pellet.

Many aspects of EM processing and ultramicrotomy can present problems; some of the common ones are highlighted in the Notes section.

2. Materials

2.1. Suppliers

EM supplies. Agar Scientific Ltd, 66a Cambridge Road, Stansted, Essex CM24 8DA.

EM supplies. TAAB Laboratories Eqpt. Ltd., 3 Minerva House, Calleva Park, Aldermaston, Berks. RG7 8NA.

Chemicals and glassware. Merck Ltd, Hunter Boulevard, Magna Park, Lutterworth, Leics. LE17 4XN.

2.2. Equipment

1. Microcentrifuge.
2. 60°C–70°C oven.
3. Rotator 2 rpm or variable speed.
4. Ultramicrotome.
5. Transmission electron microscope.
6. Horizontal rotator/mixer.
7. Knifemaker.
8. Automatic staining machine.
9. Carbon coater.

2.3. Buffers

1. Analar reagents must be used.
2. Sorensons 0.2 M phosphate buffer comprising 0.2 M sodium dihydrogen orthophosphate (NaH_2PO_4) and 0.2 M disodium hydrogen orthophosphate (Na_2HPO_4).
3. 0.2 M sodium cacodylate.
4. For the preparation of buffers, *see* **Subheading 3.1.**

2.4. Support Films

1. Copper grids (usually square or hexagonal mesh and ocassionally slots)—washed with acetone and dried prior to use.
2. 1% pioloform, butvar or formvar in chloroform. Stock solution is stored in an amber stoppered bottle. Prepare the day before by sprinkling the solid onto the surface of the chloroform and leaving to dissolve. Keeps for 6–8 mo. HAZARD— store separately away from base and alkalis.
3. Dispensing cylinder (100 mL) with tap.
4. 2 L flat beaker.
5. Lidded box.
6. Carbon rods.

2.5. Fixatives and Fixation

1. EM grade 25% glutaraldehyde HAZARD.
2. 0.2 M phosphate buffer. 0.2 M cacodylate buffer HAZARD.
3. Paraformaldehyde powder.
4. 2% aqueous osmium tetroxide HAZARD.
5. 1.5 mL Eppendorf tubes.
6. 7 mL glass processing bottles with lids.

2.6. Epoxy Resins (Epon)

1. Epon 812, e.g., Agar 100 resin, TAAB resin. HAZARD
2. DDSA—dodecenyl succinic anhydride. HAZARD
3. MNA—methyl nadic anhydride. HAZARD
4. BDMA—benzyl dimethylamine. HAZARD.
5. Low density polyethylene bottles 50 mL.

2.7. Dehydration Solvents

Use Analar reagents.

1. 30, 50 ,70, 90% ethanol or acetone in distilled water.
2. 100% ethanol or acetone.
3. 100% ethanol or acetone with added molecular sieve 5a. Do not disturb the molecular sieve.

2.8. Infiltration and Embedding

1. 50% resin: 50% solvent. Make fresh.
2. Complete resin (**Subheading 2.6.**).
3. Polythene BEEM capsules size 00 or 3.
4. Block holders.
5. Green rubber flat embedding molds.
6. Small paper labels (2 mm × 15 mm) with block numbers written in pencil.

Table 1
Preparation of 0.2 *M* Sorensons Phosphate Buffer

A (mL)[a]	B (mL)[b]	pH
90	10	5.9
85	15	6.1
77	23	6.3
68	32	6.5
57	43	6.7
45	55	6.9
33	67	7.1
23	77	7.3
19	81	7.4
16	84	7.5
10	90	7.7

[a]Solution A: $NaH_2PO_4.2H_2O$, mol wt 156.01, 3.12 g in 100 mL.
[b]Solution B: Na_2HPO_4, mol wt 141.96, 2.84 g in 100 mL.

2.9. Ultramicrotomy

1. Glass strips, 6 mm.
2. Plastic boats or tape for boats.
3. Toluidene blue stain: 1% toluidene blue in 5% borax, filtered. Microfilter each time used.

2.10. Stains and Staining

1. 0.5–1% aqueous uranyl acetate or saturated uranyl acetate in 50% ethanol. HAZARD: RADIOCHEMICAL.
2. Analar lead nitrate.
3. Analar trisodium citrate.
4. Carbonate-free 1 *M* NaOH. (Use a volumetric solution.)
5. Carbonate-free distilled water. (Use boiled or autoclaved water.)
6. NaOH pellets.

3. Methods

3.1. Choice and Preparation of Buffers

Two widely used buffers are Sorensons phosphate buffer and cacodylate buffer; they are not compatible with each other.

1. Phosphate buffer. Different pH's can be made by varying the volumes of the two constituents (**Table 1**).
   Or for each 100 mL of 0.2 *M* buffer use 0.497 g $NaH_2PO_4.2H_2O$ and 2.328 g Na_2HPO_4. Keep refrigerated.

2. 0.2 M cacodylate buffer. 21.4 g Na cacodylate in 250 mL distilled water. Adjust the pH to 7.4 with approx 8 mL of 1 M HCl and make up to a final volume of 500 mL.

3.2. Preparation of Support Films

To be performed in a fume cupboard, this method is very reliable, particularly in humid conditions.

1. Pour approx 70 mL of pioloform solution into the cylinder and drop a clean glass microscope slide, wiped with velin tissue to remove the dust, into it.
2. Cover the top with foil and open the tap fully, draining the solution into the stock bottle. Leave until the chloroform has evaporated (this is important in humid conditions as it prevents the production of holes in the film because of condensation of water vapor on the film as it dries).
3. Score 2 mm from the edge using a stout scalpel blade, breathe on the slide, and lower it at an angle of 45° into a 2 L beaker of distilled water. As the film floats off, the thickness can be judged by the interference color. Adjust the concentration of the stock solution if necessary by adding chloroform.
4. Place 20–30 grids onto the film, etched surface (matt side) in contact with the film.
5. Place a piece of paper cut from "Yellow Pages" with small print on both sides onto the film plus grids, allow it to become wet and slowly lift from the surface of the water. "Yellow Pages" paper is chosen because the quality of both the paper and printing ink allow even uptake of water at a fairly fast rate. The film will adhere to the paper, covering the grids.
6. For slot grids, remove the paper from the water when 2 mm of paper from the edge has become wet.
7. Allow to dry at room temperature in a lidded box to exclude dust.
8. The filmed grids can be carbon-coated for beam stability if a carbon coater is available and glow-discharged to improve the hydrophilicity.

3.3. Fixation of Tissue

There are two methods of fixation of tissue from organs: cardiovascular perfusion and immersion fixation. Paraformaldehyde is a monoaldehyde and penetrates faster than glutaraldehyde, but results in poorer ultrastructure. A solution is to use a mixture of both aldehydes as in perfusion fixation.

1. For immersion fixation, use 2.5% glutaraldehyde in 0.1 M buffer. The time of fixation is dependent upon the dimensions of the sample to be fixed. The largest recommended size is 1 mm^3, when there is optimal penetration. Proceed to 4.
2. For perfusion fixation, use 2% glutaraldehyde and 2% paraformaldehyde in 0.1 M buffer. The conditions depend upon the animal, its age and the organ required.
3. To prepare 100 mL of glutaraldehyde/paraformaldehyde:
   a. Add 2 g paraformaldehyde to approx 35 mL distilled water + 0.5 mL of approx. 1 M NaOH (make this each time by dissolving 5 pellets of NaOH in approx 5 mL distilled water).

b. Heat the parafomaldehyde solution in a fume cupboard to 60°C when the paraformaldehyde dissolves (it is unnecessary to use a thermometer).
c. Cool and add 8 mL of EM grade 25% glutaraldehyde.
d. Make up to 50 mL with distilled water.
e. Make up to 100 mL with 0.2 M phosphate buffer pH 7.4.
f. Filter before use in animals.
4. Once the tissue is fixed and disected, it is washed by aspiration 3 × 5 min and cut into smaller blocks of 1 mm^3. All the remaining procedures are carried out by aspiration.
5. Add 1% osmium tetroxide in 0.1 M buffer for 1 h at RT and wash in buffer 3 × 5 min. The blocks can be stored in buffer at 4°C for 1–2 wk before subsequent processing. *See* **Notes 1** and **5**.

3.4. Fixation of Suspensions, e.g., Viruses, Bacteria, Dissociated Cells

1. Centrifuge the suspension at a speed that will yield a solid pellet of the material under study.
2. Add the fixative slowly down the wall of the tube taking care not to dislodge the pellet.
3. Allow to fix for 10 min at RT and then release the pellet using a wooden cocktail stick and leave for a further 20 min. The material can now be treated as tissue blocks (*see* **Subheading 3.3.4.**).
4. If the pellet resuspends, the pellet can be recentrifuged after each part of the process. Or:
5. Resuspend in 1% low-gelling temperature agarose (37°C) in buffer, centrifuge to pellet, cool and cut into blocks and then proceed with **Subheading 3.3.4**.

3.5. Fixation of Cell Monolayers

1. Remove the culture medium, wash in appropriate buffer to remove the excess protein derived from the culture medium and flood with fixative in buffer (*see* **Subheading 3.3.**).
2. Wash and osmicate *in situ* as in **Subheading 3.3**.
3. Remove cells by scraping from the support using Parafilm-coated spatula or other appropriately shaped implement. Treat as for suspensions (*see* **Subheading 3.3.** and **Notes 1** and **5**).

3.6. Preparation of Epoxy Resins (Epon)

There are several epoxy resins to choose from that have different viscosities. The less viscous epoxy resins, e.g., Spurr resin have a carcinogenic component and are useful for hard material like bone but should be used and disposed of with care. For routine work, Epon is recommended by the author whose lab has solved many of the problems encountered by new users.

1. Hardness of Epon resin can be varied to suit the material that is to be embedded as shown in **Table 2**.

Table 2
Preparation of Epon Resin

	Soft	Medium	Hard
Epon 812[a]	20 mL (24 g)	20 mL (24 g)	20 mL (24 g)
DDSA	22 mL (22 g)	16 mL (16 g)	9 mL (9 g)
MNA	5 mL (6 g)	8 mL (10 g)	12 mL (15 g)
BDMA (approx 3%)	1.4 mL (1.5 g)	1.3 mL (1.5 g)	1.2 mL (1.4 g)

[a]Epon 812 is commercially available as: Agar 100, Polarbed, TAAB resin.

2. Warm the following items to 60°C for not less than 10 min.
   a. Glass cylinder.
   b. 50 mL bottle.
   c. Epon 812 resin, DDSA, and MNA (not the BDMA). The stock components may be warmed many times over.
3. Pour the required volume of Epon 812 resin into the cylinder, add the DDSA and MNA and pour into the 50 mL bottle. Mix gently by hand and place on rotator/mixer for 10 min.
4. Add BDMA accelerator and mix as before.
5. Complete resin can be frozen if necessary.

3.7. Dehydration

Acetone is preferred as there is less lipid loss than with ethanol dehydration. Maximum dehydration times are given below. These can be reduced for smaller or thinner samples. Again, all procedures are carried out by aspiration.

1. 30% Acetone or Ethanol 10 min.
2. 50% Acetone or Ethanol 20 min.
3. 70% Acetone or Ethanol 20 min.
4. 90% Acetone or Ethanol 20 min.
5. 100% Acetone, 3 × 20 min.
6. 100% Acetone (+molecular sieve) 20 min.

3.8. Infiltration and Embedding in Epoxy Resin (e.g., Epon)

The epoxy resin used for the 50:50 mixture can be from the frozen resin stock (*see* **Subheading 3.6.**).

1. 50:50 epoxy resin:acetone, overnight on a rotating mixer.
2. Fresh epoxy resin for 2–4 h on a rotating mixer with the caps off to allow excess acetone to evaporate.
3. Fresh epoxy resin for a further 2–4 h on a rotating mixer.
4. Embed in fresh epoxy resin.
5. The embedding molds are prepared by placing small paper labels (with numbered codes in pencil) at the top of capsules or in the base of the rubber molds.

6. The blocks are transferred from the processing bottles to the capsule or mold using a cocktail stick, orientated and then resin pipeted to fill the capsule or mold.
7. The blocks are polymerized at 60°C for 24–48 h. (*See* **Notes 2–4**).

3.9. Ultramicrotomy

Ultrathin sections are cut on an ultramicrotome using glass knives made from glass strips on a knifemaker. Examine the knives in the ultramicrotome using the back light (if available) to ensure the edge is sharp and dustfree.

1. In the ultramicrotome, trim the excess resin from the block face using the glass knife and from the edge of the block using a single-edged razor blade.
2. Cut a semithin section of 1 μm, collect onto water and use a sable paint brush to transfer to a drop of water on a microscope slide. Dry on a hotplate and stain at approx 70°C with toluidine blue for 10 s until a gold rim is visible around the drop of stain. Wash off with distilled water and dry on the hotplate.
3. Select an area and trim the face to a trapezium shape of approx. 0.5 mm^2 for ultrathin sectioning.
4. Attach a boat to the knife, fill with water, and collect ultrathin sections 70 nm thick (silver interference color) in a ribbon on the surface of the water (*see* **Notes 3, 4, 6**, and **7**).

3.10. Collection of Ultrathin Sections

The ultrathin sections may or may not form a ribbon on the surface of the water; there are different techniques for collecting the sections. An eyelash mounted on a cocktail stick with nail varnish is used for moving the sections around on the water. Touch them only on the edges and ensure the eyelash is clean with no adherent resin.

Ultrathin sections can be collected on naked grids if the sections have no holes or filmed grids for extra support if they do. Slot grids may be used if the whole section needs to be viewed and in this case the grids must be filmed to provide support.

1. If sections are in a ribbon:
   a. Place the grid in the water beneath them and raise it at a slight angle so the first section of the ribbon sticks to the edge of the grid.
   b. Slowly raise the grid out of the water and the rest of the ribbon will adhere to the grid.
   c. Blot the edge to remove excess water. Do not blot the flat surfaces of the grid.
2. If the sections are in ones, twos, or threes:
   a. Touch the grid onto the section(s) from above. This does introduce creases into the sections but is far easier than trying to collect from beneath.
   b. Blot the edge to remove excess water.
3. Place grids in a filter paper-lined Petri dish before staining.

3.11. Preparation of Stains

Uranyl acetate: The uranyl acetate stains must be made fresh before use.

1. For aqueous solutions, add 0.05 g–0.01 g of uranyl acetate powder to 10 mL distilled water and allow to dissolve. This is a radiochemical and must be handled appropriately.
2. For ethanolic solutions, prepare a saturated solution in 50% ethanol.

Reynolds lead citrate

1. Place 1.33 g of Analar lead nitrate and 1.76 g trisodium citrate in a 50-mL volumetric flask, add approx 30 mL of freshly boiled or autoclaved water (carbonate-free).
2. Stopper the flask and shake intermittently for 30 min.
3. Add 8 mL 1 M NaOH (made from carbonate-free volumetric solution), and shake to dissolve the precipitate.
4. Make up to 50 mL with carbonate-free water .
5. Allow solution to stand for 1–2 h before use. The stain may be kept at 4°C for 4–6 wk.

3.12. Staining Ultrathin Sections

This can be performed manually but there is an automatic machine commercially available. The manual method is detailed below.

1. Place the required number of drops of uranyl acetate onto wax in Petri dish, one drop per section. For aqueous stain, use 30 min at temperatures between 20°C and 60°C and for ethanolic stain, use 30 min at 37°C. If staining at temperatures higher than 20°C, place a small piece of moistened tissue in the dish to prevent drying.
2. Place the grids, section down, onto the drop. Cover the Petri dish and leave for the required time (e.g., 20 min for aqueous UAc).
3. Blot the edge of the grid and stream-wash with distilled water from a wash bottle. Touch the edge of the grid with filter paper and blot between the forceps.
4. Cover the base of a 90-mm Petri dish with NaOH pellets and put the base of a 30-mm Petri dish in the center. Moisten the NaOH pellets with distilled water and pipet approx. 2 mL of Reynolds lead citrate (*see* **Subheading 3.11.**) into the smaller Petri dish.
5. Submerge the grids (sections uppermost), cover the large Petri dish and leave for 5–10 min. **Do not breathe over the lead citrate stain as this will cause CO_2 contaminaton**.
6. With forceps, pick the grids out and stream-wash as before (*see* **Note 5**).

4. Notes

Below are some of the problems that are encountered regularly together with solutions to these problems.

1. Poor fixation can be established by examining the mitochondria for dilated, misshapen christae, distension of the space between the nuclear membranes and the presence of extracellular spaces.
   a. Check the fixation protocol, particularly the timing between acquisition of the material and immersion fixation where postmortem changes may have occurred.
   b. Check the buffer pH and osmolarity.
   c. Check the purity of the glutaraldehyde by spectrophotometer *(4)*.
2. Soft blocks—where an impression in the resin remains when a fingernail is pressed into it. This is because poor polymerization and is usually caused by out-of-date accelerator. Replace accelerator every 6 mo. If buying epoxy resin in kit form, purchase accelerator separately and renew as above. If soft blocks persist, increase infiltration times (**Subheading 3.8.**) and/or gradually increase the percentage of resin during the infiltration phase, e.g., 20:80, 40:60, 60:40, 80:20 resin:solvent.

   Attempt to section soft blocks—it is often possible. If impossible, then proceed to item 3 below.
3. Unsectionable soft blocks must be reinfiltrated and embedded. Excess soft resin should be cut away from the tissue block and the block soaked in sodium ethoxide on a rotator for 24 h at RT. (Sodium ethoxide is prepared by adding sodium hydroxide pellets to ethanol until saturated. The solution is stored overnight before use and can be kept for months.) The tissue block is rinsed in four changes of ethanol, infiltrated and embedded as normal (**Subheading 3.8.**).
4. Polymerized blocks are brittle in the center. This may occur in one block of a batch due to poor infiltration of that solitary block (*see* **Note 2**) or it may occur in a complete batch of blocks. If a whole batch of blocks are brittle, this may be due to poor dehydration caused by one of several things:
   a. analytical grade solvents were not used;
   b. incorrect concentrations of solvents were used; or
   c. dehydration times were too short.
   If the material is hard, e.g., bone, bacteria then brittle blocks may be due to poor infiltration.
   a. Omit the accelerator from the resin during all infiltration steps (**Subheading 3.8.**, **steps 1–3**) and infiltrate at 60°C before final embedding in complete resin as normal.
   b. Use a resin with a lower viscosity, e.g., the epoxy resin Spurr *(2)* or an acrylic resin such as LR White *(5)*.
   c. Digest hard walls, e.g., plant material but this may damage the ultrastucture.
5. Presence of small (approx 10 nm) electron dense material throughout the tissue. This may be from the lead citrate stain.

   Check the support film or resin without tissue for the material. If present, there is a staining problem so repeat the staining procedure on unstained grids and take more care not to introduce CO_2 from breath. If absent, a precipitate has been formed from a contaminant in glutaraldehyde, osmium, and phosphate.

Use a higher quality glutaraldehyde or a newly purchased glutaraldehyde and/or cacodylate buffer.

The precipitate can be etched from unstained sections with 1% sodium metaperiodate for 20 min at RT, washed in a stream of distilled water and the sections stained (**Subheading 3.12.**).

6. The most common ultramicrotomy problems are:
   a. Sections will not cut. Try solutions i–v and viii.
   b. Sections flow over the back of the knife after cutting. Try solutions i–iii.
   c. Only a small part of the section cuts. Solution ix.
   d. Section is chattered, i.e., cuts in thick and thin horizontal lines. Try solutions iii–vii and x.
   e. Section separates into vertical strips. Solution iv.
7. Solutions to ultramicrotomy problems:
   i. Check the knife angle.
   ii. Check the speed of the arm-drop.
   iii. Check the height of the water.
   iv. Make a new glass knife.
   v. Make the block face smaller.
   vi. Collect only 20 ultrathin sections from one part of the glass knife.
   vii. Do not cut more than one 0.5 mm section before starting ultrathin sectioning.
   viii. Check the position of the cutting stroke.
   ix. Check the alignment of the block.
   x. Ensure that the sides of the block are not jagged; recut them with a fresh razor blade.

Acknowledgments

The author would like to extend her thanks to Lisa Bamber for reading through this manuscript.

References

1. Hall, J. L. and Hawes, C. (eds.) (1991) *Electron Microscopy of Plant Cells*, Academic, London, England.
2. Glauert, A. M. (1974) Fixation, dehydration and embedding of biological specimens, in *Practical Methods in Electron Microscopy*, vol. 3, part I, North-Holland, Amsterdam.
3. Reid, N. (1974) Ultramicrotomy, in *Practical Methods in Electron Microscopy*, vol. 3, part II, North-Holland, Amsterdam.
4. Robards, A. W. and Wilson, A. J. (1993) Basic Biological Preparation Techniques for TEM, in *Procedures in Electron Microscopy*, Wiley, England, ch. 5.
5. Newman, G. R. and Hobot, J. A. (1993) *Resin Microscopy and On-Section Immunocytochemistry*, Springer-Verlag, Berlin, Germany.

2

Negative Staining of Thinly Spread Biological Particulates

J. Robin Harris

1. Introduction

The negative staining of virus particles for TEM study was introduced in the late 1950s, following the establishment of a standardized procedure by Brenner and Horne in 1959 *(1)*. Rapidly, this staining technique was applied to other biological particulates, usually when a purified or semipurified aqueous suspension of freely suspended (i.e., nonaggregated) material was available. In addition to viruses, this material ranged from purified enzymes and other soluble protein molecules and components of molecular mass in the range of 200 kDa up to several MDa (such as the molluscan hemocyanins and ribosomes), to isolated cellular organelles, membrane fractions, bacterial cell walls and membranes, and filamenous protein structures of many types, also liposomal and reconstituted membrane systems, and even synthetic polymers.

The physical principle behind negative staining may at first glance appear to be rather simple, but on closer inspection, it is somewhat more complex. Essentially, a soluble heavy metal-containing negative staining salt is used to surround, support and permeate within any aqueous compartment of a biological particle. After air-drying a thin amorphous film or vitreous glass of stain supports and embeds the biological material, and at the same time generates differential electron scattering between the relatively electron-transparent biological material and the electron-dense negative stain. Simple air-drying undoubtedly leaves a good quantity of water bound to the biological material and to the surrounding negative stain, but this will be rapidly removed in vacuo within the electron microscope, unless the specimen is cooled in liquid nitrogen prior to cryotransfer of the specimen. Some of the subtleties and hazards of this oversimplified description will be expanded upon below, and have recently

From: *Methods in Molecular Biology*, vol. 117: *Electron Microscopy Methods and Protocols*
Edited by: N. Hajibagheri © Humana Press Inc., Totowa, NJ

been dealt with in some detail *(2)*. After many years of relative stability and lack of progress, negative staining is currently undergoing significant technical development and hopefully improvement, in order to overcome some of the inherent undesirable aspects, such as excessive particle flattening and drying artefacts *(3)*. This should, in turn, lead to a better understanding of the hazards that can sometimes be generated following particle-stain interaction. In addition, the combination of the established techniques of cryoelectron microscopy for the study of unstained biological materials with those of negative staining has opened up new and exiting possibilities *(4; see also* Chapter 3).

It is the aim of this Chapter to present some of the well-established and newer procedures *(5,6)* for air-dry negative staining on continuous thin carbon support films and across small holes in carbon films, with indication of the varying possibilities, applications and technical limitations. Specimens prepared in this manner can be studied at room temperature in any conventional TEM, with or without low-electron dose. Best quality high-resolution data will, however, only be obtained by following specimen cooling (e.g., to −175°C) with image recording from mechanically stable specimens (i.e., with no specimen movement/drift), with minimal image astigmatism and under low electron dose conditions.

As with many electron microscopical preparative procedures, there can be a number of alternative ways to achieve the same end result from negative staining. For instance, sample and stain can be applied to a thin support film by a fine spray (nebulizer) individually or mixed, sample and stain can be applied directly to a support film from a pipet tip, or sample and stain can be transferred from small droplets on a parffin wax (Parafilm) or clean plastic surface. All these approaches and others can work, but it is the opinion of the author that the last, the *single droplet technique*, is the simplest and most reliable. Thus, this procedure will be presented in full, and followed by a protocol for the negative-staining carbon film procedure, which is useful for two-dimensional (2D) crystal production. A number of alternative and more specialized negative-staining techniques have some value for the investigation of specific biological or molecular systems (for full coverage, *see* **ref. 2**). The possibility of performing immunolabeling experiments in combination with negative staining will be given some emphasis. Because a range of different negative stain salts are available, each with slightly different chemical properties and varying interaction with biological material, individuals often tend to favor the routine use of one or two of these, to the exclusion of others that may be perfectly suitable and usable. Thus, although several useful negative stains are mentioned in **Table 1**, emphasis will be placed upon the use of ammonium molybdate, the stain that the author currently finds to be most reliable for studies with biological and artificial membranes, protein molecules, and virus particles. The reader

Table 1
Negative Stain Solutions

Commonly Use Negative Stain Solutions

These stains are generally prepared as 1% or 2% w/v aqueous solutions[a,b]
Uranyl acetate
Uranyl formate
Sodium/potassium phosphotungstate
Sodium/potassium silicotungstate
Ammonium molybdate

Negative Stain-Carbohydrate Combinations

All of the above negative stains can be prepared as 2–6% w/v aqueous solutions
containing 1% w/v carbohydrate (glucose or trehalose).[c]

Negative Stain-PEG Combinations

The inclusion of 0.1–0.5% w/v polyethylene glycol (PEG) Mr 1000 in 2% w/v ammonium molybdate creates a solution that potentiates 2-D crystal formation.[d]

[a]A low concentration (e.g., 0.1 mM to 1.0 mM) of the neutral surfactant n-octyl-β-D-gluco-
pyranoside (OG) can be added to any of the above negative-stain solutions to improve the
spreading properties and assist permeation within biological structures.

[b]The pH of negative stain solutions can usually be adjusted over a wide range; this does not
apply to the uranyl negative stains, which readily precipitate if the pH is significantly increased
above pH 5.0. By complexing uranyl acetate with oxalic acid, an ammonium hydroxide
neutralizable soluble anionic uranyl-oxalate stain can be created, but this possesses an undesirable granularity after drying.

[c]Glucose and trehalose provide vitreous protectection to the biological material during air
drying. They also create a thicker supportive layer around the sample, thereby reducing flattening. Electron beam instability of these carbohydrates necessitates minimal routine or low-electron dose irradiation conditions, assisted by specimen cooling where possible. The inclusion of
1% trehalose reduces the net electron density of the negative-stain solution; this is why a higher
stain concentration is used.

[d]When mixed with a purified viral or protein solution and spread as a thin layer on mica *or*
across the holes of a holey carbon support film (also with trehalose present), this AM-PEG
solution can induce 2-D crystal formation (*see* text).Variation of the concentration of the PEG
and the pH of the solution is always required, to obtain the optimal conditions for 2-D crystal
formation *(2,4)*.

should, however, bear in mind that slightly different opinions may well be expressed elsewhere. Also, it must be emphasized that even with ammonium
molybdate, care should be taken to assess the possibility of deleterious effects
resulting from sample-stain interaction *(2,3)*. Once appreciated and understood,
such interactions can, however, be of biochemical value in some instances.

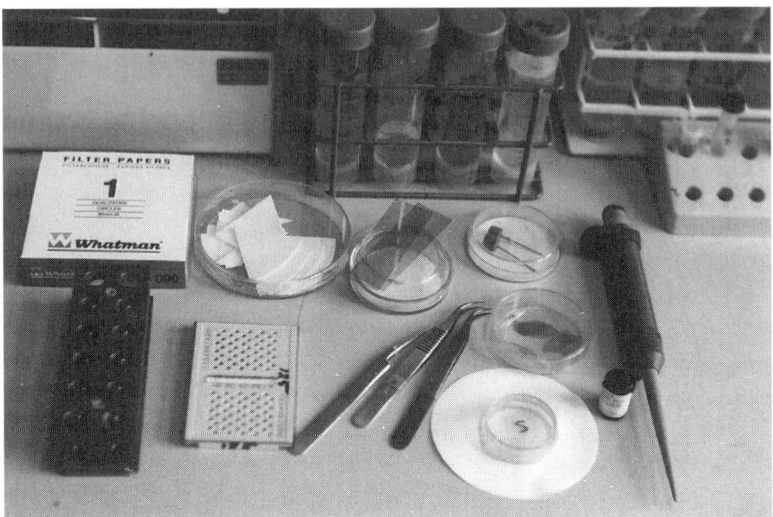

Fig. 1. An example of some of the small equipment needed for the production of negatively stained specimens.

2. Materials

2.1. Equipment

1. The principal large item of equipment that is needed in order to prepare negatively stained EM specimens is a vacuum coating apparatus (e.g., the Edwards model Auto 306 or the Bal-Tec model BAE 080 T), together with facility to perform glow-discharge treatment. This latter may be an attachment within the vacuum coating apparatus or a separate item of equipment. Carbon coating, to produce thin continuous carbon, carbon-plastic, or perforated (holey) carbon support films (*see* **Subheading 2.2.**; *see also* **ref.** *2*) can be performed using carbon rods, carbon fiber, or an electron-beam source, as described by the equipment manufacturer concerned. Glow discharge treatment of support films is particularly useful to combat the natural hydrophobicity of carbon, which interferes with the spreading and attachment of biological materials and the smooth spreading of a thin film of aqueous negative stain over and around the biological particles (i.e., embedment in a high-contrast medium).

2. Numerous smaller items of equipment are needed (*see* **Fig. 1**), such as Parafilm, fine curved and straight forceps (with rubber or plastic sliding closing ring), a range of fixed volume automatic pipets (e.g., 5 µL, 10 µL, 20 µL; and variable volume pipets, up to 1000 µL), plastic tips, scissors, metal needle/finely pointed probe, mica strips, filter paper wedges (e.g., Whatman No. 1), Petri dishes with filter paper insert, 300 or 400 mesh electron microscopy (EM) specimen grids (usually copper, but nickel or gold for immunonegative staining), grid storage boxes, a microcentrifuge, tubes, and tube racks. Last and very important, small

Kleenex or other adsorbant tissues need to be available to regularly wipe the tips of the forceps, immediately after use, to avoid cross-contamination.

2.2. Support Films

Although support films can be purchased as consumables from the various EM supplies companies, it is more usual for individuals to prepare their own. Some time needs to be devoted to the perfection of these ancillary techniques, to make available a ready supply of the necessary supports for negative staining and for the preparation of thin frozen-hydrated/vitrified specimens for cryoelectron microscopy (*see* Chapter 3).

Thin carbon support films can routinely be prepared by in vacuo carbon deposition onto the clean surface of freshly cleaved mica, with subsequent floatation onto a distilled water surface followed by lowering on to a batch of EM grids (300 or 400 mesh) positioned beneath, i.e., in a Buchner funnel or glass trough with controlled outflow of the water *(2)*. The thickness of the carbon can be assessed by a crystal thickness monitor during continuous carbon evaporation, but this is not esential. With a little experience, repeated short periods of evaporation from pointed carbon rods readily enable the desired thickness (e.g., 10 nm) to be achieved, based upon the faint gray color of a piece of white paper placed alongside the mica.

Carbon-plastic (e.g., colloidon, formvar, and butvar) support films can be produced by first making a thin plastic film on the surface of a clean glass microscope slide, from a chloroform solution (0.1–0.5% w/v). This plastic film is then released from the glass slide following scoring of the edges with a metal blade, with floatation onto a distilled water surface. An array EM grids can then be positioned individually on the floating plastic sheet, or the plastic sheet can be lowered on to an array of prepositioned EM grids at the bottom of the funnel or trough, when the water level is reduced. After drying, the batch of grids can be carbon-coated in vacuo.

Most would agree that the production of holey/perforated carbon-plastic or carbon support films (also termed microgrids) still remains something of an art. The simplest procedure is to initially perforate a drying film of plastic on the surface of a precooled (4°C) clean glass microscope slide by heavily breathing on to the surface. The small water droplets in the breath make holes in the plastic film at the zone of drying. A more reproducible way of producing a perforated plastic film is to use a glycerol-water (0.5% v/v each) emulsion in chloroform containing 0.1% or 0.2% w/v formvar. With vigorous shaking, an emulsion of small droplets is readily produced. On dipping a clean glass microscope slide into the emulsion and allowing it to dry, the plastic film so produced will be found to contain an array of small holes of varying size. The room temperature and humidity will vary, and the supplementation of moisture, by breathing gently on to the drying plastic film, may also assist the production of the perforated plastic film. A phase contrast light microscope can be used to assess the production of holes in the drying plastic film. Following scoring of the edges of the glass slide with a sharp blade, the perforated plastic film can then released from the glass slide to a water surface and lowered on to a batch of EM grids. After drying, the batch of grids should be carbon-coated, with deposition of an additional layer of gold or gold/pladium, if desired. The inclusion of the metal layer enables the quality of holey carbon grids to be readily assessed by bright field or phase contrast light microscopy. Attempts have been made by some researchers to standardize this procedure, but most tend to include their own individual variations *(2)*.

The presence of a thin plastic layer provides considerable strength to the support, which is desirable for the sequence of steps required for immunolabeling (*see* **Subheading 3.3.**), but the extra support thickness does inevitably reduce image detail and the maximal level of resolution. Thus, if desired, for both continuous carbon plastic and perforated carbon plastic films, the plastic can be dissolved by washing grids singly in an appropriate solvent such as chloroform or amyl acetate, before use.

2.3. Reagents and Solutions

Although prefixation/chemical crosslinking is not generally required for the negative staining of biological particulates, if it emerges that a sample is exceptionally unstable in the available negative stains, prior fixation with a low concentration of buffered glutaraldehyde (e.g., 0.05% or 0.1% v/v) may be included. This can be performed in solution or by direct on-grid *droplet* treatment of material already adsorbed to a carbon support film. It must, however, be borne in mind that the chemical attachment of glutaraldehyde to the available basic amino acid side groups, producing protein crosslinkage and stablization, may at the same time produce structural alterations at the higher levels of resolution. The more commonly used negative-staining salts are listed in **Table 1** (for a more detailed listing, including some of the less commonly used negative stains, *see* **ref. 2**).

These negative stains are generally used as 2% w/v aqueous solutions, but there is always the possiblility of increasing the concentration to provide greater electron density for small proteins or reducing the concentration for excessively thick biological samples that retain a greater volume of surrounding fluid. If the stain concentration is too low, up on air drying, the thin layer of fluid that surrounds a biological particle may not leave a sufficiently thick layer of amorphous salt to completely embed and support the particle, thereby resulting in partial-depth staining and in undesirable sample flattening. The adjustment of negative stain pH, to a value close to that of the sample buffer is standard. This is not usually possible for the uranyl negative stains, which precipitate at pH values above ca 5.5. It should also be borne in mind that the presence of traces of phosphate buffer is incompatible with the use of the uranyl negative staining.

Addition of a carbohydrate such as glucose or trehalose (e.g., 1% w/v) to the negative-stain solution has the advantage of creating a thicker supportive layer of dried stain, whereas at the same time helping to preserve the biogical sample during air-drying, because of the retention of sample hydration within a vitreous carbohydrate-negative stain-water layer. Inclusion of a higher than usual concentration of negative stain (e.g., 5 or 6%) is then required *(5,6)*. Specimen water, with or without the presence of a carbohydrate, will be greatly reduced once a specimen is inserted into the high vacuum of the TEM unless direct cooling in liquid nitrogen is performed first and followed by cryotransfer to the electron microscope (with the specimen maintained and studied at low temperature). As specimen lability within the electron beam is directly related to the presence of vitreous water, the early and continued success of conventional negative staining would appear to be because the rapid in vacuo removal of almost all loosely bound water from the biological material and amorphous stain, prior to electron irradiation. With the current widespread availability of TEM low-dose systems, image recording from frozen-hydrated negatively stained specimens (produced by air-drying or by rapid plunge freezing; *see* Chapter 3), can be successfuly pursued, thereby creating the possibility of improved image resolution because of sample hydration maintained at low temperature. Although negative staining and cryoelectron microscopy now appear to have some significant overlap *(2,7)*, it is likely that the separate technical approaches will be maintained for the forseeable future. Indeed, for high-resolution low-temperature negative-stain studies the phrase *high-contrast embedding media* has been introduced to avoid the *negative* connotations and undesirable limitations of conventional room temperature air-dry negative staining *(7)*.

2.4. Sample Material

For conventional on-grid negative staining, it is desirable to have purified sample material in the form of a free suspension (i.e., without large aggregates) in water or an aqueous buffer solution, at a concentration of 0.1 mg protein/mL. For the negative-staining carbon-film technique (used to produce 2-D crystals) the optimal concentration will be higher, in the order of 0.5 to 2.0 mg/mL protein or virus. For lipid suspensions, lipoproteins, nucleic acids, nucleoproteins, and viral particles, these concentration figures provide only a general guide; the main aim in all cases is to avoid overloading the specimen with sample material since particle superimposition will always obliterate structural detail. The presence of a high concentration of sucrose, urea, or other solute in

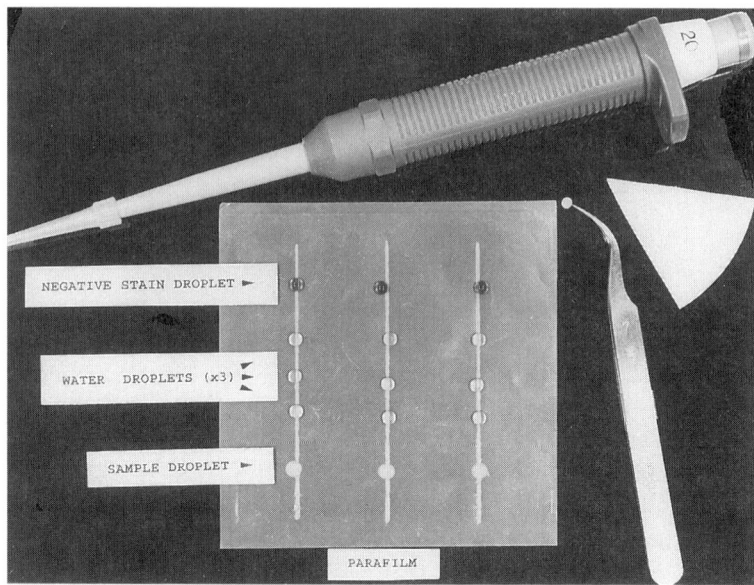

Fig. 2. A representative example of the layout of material for the single-droplet negative-staining procedure.

the sample suspension will introduce some problem for negative staining and must be removed. This can be done by prior dialysis or gel filtration using a dilute buffer solution, or by carbon adsorption and on-grid washing with distilled water or a dilute buffer solution, immediately before negative staining.

3. Methods

Protocols will be presented below for negative staining using the single droplet procedure (with the sample attached to a continuous carbon film or spread across the small holes in a perforated carbon film) and the negative-staining carbon-film (NS-CF) 2-D crystallization procedure *(2,8)*. Despite the listings below, the user should, with some limited experience, be prepared to freely introduce small technical variations to suit any local requirements determined by the biological sample, the equipment available and the aims of any individual study. Thus, an overall scientific awareness that improvements at the *grid level* can continually be sought and included is highly desirable *(2,3)*.

3.1. The Single Droplet Negative Staining Technique

1. Cut off a piece of Parafilm from a roll (length depending upon the number of samples and grids to be prepared), so that individual samples can be spaced by

1.5 cm, as shown in **Fig. 2**. (In general, it is best not to to attempt to prepare more than 10 specimens grids as a single batch).

2. Place the Parafilm, parafin-wax down, on to a clean bench surface, and before removing the paper overlay, produce a number of parallel lines by scoring with a blunt object across the paper. Then remove the paper, leaving the Parafilm loosly attached to the bench surface.

3. Place 20 μL droplets of sample suspension, distilled water (or dilute, e.g., 5 m*M* buffer solution), and negative stain solution on to the Parafilm, as in **Fig. 2**. The number of water droplets can vary between 1–5, depending upon the concentration of solute to be removed from the sample, with the proviso that each successive wash may introduce additional breakage of the fragile carbon support film.

4. Take a pair of curved forceps with an individual specimen grid coated with a) a thin continuous carbon support film or b) a holey carbon support film strengthend by in vacuo deposition of gold or gold/pladium. In the case of the former, a brief glow discharge treatment should be applied to increase hydrophilicity and thereby improve the sample and stain spreading. For holey carbon or carbon-plastic films, glow discharge is not usually necessary when a thin particulate layer of metal has been deposited. Touch the carbon or carbon-gold surface to the sample droplet. After a period of a time, ranging from 5–60 s (*see* **Note 1**), remove almost all the fluid by touching the edge of the grid carefully to a filter paper wedge.

5. Before the sample has time to dry wash the attached sample with one or more droplets of distilled water, with careful removal each time, as in Step 4.

6. Touch the grid surface to the droplet of negative stain (*see* **Note 2**), and likewise remove. Then allow the thin film of sample + negative stain to air dry before positioning the grid in a suitable container (Petri dish or commercially available grid storage box; *see* **Note 3**).

7. After drying, grids are immediately ready for TEM study, either under conventional (high-electron dose) conditions at ambient temperature or low electron dose conditions at either ambient temperature or after specimen cooling (e.g., to –180°C) (*see* **Note 4**).

Two examples of specimens prepared by the droplet negative-staining procedure are now shown. In **Fig. 3**, a sample of lobster hemocyanin dihexamer is shown, negatively stained with 2% uranyl acetate. In **Fig. 4**, a bundle of collagen fibers from keyhole limpet (*Megathura* crenulata) is shown, negatively stained with 5% ammonium molybdate containing 1% trehalose. For further explanatory comment, *see* Fig. legends.

3.2. The Negative-Staining Carbon-Film Technique

1. Prepare small pieces of mica, ca 1.5 × 0.5 cm, with one end pointed (*see* procedural outline, **Fig. 5**). Cleave the mica with a needle point to expose two untouched perfectly clean inner surfaces.

2. On a piece of Parafilm, mix 10 μL vol of sample (e.g., purified virus or protein solution, ca 1.0–2.0 mg/mL) with and equal volume of 2% ammonium molyb-

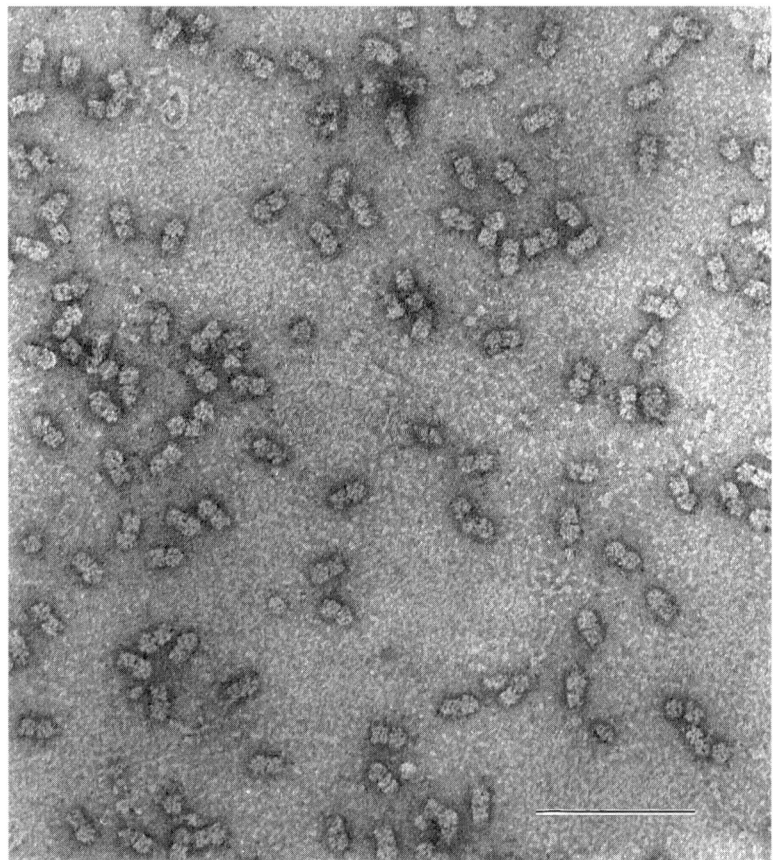

Fig. 3. The dihexamer hemocyanin molecule from the lobster *Homarus americanus*, negatively stained with 2% uranyl acetate using the droplet negative-staining procedure. The scale bar indicates 100 nm.

date (AM) containing 0.1% or 0.2% PEG *Mr* 1000 (*see* **Note 5**). The pH of the ammonium molybdate solution can be varied between pH 5.5–pH 9.0. (Some ammonium molybdate precipitation will be encountered during storage at pH 6.5 and lower, over a period of months.)

3. Apply 10 µL quantities of the sample-AM-PEG to the clean surface of two pieces of mica (held by fine forceps) and spread the fluid evenly with the edge of a plastic pipet tip. Hold each piece of mica vertically for 2 s to allow the fluid to drain towards one end, remove most of the pooled fluid and then hold horizontally, creating an even very thin film of fluid. Allow the fluid to dry slowly at room temperature within a covered Petri dish. A clearly visible zone of progressive drying towards a final deeper pool can usually be defined. 2-D crystal forma-

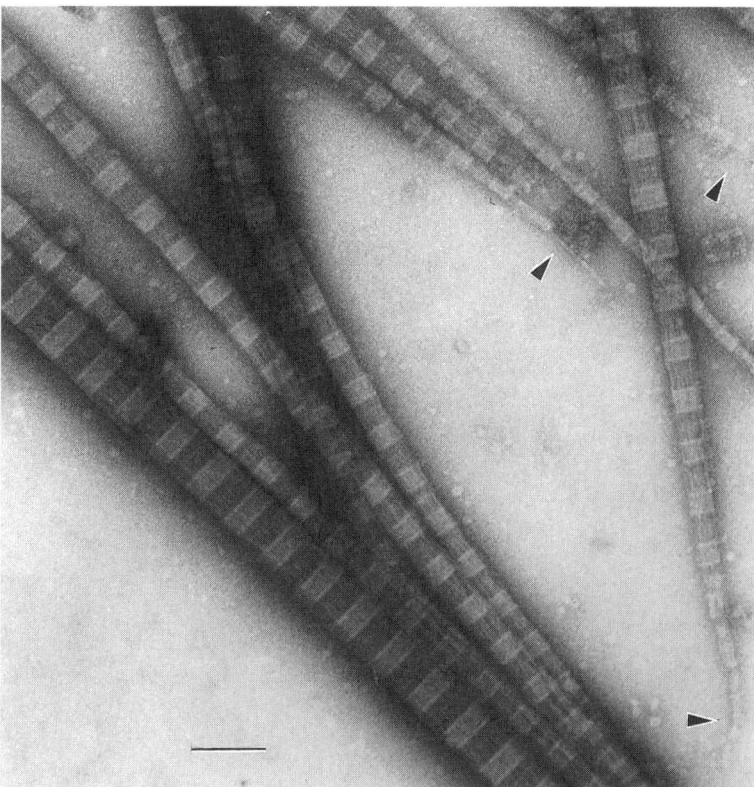

Fig. 4. Collagen fibres from the giant keyhole limpet *Megathura crenulata*, negatively stained with 5% ammonium molybdate containing 1% trehalose, using the droplet negative-staining procedure. Note the tapering of the spindle-shaped collagen fibers to very few fibrils (arrowheads). The scale bar indicates 100 nm.

tion will occur at this stage of the procedure, in all probability at the fluid/air interface, since adsorption to the untreated mica surface does not occur.

4. Coat the layer of dried biological sample on the mica surface in vacuo with a thin film of carbon (5–10 nm).
5. Float off the carbon film + attached biological material (randomly dispersed and as 2-D arrays or crystals) onto the surface of a negative-stain solutution in a small Petri dish (*see* **Note 6**). Recover pieces of the floating film directly onto uncoated 400 mesh EM grids from beneath, with careful wiping on a filter paper to remove excess stain and any carbon that folds around the edge of the grid. Often, the freshly deposited carbon film does tend to repel the aqueous negative stain, leading to understaining rather than overstaining. Allow the grid to air dry before positioning on a filter paper in a Petri dish or placing into a grid storage box. An

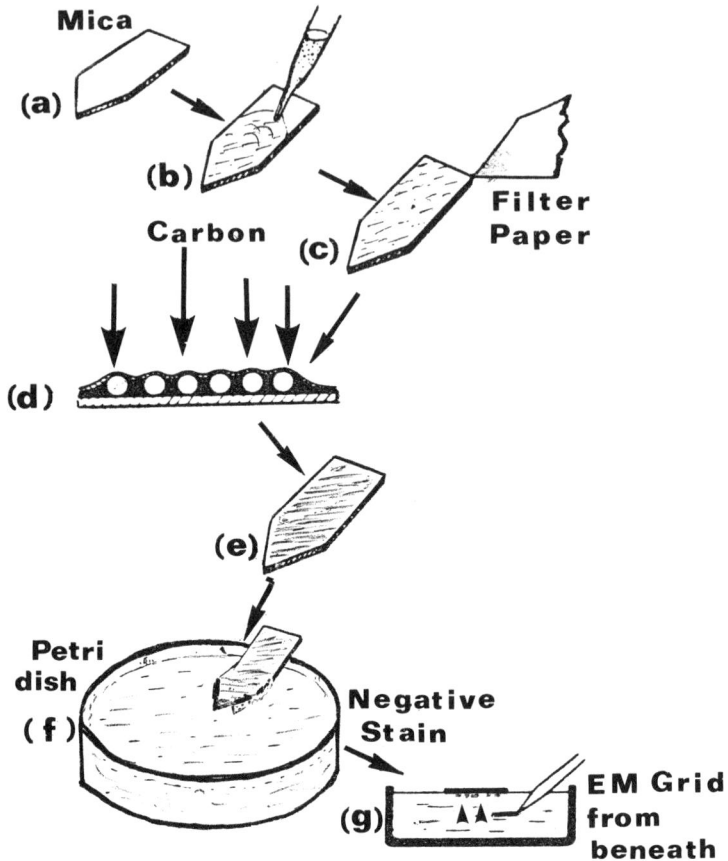

Fig. 5. A diagrammatic presentation of the succesive stages of the negative-staining carbon-film procedure *(8)*.

example of 2-D crystal formation by the *E. coli* chaperone GroEL, induced during the NS-CF procedure, is shown in **Fig. 6**. In the region shown, partial 2-D crystal formation by the cylindrical GroEL molecule (cpn60 2×7-mer) has characteristically occurred in the side-on (left hand side) and end-on (right hand side) orientations.

3.3. Immunonegative Staining

The combination of immunological labeling of protein molecules, viruses, cellular membrane fractions, intact cytoskeleton, and cytoskeletal proteins with negative staining offers considerable possibilities for antigen/epitope. Two approaches can be followed. The first requires prior preparation of the biological material in combination with a defined monoclonal antibody (IgG or Fab'

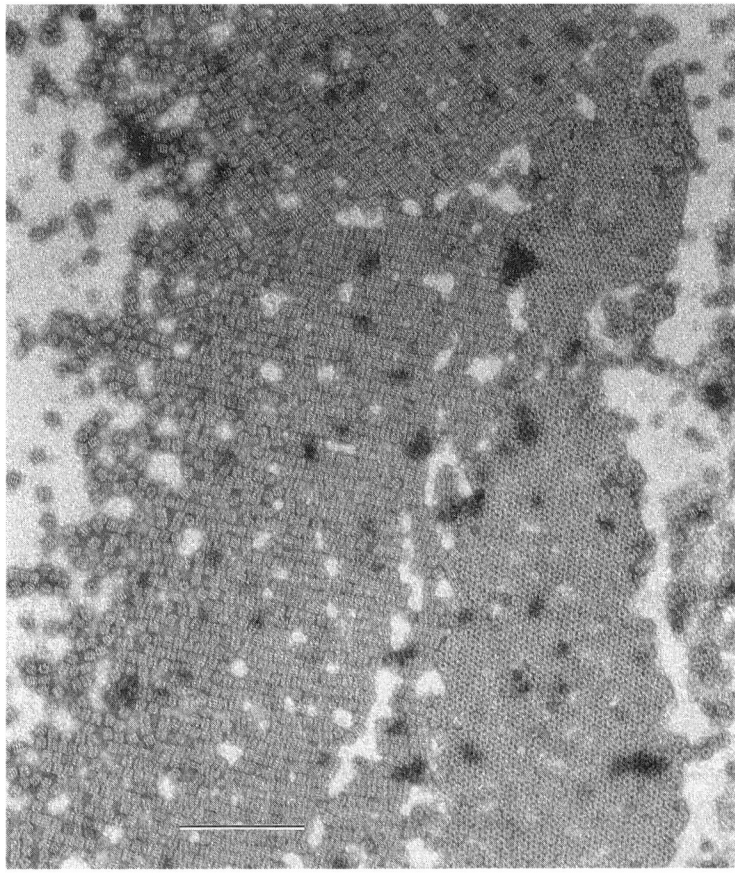

Fig. 6. The 2-D crystals of the *E. coli* chaperone GroEL (cpn60) produced by the negative-staining carbon-film (NS-CF) procedure. The final negative stain used was 2% uranyl acetate, but the 2-D crystals were induced by drying on mica in the presence of 1% ammonium molybdate containing 0.05% PEG (*Mr* 1500) at pH 8.0. Note the formation of several discrete 2-D crystals, with the molecules in two specific orientations. Free molecules can be seen at the edge of the crystals at the left hand-side. The scale bar indicates 100 nm.

fragment) in solution, alone or conjugated to colloidal gold or a smaller gold cluster probe. IgG molecules will crosslink soluble protein molecules, often creating specific linkage patterns that assist the definition of epitope location. The use of Fab' fragments does in theory present the possibility for higher resolution definition of epitope localization, but TEM study is inherently more difficult and does require some confidence that all available epitopes have a bound Fab'. With a satisfactory biological preparation, held in suspension as small immune complexes (IgG) or soluble particles (Fab'), the standard single droplet negative-staining procedure can be followed, as described in **Subheading 3.1.** (*see* **ref. 9**).

The second approach utilizes the fact that particulate biological material can be adsorbed to a carbon-coated plastic support film to create an immobilized thinly spread layer, which can then be allowed to interact successively with blocking solutions, colloidal gold-labeled (usually, 5 or 10 nm colloidal gold) or unlabeled primary monoclonal antibody solutions, wash solutions, gold-labeled secondary polyclonal antibody, distilled water, and finally a negative-stain solution. This does, in general, follow the pattern of on-grid postembedding immunolabeling of thin sectioned material (*see* Chapters 4–7), with the caution that somewhat greater handling precautions need to be adopted, particularly if the support film is carbon alone, whereas carbon-plastic support films have superior handling properties. The location of primary or secondary antibody-gold conjugates, rather than unlabeled primary antibodies alone, has been found to be most successfully pursued by this approach. Thus, with an unlabeled primary monoclonal antibody the reaction can be performed in combination with a labeled secondary antibody and/or the streptavidin/biotin reaction (with biotinylated primary or secondary antibody, followed by streptavidin-conjugated gold probe). Site-specific labeling of functional groups using derivatized gold cluster probes (Nanogold and Undecagold) also presents increasing possibilities for the future. In this instance, because of the small size of the gold probe, the use of a low-contrast negative stain such as sodium vanadate or 3% ammonium molybdate containing 1% trehalose is desirable. In general, after colloidal gold labeling, the negative staining can be performed with any appropriate negative stain. As with immunolabeling of thin sectioned material, gold or nickel grids must be used, because of the reactivity of copper in saline-containing blocking, antibody, label, and washing solutions.

An experimental layout of sequential 20 µL droplets on Parafilm, for example, sample solution, blocking solution, primary monoclonal antibody (can be biotinylated), washing solution, secondary gold-labeled polyclonal antibody (streptavidin-gold, if the first step uses a biotinylated antibody), washing solution, distilled water and negative stain, essentially identical to that shown in **Fig. 2** can be adopted. The specimen grid plus adsorbed biological sample is allowed to float on the surface of successive droplets for the required incubation times. For prolonged incubations with individual blocking or antibody droplets, or whole Parafilm sheet, can be covered with a Petri dish or other container, containing a piece of water-moistened tissue paper to reduce evaporation. Finally, the negative stain is applied, as for the single droplet procedure (**Subheading 3.1.**).

The necessary dilution of antibody and gold probe solutions and adequacy of blocking can only be determined by the individual experimenter, with the comment that the higher the antibody dilution and more efficient the blocking, the greater the labeling specificity is likely to be and the lower background/nonspecific labeling. Fixation of the biological sample should not be necessary for immunonegative staining. Thus, loss of sample antigenicity should not apply within this system, unless an on-grid brief fixation with a low concentration of glutaraldehyde (e.g., 0.1%) has had to be included, because of sample instability during the blocking, labeling, washing and negative staining sequence. A more detailed presentation of immunogold labeling techniques in relation to negative staining can be found in *(10)* and useful information on the use of gold cluster probes can be obtained in the promotional literature from Nanoprobes Inc.

Examples of droplet negative staining of immune complexes of keyhole limpet hemocyanin type 2 (KLH2), produced with a monoclonal antibody directed against a subunit domain located at the ends of the oligomeric, didecameric, and multidecameric complexes, are shown in **Fig. 7**. The use of colloidal gold labeling in conjunction with droplet negative staining is demonstrated in **Fig. 8**. Here, a suspension of cholesterol microcrystals had been incubated with biotinylated cytolytically inactive streptolysin-O mutant (N402C). The toxin mutant has bound stereospecifically to the surface of the cholesterol, as a monomolecular layer array, which has then been defined by labeling with streptavidin conjugated to 5 nm colloidal gold particles. The surface labeling of the cholesterol is clearly demonstrated, owing to the strong affinity of streptavidin for the biotinylated toxin *(11)*.

4. Notes

1. Samples with a high concentration of material require shorter carbon adsorption times. For specimens prepared across holey carbon films, higher concentrations of sample (e.g., 1.0–2.0 mg/mL) are desirable, as a greater quantity of free material is lost from the holes during the washing and staining steps. Material is, however, *held* at the opposite fluid/air interface.
2. For specimens prepared on holey carbon support films, it is necessary to use one of the negative stain-carbohydrate combinations (e.g., ammonium molybdate- or sodium phosphotungstate-trehalose) since an unsupported air-dried film of sample and stain alone breaks very readily. The situation is considerably improved by the presence of carbohydrate. Inclusion of PEG *Mr* 1000 (e.g., 0.1%–1.0%) in the specimen and wash solution, and/or in the negative-stain trehalose solution can promote 2-D crystal formation of isometric viruses and macromolecules across the holes. In this case, longer times of incubation prior to removal of sample and stain with the filter paper are desirable, to allow time for 2-D crystal formation at the air/fluid interface, before air drying.

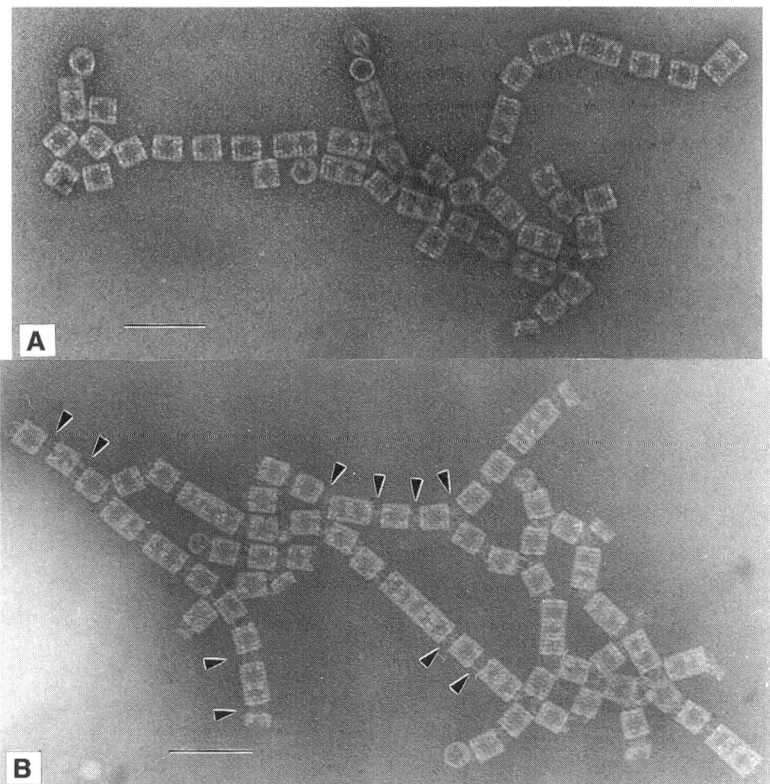

Fig. 7. Droplet negative staining of immune complexes of KLH2. (**A**) Negatively stained with 2% uranyl acetate (pH 4.5) and (**B**) negatively stained with 5% ammonium molybdate containing 1% trehalose (pH 7.0). In both cases, note the characteristic end-to-end linkage of the molecules by one or more monoclonal IgG molecules. Although the image contrast imparted by uranyl acetate is somewhat greater than ammonium molybdate, the image granularity generated by uranyl acetate is greater and the low pH is not desirable. The IgG molecules linking the hemocyanin molecules are more clearly defined in the presence of ammonium molybdate-trehalose (arrowheads, b). The scale bars indicate 100 nm.

3. Negatively stained specimens can usually be stored for many weeks. However, for specimens containing a mixture of trehalose and negative stain, after a period of approx 4–6 wk, the initially amorphous/vitreous glass from the air-dried mixture of stain and carbohydrate, which still contains a considerable quantity of bound water, will exhibit signs of undergoing crystallization of the carbohydrate and negative stain. Such specimens should, therefore, be studied in TEM within a period of a few days of preparation.

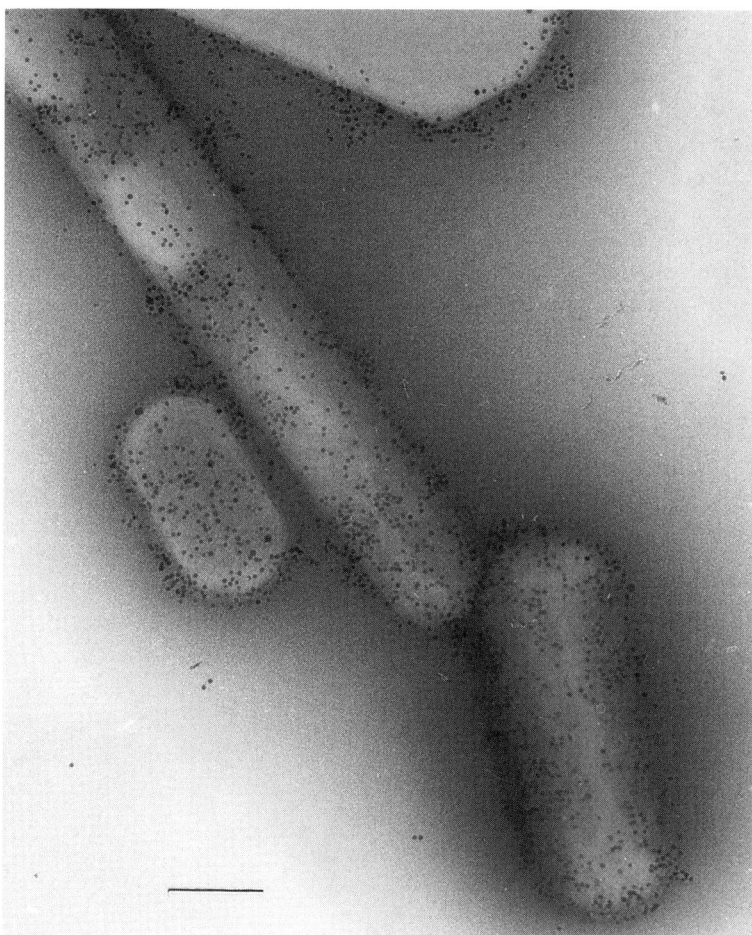

Fig. 8. Cholesterol microcystals with bound biotinylated mutant (N402C) streptol-ysin-O, following labeling in solution with streptavidin conjugated to 5 nm colloidal gold. The sample was negatively stained with 5% ammonium molybdate containing 1% trehalose, using the single-droplet technique. *See* **ref. *11.*** The scale bar indicates 100 nm.

4. If air-dried negatively stained specimens are subjected to rapid freezing and cryotransfer the remaining specimen water will be retained within the thin film of sample + stain, which must then maintained at low temperature (–180°C) and require low-electron dose study. Often, specimens will be cooled following room temperature insertion into the TEM; this produces maximal dehydration of the specimen prior to cooling to a low temperature. In general, for all specimen conditions, a low-dose study is desirable to obtain the best quality data, particularly for those specimens containing carbohydrate and is essential for those containing vitreous water.

5. The concentration and molecular weight of the PEG can be varied within a reasonable range. The concentration can range from 0.05% to 0.5% w/v and the molecular weight of the PEG from *Mr* 1000 to *Mr* 10,000. If aggregation is encountered, this will usually indicate that the PEG concentration is excessive or that the pH is too low. At higher PEG concentrations, individual micro-crystals of PEG will be produced during air drying.

6. Any of the conventional negative stain solutions, with or without the carbohydrate, glucose or trehalose can be used. Also, it is possible to float the film on to distilled water, with recovery and removal of excess water, followed by rapid plunge-freezing, as for the preparation of unstained vitreous specimens (*see* Chapter 3).

References

1. Brenner, S. and Horne, R. W. (1959) A negative staining method for high resolution electron microscopy of viruses. *Biochim. Biophys. Acta* **34,** 60–71.
2. Harris, J. R. (1997) Negative staining and cryoelectron microscopy, *RMS Microscopy Handbook*, Number 35, BIOS Scientific Publishers Ltd., Oxford, England.
3. Harris, J. R. and Horne, R. W. (1994) Negative staining: A brief assessment of current technical benefits, limitations and future possibilities. *Micron* **26,** 5–13.
4. Adrian, M., Dubochet, J., Fuller, S. D., and Harris, J. R. (1998) Cryo-negative staining. *Micron* **29,** 145–160.
5. Harris, J. R., Gebauer, W., and Markl, J. (1995) Keyhole limpet hemocyanin: Negative staining in the presence of trehalose. *Micron* **26,** 25–33.
6. Harris, J. R., Gerber, M., Gebauer, W., Wernicke, W., and Markl, J. (1996) Negative stains containing trehalose: Application to tubular and filamentous structures. *J. Microsc. Soc. Amer.* **2,** 43–52.
7. Orlova, E. V., Dube, P., Harris, J. R., Beckman, E., Zemlin, F., Markl, J., and van Heel, M. (1997) Structure of keyhole limpet hemocyanin type 1 (KLH1) at 15 Å resolution by electron cryomicroscopy and angular reconstitution. *J. Mol. Biol.* **272,** 417–437.
8. Horne, R. W. and Pasquali-Ronchetti, I. (1974) A negative staining-carbon film technique for studying viruses in the electron microscope. I. Preparation procedures for examining isocahedral and filamentous viruses. *J. Ultrastruct. Res.* **47,** 361–383.
9. Harris, J. R. (1996) Immunonegative staining: Epitope localization on macromolecules. *Methods* **10,** 234–246.
10. Hyatt, A. D. (1991) Immunogold labelling techniques, in *Electron Microscopy in Biology: A Practical Approach* (Harris, J. R., ed.) IRL Press, Oxford, pp. 59–81.
11. Harris, J. R., Adrian, M., Bhakdi, S., and Palmer, M. (1998) Cholesterol-streptolysin O interation: An EM Study of wild-type and mutant streptolysin O. *J. Struct. Biol.* **121,** 343–355.

3

Preparation of Thin-Film Frozen-Hydrated/ Vitrified Biological Specimens for Cryoelectron Microscopy

J. Robin Harris and Marc Adrian

1. Introduction

The production of rapidly frozen thin-film unstained vitrified specimens for cryoelectron microscopy was established as a standard procedure in the early 1980s, primarily because of the persistent efforts of Jacques Dubochet and his colleagues *(1–3)*. Many scientists have now used this approach to study thin unstained films of numerous biological samples, ranging from DNA to protein molecules and macromolecular assemblies, ribosomes, protein filaments, and microtubules, membrane systems, liposomes, and many different virus particles. Detailed surveys of the technical, methodological considerations, and applications of cryoelectron microscopy have been presented *(3,4)*.

The principal advantages seen to be gained from the study of unstained vitrified biological specimens are as follows.

1. The maintenance of the frozen-hydrated state throughout specimen preparation and study in the TEM.
2. The complete absence of any stain-particle interaction.
3. The presence of electron density information within the images that is derived from the biological material alone (that is, from the atomic composition of the biological material versus the lower electron density of the surrounding water).
4. The theoretical possibility of near atomic-level image resolution, from single particle averaging and crystallographic analysis.
5. The possibility to perform dynamic biophysical and biochemical experiments where specific metabolic or physically induced states with varying structural conformations of the biological material are instantly *trapped* by the rapid vitrification process.

From: *Methods in Molecular Biology*, vol. 117: *Electron Microscopy Methods and Protocols*
Edited by: N. Hajibagheri © Humana Press Inc., Totowa, NJ

To produce high-quality electron microscopical data from thin vitrified specimens, following successful cryotransfer of a specimen grid to the TEM, is not as straight forward as for air-dried specimens that have been inserted into the TEM at room temperature (with or without subsequently cooling of the specimen grid inside the TEM). The sensitivity of frozen-hydrated specimens in the electron beam is such that study and image recording is only possible under strict low-electron dose irradiation conditions, because rapid bubbling of vitreous water within and around the biological material occurs at higher levels of irradiation, thereby destroying all image detail. The image contrast produced from unstained biological samples in vitreous water is extremely low, because of the relatively small difference in net electron density of the biological material (C, N, O, H, P, S), and consequent differential electron scattering, versus water. Indeed, if the layer of vitreous water is too thick, this further reduces the contrast because of excessive overall density, but during specimen preparation, this situation can allow the sample greater mobility and therefore a greater range of angular orientation of the biological particles *(4)*. Even with an optimal thickness of vitreous water many biological particles are almost invisible, because of the low amplitude contrast, when images are recorded close to instrumental focus. To overcome this limitation, images are routinely recorded with considerble underfocus (1.2–2.0 μm), which enhances the phase contrast content of the image, whereas at the same time introducing density variations because of the pronounced oscillations of the contrast transfer function (CTF) with increasing defocus. Modern image processing techniques now enable correction for defocus and the CTF to be introduced, when considered appropriate, in order to retrieve the highest possible resolution. It is not the place of this short technical chapter to deal further with such electron optical and image processing considerations and the reader is referred to the available literature *(5,6)*.

Recent technical innovations in the area of negative staining *(4)* have now been found to integrate well with cryoelectron microscopy, in that it has been found that vitreous specimens can be prepared directly from thin aqeous films containing a high concentration (i.e., 16% w/v) negative stain *(7)*. With cryotransfer and low-electron dose study, images from such specimens yield high constrast images, similar to those from conventional air dry negative staining, but with a markedly reduced stain granularity. Image recording with vitreous negatively stained specimens can be performed substantially closer to image focus (theoretically, with higher resolution) than is possible with unstained vitreous specimens, thereby avoiding the requirement for CTF correction. Subnanometer resolutions have been obtained from these cryo-negatively stained images *(7)*, which are in agreement with the improvement obtained from air-dried negativcly stained specimens in the presence of glucose, following specimen cooling within the TEM *(8)*. It should, however, be born in mind that to date the highest resolution cryo-images have been derived from unstained two-dimensional (2-D) crystalline material, in the presence of glucose, trehalose or tannic acid, combined with electron crystallographic analysis.

2. Materials

The equipment required for the production of vitreous unstained and cryonegatively stained specimens is relatively simple and inexpensive. The necessary cryospecimen holder and cryotransfer system is somewhat more expensive, as is a suitable cryo-TEM with an efficient double-blade anticontaminator system located close to the specimen. Also required is access to a vacuum coating unit, for the preparation of perforated (holey) carbon support films, and for gold or gold-paladium coating of these films to render them more stable and hydrophilic. Apparatus for glow-discharge treatment of holey carbon support films may also be required. Emphasis here will be placed only upon specimen preparation, with some limited comment on the cryotransfer procedure and low-temperature EM study.

2.1. Equipment and Materials

The main item required is a free-fall (*guillotine-type*) rapid plunge-freezing apparatus, of the type shown in **Fig. 1** (as designed by the EMBL, Heidelberg, Germany). The following additional items are also needed: a supply of liquid nitrogen, large and small dewar flasks, a polystyrene container (**Fig. 2**), a small double-wall aluminium freezing chamber, a cylinder of liquid ethane with outlet tube connected to a pipet tip, straight fine forceps, holey carbon support films (*see* Chapter 2, **Subheading 2.2.**), filter paper wedges/arcs, automatic pipets and tips (5 μL through to 1000 μL), the small accessory items (e.g., small circular plastic grid box and aluminium base, and holder for the specimen grid clip/ring) provided by manufacturers of cryospecimen holders and cryotransfer systems (e.g., Gatan Inc., Pleasanton, CA; Oxford Instruments Ltd., Eynesham, Oxford, UK), *see* **Fig. 3**.

2.2. Solutions

Sample solution (with known protein/nucleic acid/lipid concentration), together with buffer or distilled water for the production of an appropriate dilution is needed. For cryonegative staining a freshly prepared 16% solution of ammonium molybdate, with pH usually adjusted to neutrality, is required. Polyethylene glycol (PEG) *Mr* 1000 can be added to the ammonium molybdate solution, usually in the concentration range from 0.1% to 1.0% (w/v), for studies directed towards the production of 2-D crystals of protein molecules or viral particles.

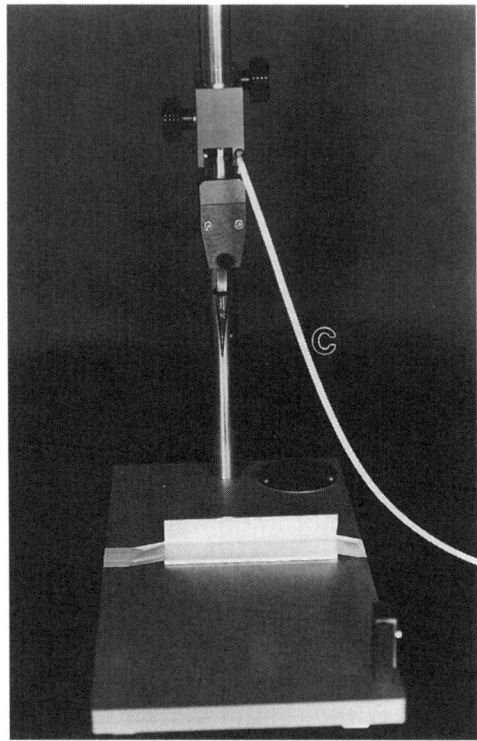

Fig. 1. A representative *guillotine* plunge freezing apparatus, of the type designed by the EMBL, Heidelberg. The rapid free-fall of the apparatus can be activated by hand via the cable-release (C).

3. Methods

The apparatus described in **Subheading 2.1.** should be conveniently arranged within a negative pressure fume extraction hood. Some personal experience is needed to be sure of having all materials immediately at hand, appropriately positioned and ready for rapid use.

3.1. Preparation of Unstained Thin-Film Vitreous Cryospecimens

1. Fill a small polystyrene chamber/box with liquid nitrogen, to a level just beneath the top of the double-walled aluminium freezing chamber. Ethane gas is then slowly passed into the center of the chamber, with liquification, until the chamber is full. After a few minutes, the liquid ethane takes on a slightly opaque appearance; it is then ready for use as a rapid freezant (at ca −180°C).
2. A 5 μL volume of sample is pipeted directly onto a holey carbon grid held by a pair of straight forceps material (usually in the order of 1.0 mg protein/mL is

Fig. 2. A suitable aluminium double-wall freezing chamber within a polystyrene liquid-nitrogen container, positioned centrally with respect to the foceps (F) held in a plunge-freezing apparatus.

Fig. 3. A plastic grid holder to take four vitrified specimens, positioned within an aluminium base. A removal rod, with screw thread (arrowhead) is used to transfer the holder from the liquid nitrogen tray to the cryotransfer system. The plastic lid is necessary for long-term storage of the holder in a suitable liquid-nitrogen container.

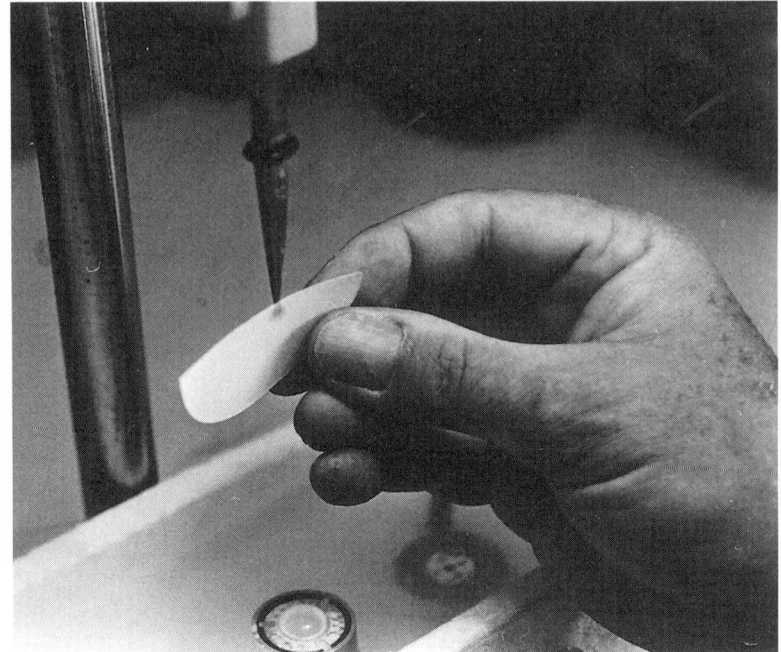

Fig. 4. The direct blotting of a sample on a holey carbon support film with a filter paper, immediately prior to plunge freezing.

satisfactory, but varying concentrations can readily be utilized by trial and error; *see* **Note 1**). Glow-discharge treatment of the holey carbon grid can be applied to improve the hydrophilicity and sample spreading, but if the holey carbon has been metal-coated (gold or gold-paladium), no glow-discharge treatment is necessary.

3. The forceps with grid and sample are then positioned within the clamp of the plunge freezing apparatus. The grid is carefully blotted face-on with a filter paper, which is held in direct contact for approx 1–2 s (**Fig. 4**), and gently lifted off. After a further 1–3 s, during which time some evaporation of water will occur to produce an optimal thickness of the aqueous film spanning the holes (achieved by personal experience only), the release catch of the *gullotine* is instantly activated (by a hand or foot mechanism) and the grid rapidly plunged into the liquid ethane. A thin layer of vitreous water is produced, containing unstained sample material, crossing the small holes and surface of the holey carbon support film.

4. The forceps are then released from the holding mechanism and very quickly lifted from the liquid ethane into the liquid nitrogen, with removal of as much adhering liquid ethane as possible (any remaining will solidify as a contaminant on the grid and can obliterate the specimen and/or interfere with the grid-holding screw or clip of the cryoholder).

Fig. 5. The rapid transfer of a newly prepared vitreous specimen from the liquid ethane cryogen of the freezing chamber to a plastic grid holder, positioned under liquid nitrogen.

5. At this stage one or more grids can be stored in the small circular plastic transfer holder (**Fig. 5**). Such holders, with a screw-on cap, can be stored in ice-free liquid nitrogen, using a modification of the system routinely used for storing aliquots of viable cultured cells. Repeat specimens can be prepared, with the precaution that the tips of the forceps holding a new holey carbon grid must not contain any water, as this will prevent the easy removal of the grid following the rapid freezing sequence (light coating of the forceps points with Teflon will help).

6. Meanwhile, the cryotransfer holder should have been precooled, either when within the electron microscope or when inserted into the cryoworkstation (**Fig. 6**). The anticontaminator system(s) of the electron microscope must also be precooled. The small holder for the cryospecimens is then rapidly moved from the polystyrene box of liquid nitrogen to the liquid nitrogen within the cryostation. From there, a grid is positioned in the space at the end of the cryoholder, the holding clip or screw inserted, the sliding protective shield moved over the specimen, and the specimen holder rapidly transferred to the electron microscope. The exact sequence of actions relating to the individual electron microscope air-lock evacuation, insertion, and rotation of the cryoholder need to be performed as descibed by the manufacturer.

7. Following insertion into the transmisison electron microscope, the small dewar flask of the cryoholder should be rapidly refilled or topped-up as necessary and time allowed for the specimen holder temperature (usually $-170°C$ to $-185°C$) to stabilize and the microscope to achieve its routine high vacuum.

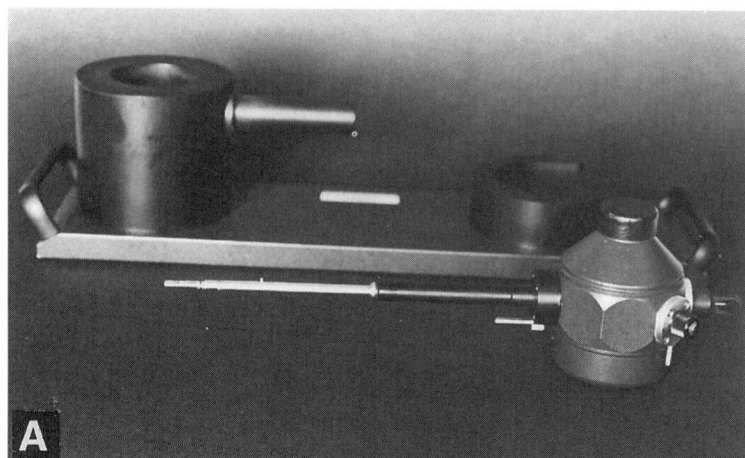

Fig. 6. A cryospecimen holder (**A**) positioned beside a cryo-workstation (Gatan). Liquid nitrogen precooling of the cryospecimen holder, can be performed within the cryoworkstation or in the TEM specimen airlock (**B**). Within the cryo-workstation, a vitreous specimen grid is transfered to the cryoholder under liquid nitrogen, followed by rapid cryotransfer to the electron microscope specimen airlock.

8. After removal of the sliding shield of the specimen holder, low-electron dose study and image recording can commence. For precise details of the optimal defocus conditions, the reader is referred to relevant journal publications from recent years, dealing with viral and macromolecular imaging, image processing and three-dimensional (3-D) reconstruction (e.g., *Journal of Structural Biology,*

vol. **116,** 1996, and **ref.** *3*). At least one visit to a laboratory where scientists already perform routine cryoelectron microscopy is strongly recommended.

Some examples of cryo-TEM images from unstained frozen-hydrated specimens will now be given; the individual legends supplement the comments made within the text. In **Fig. 7**, the characteristic rods of tobacco mosaic virus (TMV) are shown, with the thickness of the vitreous water film increasing from the bottom to the top of the figure. The mixed population of keyhole limpet hemocyanin types 1 and 2 (KLH1/KLH2) is shown in **Fig. 8**. The molecules shown here are located on a thin carbon support film and across a hole. Note the superior detail of the molecules dispersed across the hole. In the presence of a high concentration of calcium and magnesium ions (e.g., 100 m*M*), the subunit of KLH1 can be reassociated into a helical tubular/polymeric form. **Figure 9** shows a paracrystalline bundle of KLH1 tubules *(10,11)*.

A mixed population of intact and partially disrupted rotavirus is shown in **Fig. 10**, along with some TMV rods. (A direct comparison can be made with the same specimen material when prepared by the cryonegative staining procedure [**Fig. 11**], shown in **Fig. 12.**)

3.2. Immunolabeling

Direct immunolabeling of viruses, macromolecules, and membranes with monoclonal IgG molecules or Fab' fragments extends the possibilites of cryoelectron microscopy of unstained material into the relm of high-resolution epitope analysis. If Fab' is conjugated with 14 Å nanogold or 8 Å undecagold atomic clusters labels (Nanoprobes, Inc., Stony Brook, NY) it provides further possibilities, when location of the antibody fragment alone proves to be difficult. In addition, site-specific labeling of functional groups using derivatized gold cluster probes, also has much potential, as does utilization of the streptavidin/biotin system. All these approaches require prior preparation and standardization of the immunological reagents (e.g., blotting or ELISA) prior to incubation with the the biological sample. Immunolabeled material can then be utilized as a sample suspension within the protocol presented in **Subheading 3.1** (for example, *see* **ref.** *12*).

3.3. Preparation of Thin-Film Vitreous Cryonegatively Stained Specimens

This procedure is a variant of that described above under **Subheading 3.1.** and is designed to produce frozen-hydrated specimens, where a thin aqueous film of *negative stain* containing the biological sample has been rapidly frozen. The materials required are essentially the same as in **Subheading 2.1**. An additional procedural stage is included, within which the sample is mixed and incubated with a high concentration of ammonium molybdate solution (pH usually

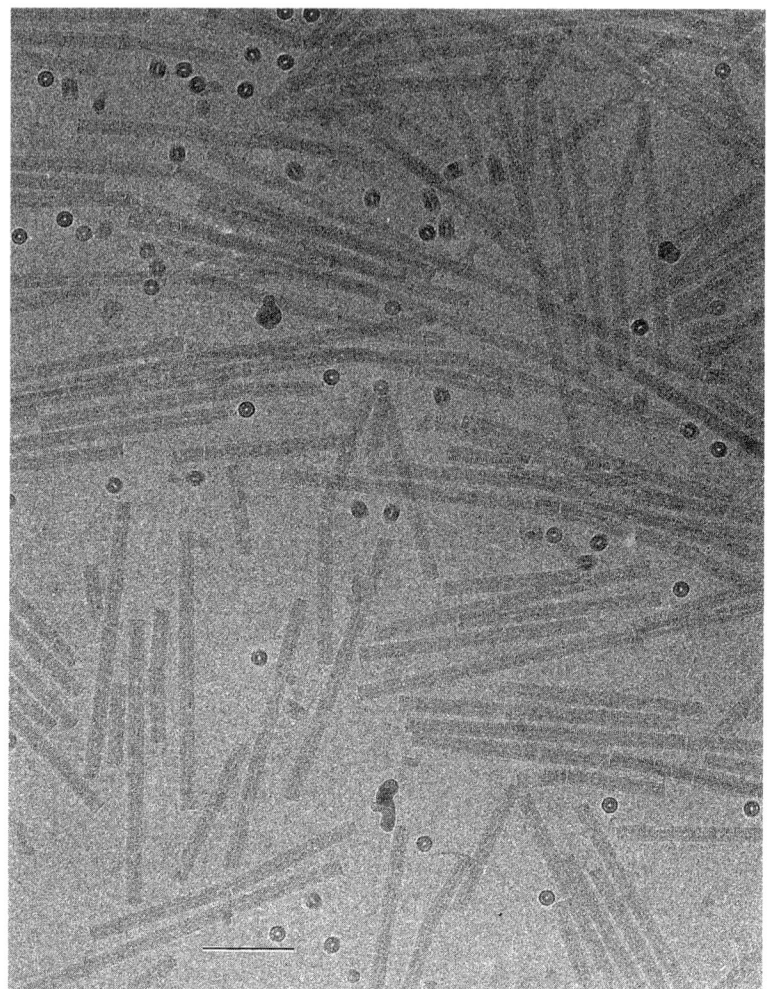

Fig. 7. A cryoelectron micrograph of unstained frozen-hydrated TMV. Note the orientation of the TMV rods parallel to the plane of the film of vitreous ice and also a number of rods (end-on) perpendicular to the plane of the ice. Towards the top of the figure, the ice is somewhat thicker, as it is closer to the edge of a hole; here there the obliquely positioned TMV rods show more clearly as does the central water-filled channel of the end-on viruses (enhanced by defocus phase contrast). The scale bar indicates 100 nm.

in the range 6.0–8.0) prior to blotting and rapid freezing. The possibility of being able to create monolayer 2-D crystals of isometric viral particles and macromolecules immediately prior to freezing, by including polyethylene glycol (PEG) in the sample and negative stain solution, is a new and as yet largely unexplored explored benefit stemming from this approach (*see* **Note 2**).

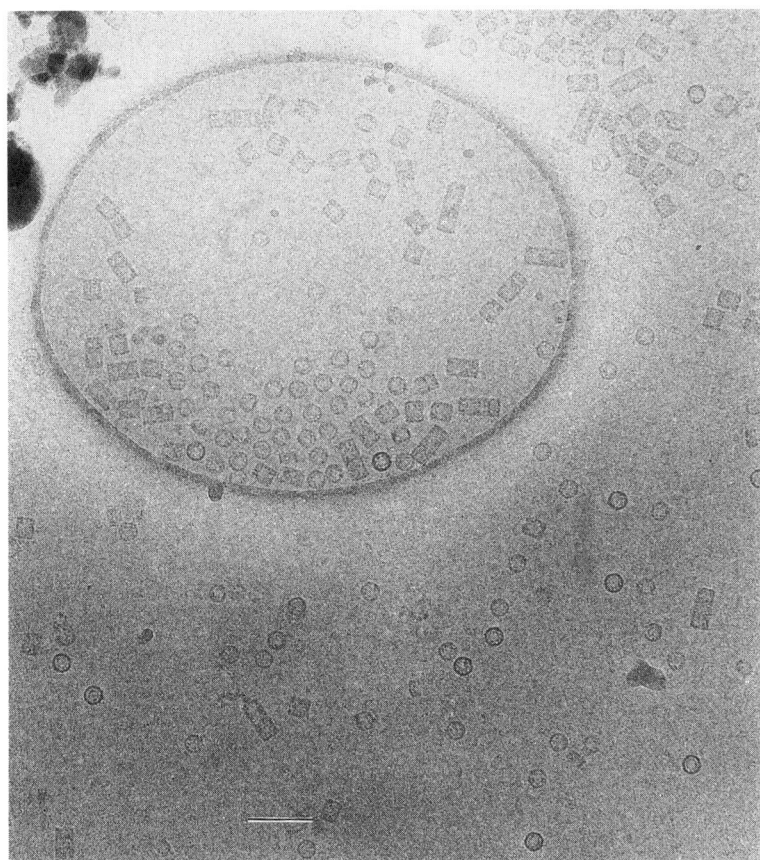

Fig. 8. A cryoelectron micrographs of unstained total keyhole limpet hemocyanin (i.e., a mixture of KLH1 and KLH2). Note the presence of hemocyanin decamers and stacked decamers (multidecamers), adsorbed to the carbon support film and spread in vitreous ice alone within the hole. The clustering of particles towards the edge of the hole is a common feature of cryospecimens. The scale bar indicates 100 nm.

As in **Subheading 3.1.**, a 5 µL of sample solution is applied to a holey carbon support film held by a pair of straight fine forceps. This is followed by short period of time, ranging from 10 s to 60 s, during which some concentration of sample at the fluid/air interface will occur.

1. *Without* blotting, the grid + sample solution is inverted onto a 100 µL droplet of 16% ammonium molybdate solution and allowed to float for a period of time, usually ranging between 10 s–3 min.
2. The forceps + grid and sample mixed with ammonium molybdate are then positioned in the holder of the plunge freeze apparatus and the procedure follows in

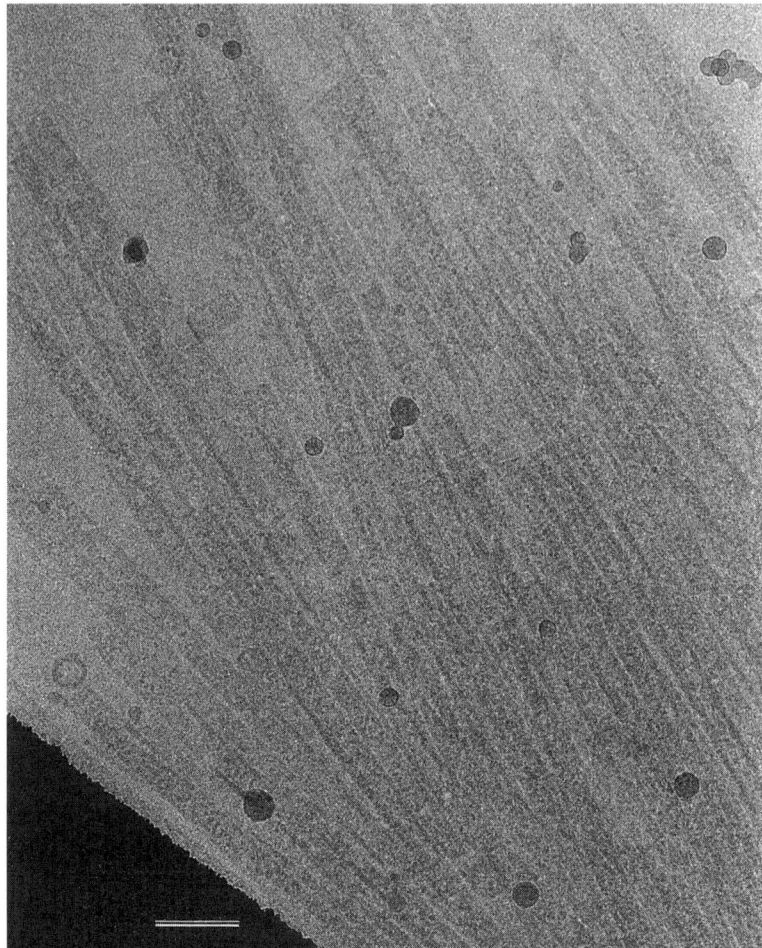

Fig. 9. A cryoelectron micrograph of a unstained bundle of KLH1 helical tubules, produced from purified KLH1 subunits by in vitro reassociation in the presence of 100 m*M* calcium and magnesium chloride *(10,11)*. Note the presence of occasional crystals of hexagonal ice. The scale bar indicates 100 nm.

an identical manner to that described in Steps 3 to 8 of **Subheading 3.1.**, with sequential blotting and rapid vitrification.

A diagrammatic description of the overall cryonegative staining procedure is given in **Fig. 11**, taken from **ref.** *(7)*. Examples of data obtained from some cryonegatively stained specimens is shown in **Figs. 12–14**.

In **Fig. 12**, the same rotavirus sample as used for the preparation of the unstained cryospecimen shown in **Fig. 10** is given. Again, note the varying

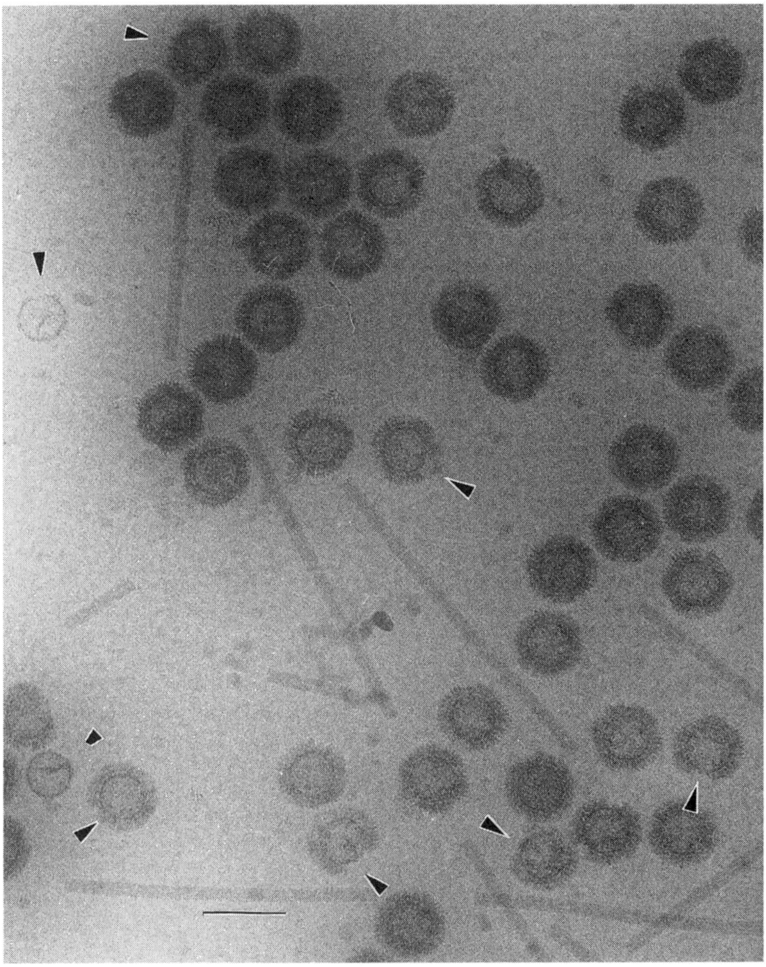

Fig. 10. A cryoelectron micrograph of unstained rotavirus particles, together with some TMV rods. Note the varying integrity of the rotavirus sample; arrowheads indicate damaged virions and isolated cores. See also the same sample frozen in the presence of ammonium molybdate, shown **Fig. 12**. The scale bar indicates 100 nm.

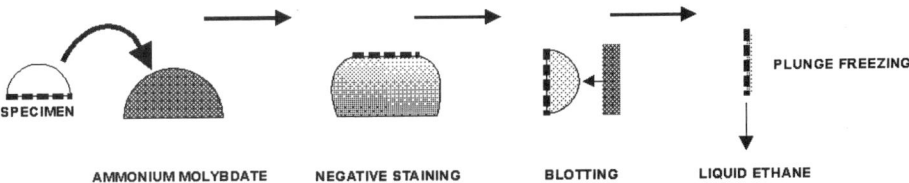

Fig. 11. A diagrammatic scheme showing the various stages during the production of a thin-film vitreous cryonegatively stained specimen (7).

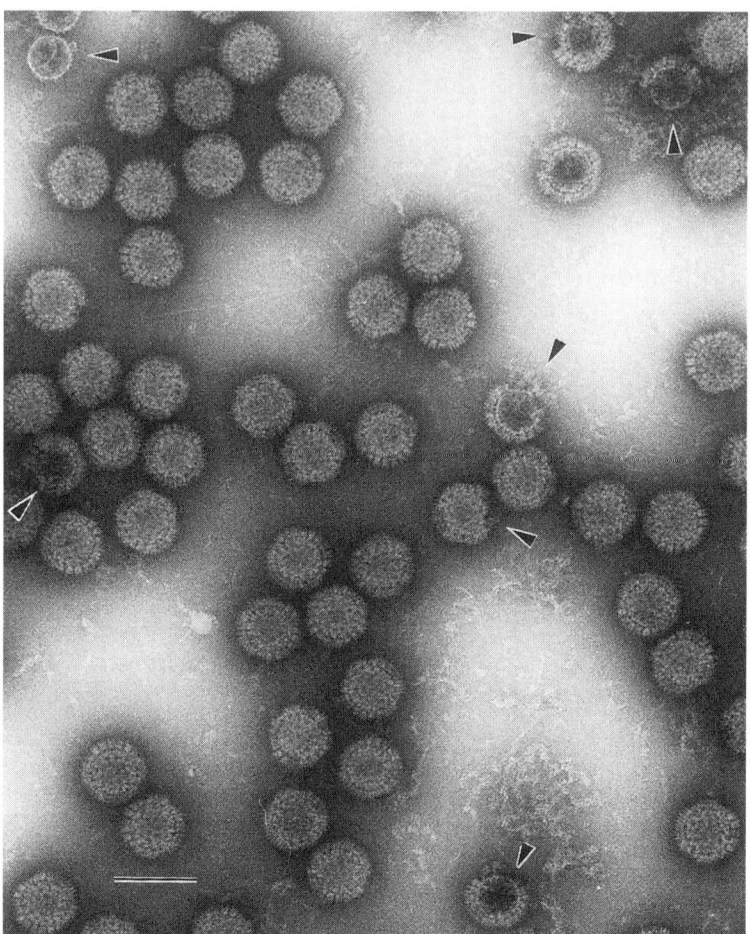

Fig. 12. A cryoelectron micrograph of negatively stained rotavirus (*see* **Fig. 10**). The capsomers of the intact virions show with considerable clarity, and disrupted virions and isolated cores are well defined (arrowheads). Within the background, strands of DNA and individual capsomers can be seen, following release from damaged virions. The scale bar indicates 100 nm.

state of integrity of the virions. From this negatively stained specimen, it is possible to define the nucleic acid strands and individual capsomers that have been released; although present, these particles cannot be so readily defined in the absence of negative stain (*see* **Fig. 10**). When PEG is included in the 16% ammonium molybdate solution used for the cryonegative staining, tomato bushy stunt virus (TBSV) has a strong tendency to produce ordered 2-D arrays *(7)* and crystals. The 2-D array of TBSV shown in **Fig. 13** was produced in the

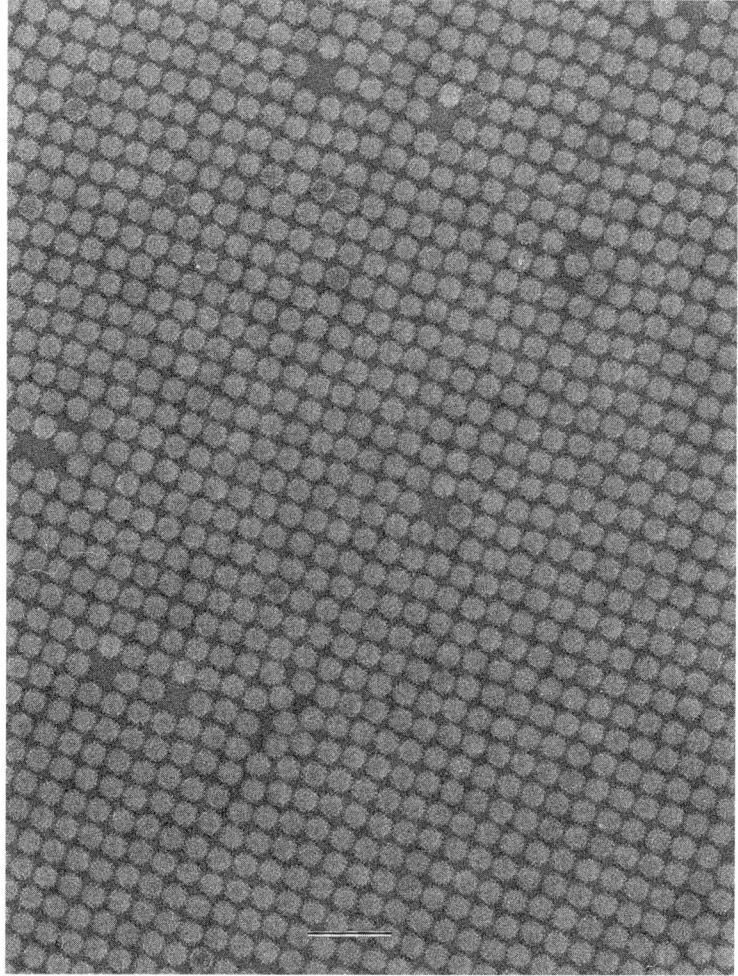

Fig. 13. A cryoelectron micrographs showing part of a large 2-D array of tomato bushy stunt virus, produced by the cryonegative staining procedure when performed in the presence of 2% PEG (*Mr* 1000). Although short-range 2-D crystalline order is present, overall there is considerable lattice disorder. The scale bar indicates 100 nm.

presence of 2% PEG. Although considerable lattice disorder is present, this figure serves to indicate the considerable potential of this approach for the production of 2-D crystals. The oligomeric and polymeric forms of KLH1 and KLH2 produced in the presence of a high concentration of calcium and magnesium *(10,11)*, are not stable in the presence of 16% ammonium molybdate *(7)*. **Figure 14** shows the break up of the tubular polymers and multidecameric

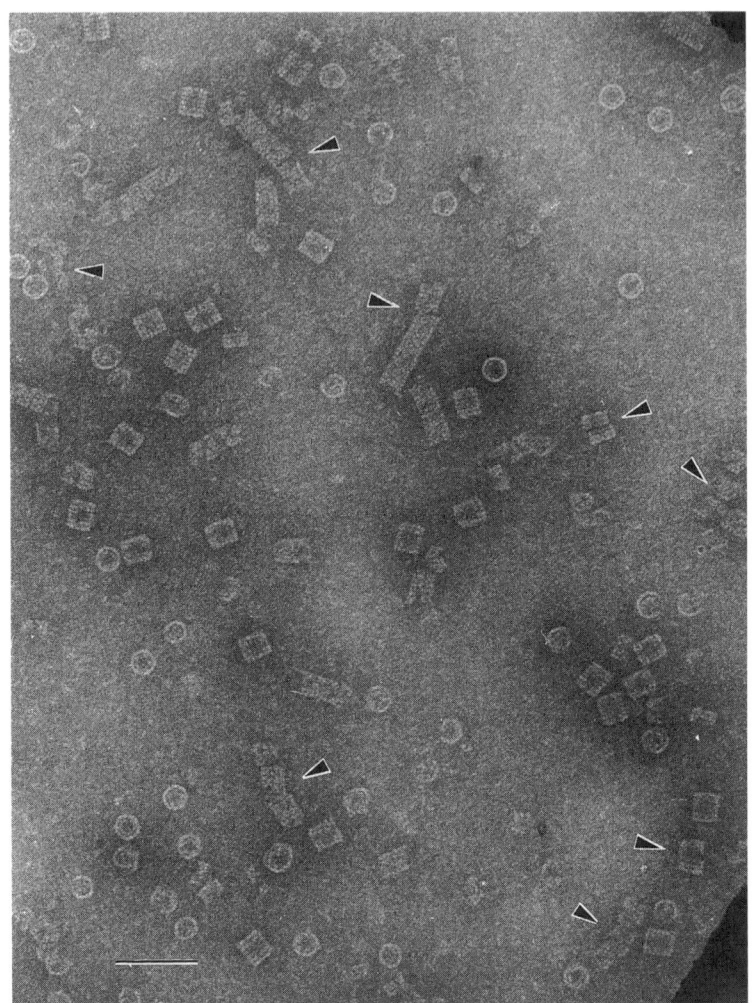

Fig. 14. A cryoelectron micrograph showing dispersed decameric oligomers and tubular polymers of reassociated KLH2 *(10,11)*. In the presence of 16% ammonium molybdate (pH 7.0), keyhole limpet hemocyanin is not stable *(7)*. Splitting of the decamers and tubules is indicated (arrowheads), and many damaged molecules and smaller subunit aggregates are present. This instability is not desirable during specimen preparation, but once appreciated and understood, it has been found to be of biochemical value for the study of KLH *(10,11)*. The scale bar indicates 100 nm.

oligomers of KLH2 under these conditions. Observations such as this indicate that considerable caution must be adopted until knowledge of the sample stability in the presence of 16% ammonium molybdate has been established. Data obtained from conventional negatively stained specimens (using a range of different stains) and from unstained frozen-hydrated specimens will usually provide reliable comparisons.

4. Notes

1. If the aqueous sample solution being used for the preparation of either unstained or negatively stained cryospecimens contains a high concentration of solute such as sucrose, glycerol, urea, or buffer salts, it will usually be necessary to remove these reagents. If prior dialysis or a column filtration cannot be done, on-grid washing/dialysis can be performed actually on the holes of the holey carbon support film, by withdrawing most of the sample by direct blotting or touching a filter paper to the edge of the grid, followed by application or one or more 5 μL quantities of water or low concentration buffer to the grid, each time with blotting, prior to direct freezing or addition of ammonium molybdate. For this on-grid microdialysis to work successfully, with entrapment of the biological material at the fluid/air interface opposite to the side of blotting *(9)*, a high initial concentration of sample is usually required (e.g., 1–2 mg/mL).

2. For 2-D crystallization of icosahedral and filamentous viral particles, macromolecules, and enzyme complexes during the cryonegative staining procedure, the optimal concentration of the start material, the pH of the ammonium molybdate solution, the best concentration of PEG, and the time of incubation of the sample with ammonium molybdate, all need to be determined by the experimenter for each sample under investigation.

3. The 16% ammonium molybdate solution may induce biochemical instability with some samples *(7)*. A pH variation should then be explored, as well as prior fixation with a low concentration of glutaraldehyde (e.g., 0.05–0.1% v/v) and there is possibility of using an alternative negative-stain solution such as sodium phosphotungstate or silicotungstate (uranyl acetate solution is not considered to be suitable in this instance). The higher than usual concentration of ammonium molybdate (e.g., 16% here, compared to 2–5% for air dry negative staining) is necessary because a very thin aqueous film of sample + stain, across the holes of the grid, is produced by direct blotting (**Fig. 4**).

References

1. Adrian, M., Dubochet, J., Lepault, J., and McDowall, A. W. (1984) Cryo-electron microscopy of viruses. *Nature* **308,** 32–36.
2. Lepault, J., Booy, F. P., and Dubochet, J. (1983) Electron microscopy of frozen biological suspensions. *J. Microsc.* **129,** 89–102.
3. Dubochet, J., Adrian, M., Chang, J.-J., Homo, J.-C., Lepault, J., McDowall, A. W., and Schultz, P. (1988) Cryo-electron microscopy of vitrified specimens. *Quart. J. Biophys.* **21,** 129–228.
4. Harris, J. R. (1997) *Negative Staining and Cryoelectron Microscopy: The Thin Film Techniques,* in *RMS Microscopy Handbook, No. 35,* BIOS Scientific Publishers Ltd., Oxford, UK.
5. *Journal of Structural Biology* (1996) vol. 116, no.1, Special Issue: Advances in Computational Image Processing for Microscopy.

6. Frank, J. (1996) *Three-Dimensional Electron Microscopy of Macromolecular Assemblies*, Academic, San Diego, CA.

7. Adrian, M., Dubochet, J., Fuller, S. D., and Harris, J. R. (1998) Cryo-negative Staining. *Micron* **29,** 145–160.

8. Orlova E. V., Dube, P., Harris, J. R., Beckman, E., Zemlin, F., Markl, J., and van Heel, M. (1997) Structure of keyhole limpet hemocyanin type 1 (KLH1) at 15 Å resolution by electron cryomicroscopy and angular reconstitution. *J. Mol. Biol.* **271,** 417–437.

9. Cyrklaff, M., Roos, N., Gross, H., and Dubochet, J. (1994) Particle-surface interaction in thin vitrified films for cryo-electron microscopy. *J. Microsc.* **175,** 135–142

10. Harris, J. R., Gebauer, W., Guderian, F. U. M., and Markl, J. (1997) Keyhole limpet hemocyanin (KLH), I: Reassociation from *Immucothel®* followed by separation of KLH1 and KLH2. *Micron* **28,** 31–41.

11. Harris, J. R., Gebauer, W., Söhngen, S. M., Nermut, M. V., and Markl, J. (1997) Keyhole limpet hemocyanin (KLH), II: Characteristic reassociation properties of purified KLH1 and KLH2. *Micron* **28,** 43–56.

12. Biosset, N., Penczek, P., Taveau, J.-C., Lamy, J., Frank, J., and Lamy, L. (1995) Three-dimensional reconstruction of *Androctonus australis* hemocyanin labeled with a monoclonal Fab fragment. *J. Struct. Biol.* **115,** 16–29.

4

The Production of Cryosections Through Fixed and Cryoprotected Biological Material and Their Use in Immunocytochemistry

Paul Webster

1. Introduction

Immunocytochemistry is the name given to methods that use antibodies to detect the location of proteins within cells using electron microscopy (EM). The antibodies bind specifically to the protein being investigated and electron-opaque markers visualize their location within the cell. For electron microscopy, it is important to have preparation methods that open the cells and allow access of antibodies to the proteins (or antigens) they bind to without destroying the normal cellular organization (or morphology). If possible, the data obtained from immunocytochemical experiments should be quantitative (*see* **Note 1**).

Successful immunocytochemistry is only possible when antibodies (or other affinity markers) are able to bind to their target ligands. For extracellular antigens, this is not a problem, but for molecules inside cells the affinity markers must gain access. At the moment, the best way to gain access into cells, for electron microscopy, is to cut thin sections through them. In this way, the morphology is retained, but antibodies gain total accessibility to the cut face of the cell. To cut sections thin enough to be examined in the electron microscope (approx 40–100 nm), biological samples must be made hard; this can usually be done by replacing the water within the material with a medium that can be hardened, either by freezing or by polymerization.

The Tokuyasu cryosection technique *(1–7)* is now one of the two most important techniques for subcellular immunocytochemistry, the other being resin embedment and sectioning (*see* Chapters 5, 6, and 7). Cryosections are used for the localization of many different antigens and have been used in many different affinity labeling experiments *(8–15)* including *in situ* hybridization *(16–18)*.

From: *Methods in Molecular Biology*, vol. 117: *Electron Microscopy Methods and Protocols*
Edited by: N. Hajibagheri © Humana Press Inc., Totowa, NJ

For some antigens, it appears to be the most sensitive detection method avail-able *(19)* For other antigens, where antibodies do not label resin sections, it appears to be the only detection method available *(14)*.

The ease and rapidity of sample preparation for cryosectioning make this a convenient way of performing routine immunocytochemical localizations in a busy laboratory environment. However, except for in a few specialized labora-tories, the cryosectioning technique has not yet been routinely adapted. The main reason for this is that it is perceived to be too complex for most research-ers or technicians to master . Poor preliminary results, usually because of poor morphology (*see* **Note 21**), also turn researchers away from further attempts at this relatively simple technique. Recent developments in the design of cryoultra-microtomes, the manufacture of cryodiamond knives, and section retrieval tech-niques *(20)* make it possible to clearly describe a method for cryosectioning for the first time. If this method is followed exactly, even inexperienced operators will be able to successfully produce, label, and contrast cryosections with good morphology for immunocytochemistry (**Table 1**).

Briefly, the biological material is fixed, usually by immersion in a buffered aldehyde solution. The material is then cut into small pieces, infused with 2.3 *M* sucrose, which acts as a cryoprotectant, and then placed onto small speci men pins. The pins, with attached specimen, are frozen by immersion in liquid nitrogen, quickly transferred to a cooled cryochamber fitted onto an ultrami-crotome, and trimmed to a suitable shape. The sections are obtained using a dry diamond knife and collected on the knife surface. The sections are retrieved from the knife by picking them up on a small drop of sucrose/methyl cellulose and transferred onto formvar/carbon-coated specimen grids. The grids are floated on drops of buffered saline with the section side in the liquid. The back of the grid is kept dry at all times and the section side is kept hydrated. The sections on the grids are immunolabeled with affinity marker and then incu bated with a gold visualization probe. Finally, the sections are dried in the presence of a thin plastic film to avoid surface tension damage and examined in a transmission electron microscope.

2. Materials

2.1. Major Equipment

1. Transmission electron microscope.
2. Ultramicrotome equipped with cryoattachment.
3. Antistatic device for ultramicrotome.
4. Vacuum evaporator.
5. Diamond cryoknife.
6. Diamond cryotrim tool.
7. Aluminium specimen mounting pins.

Table 1
Flow Chart for Cryosectioning and Immunolabeling Biological Material

1. Fix fresh, living material with 4% formaldehyde/0.2% glutaraldehyde in 100 mM phosphate buffer. Maximum for 2 h fixation.
2. For tissues, perfusion fix if possible and dissect out carefully to avoid mechanical trauma.
3. For cells carefully scrape from solid substrate, pellet by centrifugation and embed in 10% gelatin.
4. Cut sample into slabs and then into small blocks no bigger than 3 mm³.
5. Transfer blocks into 2.3 M sucrose and leave overnight on a mixing wheel at 4°C.
6. Mount individual blocks onto specimen pins (supplied with cryoultramicrotome) and freeze by immersion in liquid nitrogen.
7. Cool cryochamber of ultramicrotome to –90°C and lock cryotrim tool into knife holder.
8. Transfer specimen pin to cryochamber of ultramicrotome and fasten securely into specimen arm.
9. Use cryotrim tool to smooth off the front face of the block; for this, use 500 nm advance.
10. Trim the sides of the block so that a mesa of approx 1 mm wide by 2 mm long and 1 mm high is formed.
11. Remove cryotrim tool and insert the cryodiamond knife in knife holder.
12. Cool the cryochamber to –120°C and adjust thickness setting for ultramicrotome to section at 80 nm.
13. Advance knife and specimen block and cut sections automatically.
14. Manipulate the sections with an eyelash probe and collect them in small groups on the surface of the knife.
15. Pick up sections on a drop of methyl cellulose/sucrose (1 part 2% methyl cellulose; 1 part 2.3 M sucrose) held in a 2-mm diameter wire loop.
16. Warm up the drop to which the sections are attached and place on a coated specimen grid.
18. Float the grid on a drop of PBS. From now on, keep the section side of the grid wet and the back of the grid dry.
19. Transfer the grid to PBS containing blocking agent (e.g., 10% fetal calf serum). Leave for 5 min.
20. Float grid on small (5 μL) drop of antibody for 20 min.
21. Wash by floating on drops of PBS. Use 5 drops for 2 min each.
22. Incubate on small (5 μL) drop of diluted protein A-gold for 20 min.
23. Wash on 5 drops of PBS for 2 min each.
24. Wash on 4 drops of water, 1 min each.
25. Transfer grid to a drop of methyl cellulose containing uranyl acetate (9 parts 2% methyl cellulose:1 part 3% uranyl acetate). Leave on ice for 10 min.
26. Lift off the grid with a 3.5 mm diameter loop of 3.5 and remove excess liquid using a hardened filter paper.
27. Dry and dislodge grid from loop and examine in the TEM.

8. Eyelashes mounted on long wooden sticks.
9. Small loops (2 mm diameter) for section retrieval, made from copper or platinum wire 0.2 mm diameter and mounted on plastic or wooden sticks. Mount the loops on sticks that are long enough to go into the cryochamber and still be held comfortably outside.
10. Large loops (3.5 mm diameter) for the final drying step. Mount onto sticks or on Eppendorf tips. Each labeled grid is dried on a loop so make at least 20 of them. Loops can be made by wrapping wire around drill bits of suitable diameter.
11. Fine forceps (e.g., Dumont Biologie #5 or #7).
12. EM specimen grids [Hexagonal 100 mesh grids which have been formvar and carbon coated and which have preferably been glow discharged] (*see* **Note 2**).
13. Hardened filter paper.
14. Stereo microscope with light source.
15. Scalpel blades and handles.
16. Liquid nitrogen Thermos flasks with open neck.
17. Liquid nitrogen storage tank (for specimen storage).
18. Parafilm (American Diagnostica, CT).

2.2. Chemicals and Solutions

1. Fixative consisting of 4% formaldehyde containing 0.1% glutaraldehyde buffered in 100 mM sodium phosphate buffer (pH 7.2) (*see* **Notes 3–8**). To prepare, dissolve 8 g of paraformaldehyde powder in 80 mL of distilled water heated to 60°C. The solution will appear cloudy. Add 1 *N* sodium hydroxide dropwise until the solution clears. Cool the solution, add 2 mL of 10% aqueous glutaraldehyde solution, add 100 mL of sodium phosphate buffer, adjust to pH 7.4, and make up to a final volume of 200 mL with distilled water. This stock solution can be kept frozen in small aliquots for long periods. When required, thaw an aliquot and use immediately.
2. Phosphate buffered saline (PBS).
3. 2.3 *M* sucrose in PBS.
4. Fetal calf serum (10% in PBS).
5. Primary antibodies with known antigen specificity.
6. Protein A-gold. Protein A-gold preparations with different particle sizes are required if multiple labeling experiments are to be performed.
7. Secondary antibodies. Rabbit IgG's which recognize mouse or rat IgG's or IgM's (if the primary antibodies are mouse monoclonals—anti-rat antibodies if the primaries are rat monoclonals, etc.) will be required if protein A-gold is to be used to detect all antibodies. This is an economical and simple alternative to purchasing immunoglobulins coupled to colloidal gold. Currently, our protein A-gold is purchased from the University of Utrecht, The Netherlands (*21,22*).
8. 1% glutaraldehyde in 100 m*M* sodium cacodylate (pH 7.2).
9. 0.12% glycine in PBS.
10. 3% uranyl acetate in water.

11. 2% methyl cellulose in water. (Use 25 centipoise viscosity methyl cellulose Sigma M6385.) Stir the powder into hot water. Methyl cellulose is slow to dissolve and dissolves better in cold water, so leave for 2 d at 4°C. When fully dissolved, centrifuge at high speed (e.g., 60,000 rpm in a Beckman 60Ti or 70Ti rotor) for 60 min. at 4°C. The methyl cellulose should be removed without disturbing the pellet at the bottom of the tube and should be kept cold at all times.

3. Methods

3.1. Formvar Coating Specimen Grids

All successful TEM experiments rely totally on the quality of the support film covering specimen grids. These films must be strong, clean, and remain attached to the specimen grids even through many manipulations. As there is still some mystery in the methods used to prepare these grids, with some very unscientific methods being used, this protocol will clearly define a reproducible method.

1. Clean 100 mesh, hexagonal specimen grids. Place them in a 200-mL glass flask, rinse with 2% aqueous acetic acid, multiple washes with distilled water, and finally with 3 changes in acetone. Dry over a warm plate.
2. Clean a glass slide by immersing it in ethanol, instantly removing it and drying with lens tissue. Use "precleaned 25 × 75 mm micro slides selected" Catalogue #48312-002 from VWR Scientific, Inc., Media, PA.
3. Dip slide into a solution of formvar dissolved in chloroform (2% w/v) and withdraw carefully. Speed of withdrawal will affect film thickness.
4. Leave slide to dry vertically.
5. Score around slide on all four sides with a razor blade to define limits of film to be dislodged from slide.
6. Fill a deep glass dish with distilled water and place a light so that it reflects on the water surface.
7. Clean the water surface by passing a piece of lens tissue over the surface.
8. Hold the formvar-coated slide at a 45° angle at the edge of the glass dish and with the water just touching the bottom of the slide. The formvar film should start to come away from the glass.
9. Slowly push the slide into the water, watching the film as it lifts off the glass and floats on the water surface.
10. Remove the glass slide from the water and examine the floating film. The incident light should show interference colors in the film. These should be silver or gray. Yellow, purple, or green colors indicate that the formvar solution should be diluted more by adding chloroform. Films that are too thin mean that the formvar solution is too dilute. To increase the thickness, either remove the glass slide from the solution more slowly or increase the concentration of formvar. Addition of a stock solution of 4% formvar in chloroform will increase the concentration of the coating solution.

11. Place cleaned specimen grids onto the floating film until it is filled. Place the grids so that the shiny surface of the grid is facing up and the dull side is on the film.
12. Dry the glass slide used for casting the film and attach a 1 in. by 2.5 in. white address label to it.
13. Hold the glass slide vertically over one edge of the floating film and push it into the water so that the film with the attached grids stick to the address label.
14. Once the glass slide has been totally immersed remove it, with the grids attached and leave to dry.
15. Coat the grids with carbon using a vacuum evaporator.
16. Although not essential, the grids can be made hydrophilic prior to use using a glow-discharge apparatus in the vacuum evaporator.

3.2. Chemical Fixation of Sample

If we are to study the location of molecules within cells, then some form of immobilization must take place prior to labeling. Molecules that are not immobilized may be displaced from their normal location resulting in either a false localization or a negative result (if the molecules are washed away, for example).

The easiest method of immobilization is by chemical fixation and aldehydes have classically been used for this purpose *(6)*. In addition to immobilizing antigens in place, the fixative will also help preserve specimen morphology (*see* **Notes 3–8**).

For immunocytochemistry, tissues or cells can be fixed by immersion in a buffered solution of aldehyde using the same methods that are used for Epoxy resin embedding. The pieces of material should be small for fast aldehyde penetration but should not be damaged by over manipulation (squeezing, pulling, cutting, etc.). Large organs are best fixed by perfusing with warm fixative.

The aldehyde to be used, and the concentration, depends on the antigen to be studied. For example, although glutaraldehyde produces better cellular morphology, some antigens are not recognized by antibodies after this fixation. Other antigens seem unaffected even by high concentrations of glutaraldehyde. So-called "insensitive" antigens can be fixed in glutaraldehyde concentrations of up to 1% for 15–60 min. For "sensitive" antigens, formaldehyde concentrations of up to 8% can be used.

3.2.1. Fixing Tissues

1. Fix tissue by perfusion with the fixative. Routes of perfusion and perfusion conditions will vary with the tissue under examination.
2. Carefully dissect out the tissue under study once the perfusion has been completed and transfer it whole in fresh fix. Leave for 1 h. Any pulling or squeezing at this stage may adversely affect the subcellular morphology of the tissue. Do not mince the tissue as is sometimes recommended.

3. Transfer the tissue to 100 mM phosphate buffer and slice into 1–2-mm thick slabs using two sharp, clean razor blades. Use one blade to immobilize the tissue by applying slight pressure and gently cut into the slice with the second blade. It is important to avoid any squeezing or pulling. This may damage or alter subcellular morphology.

4. Use similar cutting techniques, with two blades, to slice the slabs into rods and then cut the rods into small cubes approx 1–3 mm^3.

3.2.2. Fixing Cells

1. Add double strength fixative (8% formaldehyde/0.4% glutaraldehyde, ratio 1:1) to the cell culture. Cell suspensions can be fixed in suspension and immediately pelleted by gentle centrifugation. If possible do not centrifuge cells prior to fixation. The presence of extracellular proteins (e.g., fetal calf serum) does not normally affect intracellular labeling.

2. For cells grown on culture dishes or in flasks, fix by the addition of double strength fixative to the culture medium. Carefully scrape the cells off the substrate using a soft wooden stick or piece of Teflon that has been beveled on one edge to form a blade (*see* **Note 9**).

3. Pellet scraped cells by centrifugation (*see* **Note 10**).

3.3. Embedding in Gelatin

If the cell pellet can be manipulated without falling apart, then ignore this step and move onto the cryoprotection step. If the pellet is loose or is falling apart, it will be necessary to embed in gelatin (*see* **Note 11**).

1. Remove the fixative supernatant from the cell pellet.

2. Resuspend the pellet in a small volume of 10% gelatin dissolved in water. Because the gelatin is still warm, repellet the cells. If the gelatin sets before the cells have pelleted, warm the tube until the gelatin becomes liquid again and recentrifuge.

3. Cool the pellet in ice for 30 min or until the gelatin has solidified.

4. Cut off the tip of the tube and remove the gelatin-embedded pellet.

5. Slice the pellet into slabs, cutting away any gelatin not containing cells.

6. Slice the slabs into rods and then into small blocks using the same method as for the tissue pieces (*see* **Note 12**).

3.4. Cryoprotection and Freezing

1. Transfer cubes of tissue or cell pellets to 1 mL of 2.3 M sucrose in an Eppendorf tube, seal the cap and leave overnight on a rotating wheel at 4°C (*see* **Note 13**).

2. Transfer each infiltrated sample block onto a clean specimen stub, remove excess sucrose with a damp filter paper and immerse the specimen stub with sample block in liquid nitrogen until freezing is complete (**Fig. 1A**; *see* **Note 14**).

3.5. Specimen Transfer to Precooled Cryochamber

1. Precool the cryochamber of the ultramicrotome to –90°C.

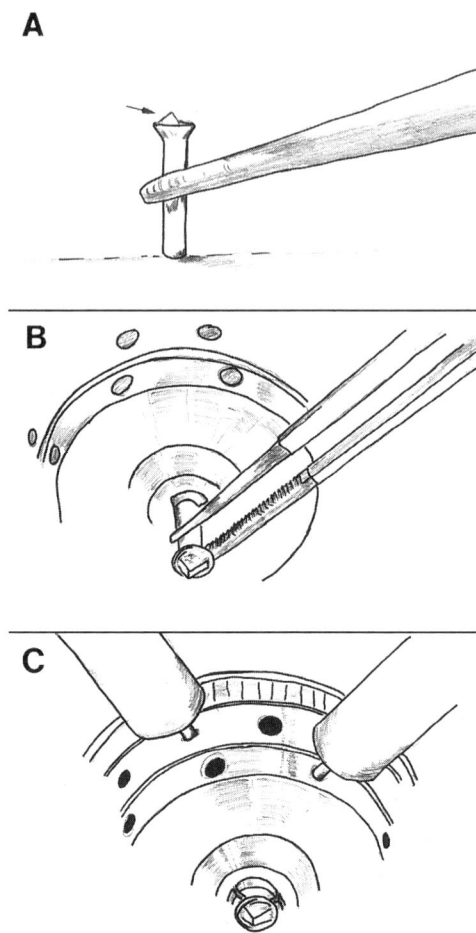

Fig. 1. Mounting the specimen and inserting it into the cryochamber. (**A**) The sample (arrow), cryoprotected by immersion in 2.3 M sucrose and trimmed to a convenient size and shape, is mounted on a metal specimen pin ready for immersion freezing in liquid nitrogen. (**B**) The specimen pin, with the frozen sample attached, is inserted into the precooled sample holder of the ultramicrotome. This is carried out without warming the sample. (**C**) The sample holder is tightened, holding the specimen pin in place (*see* **Note 22**).

2. Insert the diamond cryotrim tool and allow to reach equilibration temperature.
3. Rapidly transfer the specimen, on the specimen stub, from the liquid nitrogen using precooled forceps and quickly place it into the precooled cryochamber (**Fig. 1B**).
4. Place the specimen stub into the specimen holder and lock it in place using minimal force (**Fig. 1C**). Any tools inserted into the cryochamber, and especially those which are used to manipulate the specimen, must be precooled in liquid nitrogen.

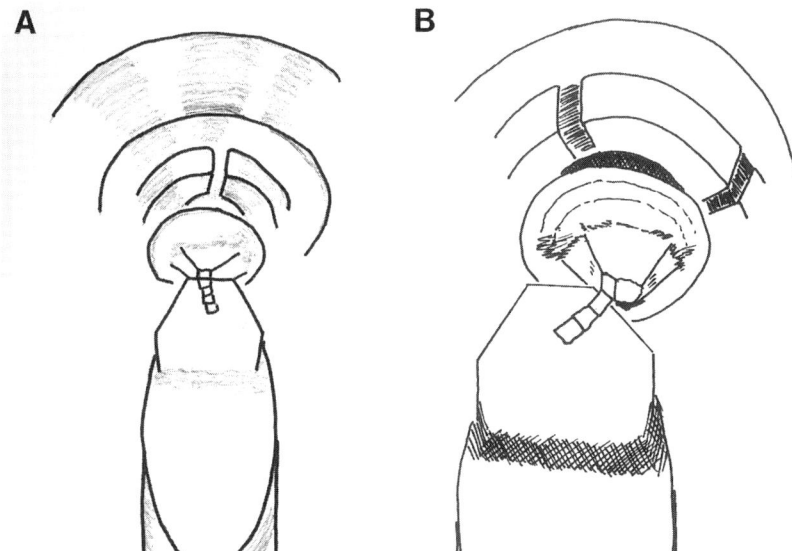

Fig. 2. Trimming the sample block. (**A**) Use a diamond trim tool to cut into the front of specimen block. (**B**) The side of the trim tool can be used to shape the edges of the block into a rectangular shape. A block with smooth horizontal sides parallel to the knife, and smooth vertical sides perpendicular to the knife, will section best.

3.6. Trimming the Block

1. Carefully advance the front of the trim tool to the surface of the block until it is just touching the front of the block. First time users may benefit from a few lessons from more experienced ultramicrotome operators at this step (*see* **Note 15**).
2. Set the ultramicrotome to cut sections of 500 nm (5 μm) and leave to section at maximum speed. The Ultracut S, which will switch off when the end of the feed is reached, can be left to trim unattended.
3. Once a substantial portion of the block (revealing an area of between 1–3 mm^2 has been smoothed away ("faced-off"), stop sectioning (**Fig. 2A**).
4. Use the edge of the trim tool to shape the front of the block into a "mesa" protruding approx 1 mm from the block surface, approx 1 mm wide and 2 mm long (**Fig. 2B**) (*see* **Note 15** for a detailed description of the trimming process).

3.7. Sectioning

1. Mount the diamond knife in its holder, place in the cryochamber and leave to cool down.
2. Cool the cryochamber to –120°C and leave to equilibrate.
3. Carefully move the trimmed block to the knife edge and set the cutting window so that the section stroke starts just above the knife and ends just below the knife.

4. Set the ultramicrotome to cut sections of 80 nm thickness.
5. Use the rough advance to bring the knife close to the block (about 1–2 mm away) then use the fine advance to either move the knife or the block. When it seems as if the knife and the block are almost touching, start sectioning by hand, using the specimen advance, to bring the knife and the block together. Switch over to automatic sectioning when sections appear on the knife (**Fig. 3A**).
6. Set the cutting speed to between 0.6 mm/s and 2 mm/s. The slow cutting speed may be combined with a fast return cycle in order to reduce the possibility of thermal changes occurring in the specimen (*see* **Notes 16–17**).

3.8. Manipulating the Sections

Once sections have been obtained the sections will lie on the knife surface close to the edge. They can be carefully moved away using an eyelash mounted on a stick (**Fig. 3B**). It is important to separate sections that are lying on top of each other and to arrange them into small separate groups ready for picking up. When using the eyelash, it may be best to hold the sections down rather than trying to move them with the eyelash positioned under them.

3.9. Section Retrieval

1. Dip a 2-mm diameter loop in a mixture of one part 2% methyl cellulose: one part 2.3 *M* sucrose and lift out a drop on the loop *(20)*. Keep the methyl cellulose/sucrose mixture in a small tube on ice (*see* **Note 18**).
2. Transfer the loop to the cryochamber and, looking through the binoculars, direct the drop so that is just above a group of sections. Move the drop to the sections and pick them up onto the drop (**Fig. 3C**).
3. Remove the loop from the cryochamber, thaw and place the drop, section side down, onto a formvar/carbon coated specimen grid. The drop can be left to dry on the grid until it is ready for labeling *(23)*.

3.10. Immunolabeling of Thin Sections for EM

Thin sections of biological material, mounted on specimen grids, can be easily labeled by floating them section-side down on small drops of antibody. This method will work for sections through resin-embedded material as well as for thin, thawed cryosections.

1. Attach a long piece of parafilm to a bench surface with a drop of water. As this will form a clean surface on which to label and wash, only expose as much of the Parafilm as needed and do not contaminate the clean parafilm surface.
2. Float specimen grid, section side down on a drop (100 μL) of PBS placed on the clean parafilm surface (**Fig. 4A**). Leave for 5 min. It is important now to always keep the section side of the grid wet and the back surface dry.
3. Transfer, and float, the grid on a drop of PBS containing 10% fetal bovine serum (blocking solutions; *see* **Note 19**). Use 3.5 mm diameter loop to transfer grid. Leave 5 min.

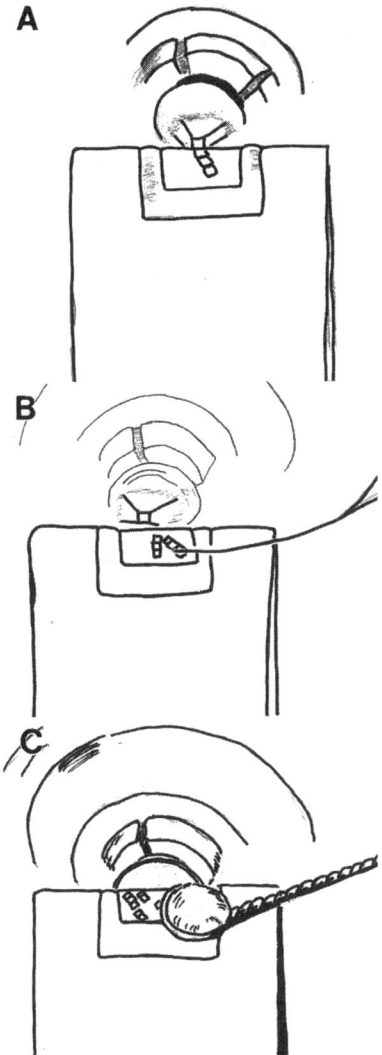

Fig. 3. Sectioning and section retrieval. (**A**) The trimmed block is carefully brought to the diamond knife and the ultramicrotome is left to section automatically (*see* **Note 22**). (**B**) Move accumulated sections from the edge of the knife using an eyelash probe and arrange them in small groups. Make sure the sections remain as a single layer and keep the number of sections in each group small enough to fit on the pick-up drop. (**C**) Use a drop of 1 part 2% methyl cellulose: 1 part 2.3 *M* sucrose loaded onto a wire loop to pick up a group of sections. Do not attempt to pick up multiple groups. Remove the loop from the cryochamber, thaw and place the drop, section side down, onto a formvar/carbon coated specimen grid.

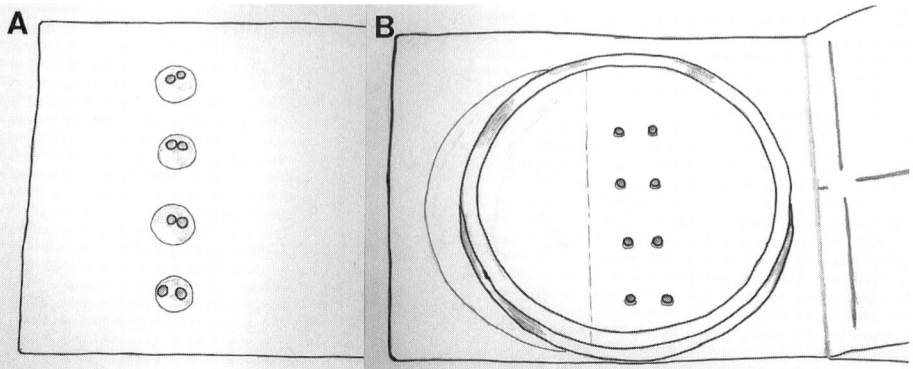

Fig. 4. Antibody incubation. (**A**) Wash away the pick-up solution from the surface of each grid by floating them, section side down, on a drop of PBS. Multiple 100 μL drops can be placed on a clean layer of parafilm attached to a level surface. Clean Parafilm is exposed only as it is needed. (**B**) After washing, transfer each grid to single 5 μL drops of diluted antibody solution on the Parafilm surface. Cover the drops with a Petri dish and place a wet filter paper under the dish to prevent the antibody drops from drying.

4. Use forceps to transfer grid to a 5-μL drop of rabbit antibody diluted in blocking solution (*see* notes for details on antibody dilution). The antibody, diluted to a suitable concentration, is centrifuged prior to use. This will remove any aggregates formed during storage.
5. Place a wet piece of filter paper close to the antibody drop, cover with a plastic dish and leave for 20 min (**Fig. 4B**) (*see* **Note 20**).
6. Remove the grid from the antibody with forceps and float on PBS. Wash by transferring the grid to five drops of PBS, leaving the grid on each drop for 2 min.
7. Use forceps to transfer the grid to a 5-μL drop of protein A-gold diluted in blocking solution.
8. Again cover with a plastic dish and leave for 20 min in a moist atmosphere.
9. Remove the grid with forceps and wash by transferring the grid to 5 drops of PBS, leaving the grid on each drop for 2 min.
10. Wash with water, 4 changes, 1 min each. Although only a short wash, this step removes salts prior to incubation with uranyl acetate. Phosphates present in the sections will precipitate the uranyl salts.

3.11. Contrasting and Drying of Cryosections

This step is essential for fine structure preservation as well as for producing enough contrast to visualize subcellular detail in the electron microscope. Many different contrasting methods exist and all can be tried using hydrated cryosections mounted on specimen grids. Here is a simple, reproducible method.

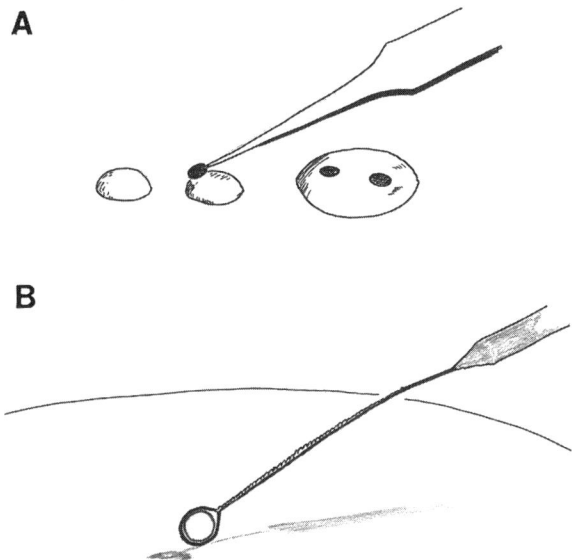

Fig. 5. Final contrasting and drying. **(A)** Place drops of cold 2% methyl cellulose containing 0.3% uranyl acetate onto a clean Parafilm surface, on ice. Transfer the labeled grids from the final water wash with forceps and touch each onto the first two drops to wash away excess water. Leave each grid on the final, larger drop, for 10 min. **(B)** Remove each grid individually from the cold methyl cellulose/uranyl acetate solution with a wire loop. Place the loop onto a hardened filter paper and let excess solution drain off. If the grid falls out of the loop it can be rapidly transferred back to the solution on ice with forceps and the process repeated.

1. Mix 2% methyl cellulose with a 3% aqueous uranyl acetate solution to give a final concentration of 0.3% uranyl acetate (nine parts methyl cellulose to one part uranyl acetate). Mix immediately prior to use and keep the solution on ice at all times.
2. Put drops of this solution onto a clean surface (Parafilm) on ice and float the grids, sections down, on the drops for 10 min. To wash off water from the previous step, transfer the grids over two small drops onto a larger drop for final incubation (**Fig. 5A**).
3. Loop off each grid individually from the methyl cellulose-uranyl acetate solution using a 3.5-mm diameter wire loop.
4. Remove excess liquid from the loop by placing the loop at a 45° angle onto a clean hardened filter paper, with the section side facing down toward the filter paper (**Fig. 5B**). Once the methyl cellulose starts to be absorbed by the paper, gently drag the loop away from the wet patch until no more liquid is removed. Leave the grid in the loop to dry. Films of optimal thickness have gold to blue interference colors. When the film is dry, the grids can be carefully removed from

the wire loop using pointed forceps and stored section-side up or immediately examined in the transmission electron microscope (*see* **Notes 20–21**).

4. Notes

1. Limitations of thin section immunocytochemistry:A good immunocytochemical protocol should be quantitative. This means that the amount of label present on the section should be directly related to the amount of antigen present in the sample. If only small amounts of antigen are present then only small amounts of signal should be present. The amount of antigen accessible to antibody on the surface of a well-fixed thin cryosection may be up to x100 less than the amount accessible in a permeabilized paraffin embedded section or a 3–10-μm thick section obtained using a cryostat. Evaluation of the signal obtained in thick sections or whole cells is recommended before progressing to a study using electron microscopy. If only small amounts of antigen are present or the antigen is located to small areas within the sample, then electron microscopy localization may not be possible without extensive sectioning and examination.

2. Specimen grids are made from many different materials. The most common ones are made of copper or nickel. Although, we routinely use copper and nickel grids, other researchers have found disadvantages with both. If the formvar grids are damaged copper grids react with salts in the buffers to produce precipitates on the carbon film (**Note 20**). Nickel grids are easily magnetized making them difficult to manipulate and potentially cause astigmatism when being examined in the electron microscope. If preferred, gold or palladium grids can be used).

3. Fixatives: All chemical fixatives used for chemical crosslinking are hazardous materials. They will crosslink any biological material they come into contact with, so they must be treated with great care. Fumes from fixatives have similar fixation properties.

4. Aldehyde fixation: Tissue pieces should be handled carefully throughout the whole process so that ultrastructural damage does not occur from squeezing, pulling, or shaking.

5. Aldehyde fixation:Whether using formaldehyde, glutaraldehyde, or a mixture of the two, it is important to note three things as follows.
   a. Unlike glutaraldehyde, when used at low concentrations (< 4%) formaldehyde-linking is partly reversible. It is important to avoid extensive washing after fixation with formaldehyde alone.
   b. Again, unlike glutaraldehyde, the crosslinking reactions of formaldehyde occur much slower. For this reason, it is best to leave the cells or tissues to be fixed for a longer time than when using glutaraldehyde.
   c. Formaldehyde exists in solution as monomers and polymers and it is believed that polymers are more active at crosslinking. The polymers are present in higher numbers in more concentrated solutions of formaldehyde. The polymers also form when formaldehyde solutions are cooled to 4°C. It is possible, therefore, that short fixation times can be used with high concentrations of formaldehyde. *See* **ref. 6** for a detailed discussion of chemical fixaiton.

6. Aldehyde fixation: Material can be fixed in formaldehyde alone. We routinely use 4% formaldehyde in sodium phosphate buffer (pH 7.4), but other buffers will work. One protocol for formaldehyde fixation that has worked is to fix in 4% formaldehyde (made from paraformaldehyde powder) for 30–60 min and then overnight in 8% formaldehyde. Formaldehyde-fixed material should be stored and transported in formaldehyde.

7. Fixation: Leaving tissue samples in glutaraldehyde for longer than 2 h will make them brittle and difficult to cryosection.

8. Fixing Tissues: If perfusion fixation is not possible, then immersion fixation is an available option. However, it is important to transfer the sample to the fixative as rapidly as possible after removal and to avoid all forms of mechanical trauma. If necessary, the tissue can be carefully sliced into small slabs as soon as it has been immersed in fixative. This will allow for rapid access of fixative to the tissue. Again, care must be taken to protect fixed and unfixed tissue from any form of mechanical trauma unless this is part of the experiment.

9. Removal of cells on solid substrates: An alternative way of removing cells from culture dishes is to treat them with proteinase K. Be aware, however, that this may result in the removal of surface antigens. The cell monolayer is treated on ice for a few minutes with 20–50 µg/mL proteinase K in PBS. For most cell types, the lower concentration works well. Cool the solution and cells on ice, mix the two and gently pipet the proteinase K solution up and down until cell free areas appear on the plastic. The solution, which will become turbid, can be removed and centrifuged. Some fixative should be added to the cell suspension prior to centrifugation (to preserve the morphology) and a protease inhibitor can be added (PMSF works well) to inhibit the proteinase K activity.

10. Cell Pelleting: Pelleting of small numbers of cells is best performed in the presence of protein (BSA or FCS will work). Add a small amount (1–2%) to the suspension of cells in fixative just prior to centrifugation. The protein will stop the cells from sticking to the sides of the tube during centrifugation.

11. Embedding in gelatin: Gelatin is supplied by many manufacturers and comes in many grades of quality. If prepared in phosphate buffer, the salt may occasionally precipitate out of solution and the crystals cause problems during sectioning. Centrifugation of warm gelatin will remove these crystals. It is for this reason that dissolving gelatin in water is preferred. However, a good source of gelatin, which can be dissolved in phosphate buffer without forming precipitates, is that used in cooking. Large amounts can be obtained at low cost from food stores.

12. Preparing small blocks of specimen: Ideally, the pieces of tissue or cell pellets should be trimmed to a pyramidal shape so that when mounted onto the specimen stub the base is wider than the top. If possible, the block face should be flat and resemble a square, rectangle, or trapezoid.

13. Sucrose Infiltration: Although an overnight infiltration in 2.3 *M* sucrose is recommended, it is possible to leave tissue pieces and cell pellets in the sucrose for much shorter times (< 30 min). However, uneven infiltration may occur that will make the sample more difficult to section.

Table 2
Effect of Temperature on Sectioning Quality

								Thickness setting (nm)										
		900	800	700	600	500	400	300	200	100	90	80	70	60	50	40	30	20	10
Sectioning	−90	X	X	X	X	X	X	X	X	X									
Temperature	−100			X	X	X	X	X	X	X	X	X							
(in °C)	−110							X	X	X	X	X	X						
	−120									X	X	X	X	X	X				
	−130										X	X	X	X	X	X			
	−140											X	X	X	X	X	X		

X—indicates that sections can be obtained.

Blocks of rat liver fixed, by perfusion, in 2% formaldehyde containing 0.2% gluttaalrdehyde and buffered in 100 mM sodium phosphate buffer (pH 7.4; phosphate buffer), were cryoprotected by immersion in 1 mL of 2.3 M sucrose in phosphate buffer and incubating on a rotating wheel for 24 h at 4°C. The blocks were frozen on specimen pins and sectioned using an Ultracut UCT with attached FCS cryosectioning system (Leica, Deerfield, IL), and a 3 mm cryodiamond knife (Harris Diamond Co. Drukker International). Similar results (not shown) were obtained using a 3 mm cryodiamond knife supplied by Diatome US (Ft. Washington, PA). This is a subjective experiment indicating settings on the ultramicrotome that will easily produce sections. The outer limits of sectioning at each temperature may require some technical care and patience. As a general rule, as the sections become thinner, the greater the folding that occurs. If the section folding is too great then the cryochamber and specimen block should be cooled futher. An antistatic line in the cryo chamber can be used to help flatten sections when very thin sections are being produced. With care, it may also be possible to section beyond these limits.

14. Freezing: Some sucrose should be left around the base of the block so as to anchor the block to the stub. It is also important not to allow the specimen to dry out. Not only will higher sucrose concentrations produce softer blocks, but there is also a possibility of damage to subcellular morphology occurring within the sample. Sometimes during extreme drying, it is possible for small crystals of sucrose to form within the tissue. These will prevent proper sectioning and may even damage the diamond knife.

15. Trimming the block: Start by positioning the block in front of the trim tool and cut into the block until a flat surface is produced. Then position one corner of the trim tool at one side of the block and cut 500 nm increments into the block. Move the trim tool to the other side of the block and repeat the trimming. Turn the block one quarter turn clockwise and repeat the side trimming process. Turn the block one quarter turn counterclockwise and examine the block. Make sure the face of the block is mirror smooth and that the sides of the block are parallel to each other and perpendicular to the knife edge. The top and bottom of the block should be parallel to the knife edge. The edges of the block should be sharp, showing no signs of chipping, and clear of debris.

16. Sectioning: The most important factor to obtaining good cryosections is the quality of the knife. Until recently, glass knives were the preferred tool for production of cryosections. Recent improvement in diamond knife production means they can now be used routinely for the production of good cryosections. However, knife-making machines are commercially available and the theory behind glass knife making has been covered in great detail *(7)*. A simple modification to an older knife-making machine, commonly found in most electron microscopy laboratories, is also available if required *(24)*. More complex modifications, requiring the use of a machine shop, are also available *(25)*.

17. Sectioning: Section thickness can be varied through a wide range by manipulating the temperature of the cryochamber (*see* **Table 2**). For any particular sucrose concentration, cooling the block will make it harder and more brittle. Warming the block will make it softer and more plastic.

18. Section Retrieval: A drop of 2.3 *M* sucrose can be used in place of the methyl cellulose/sucrose mixture. The drop, on a small wire loop is placed into the cryochamber and either used immediately to pick up the sections or left to cool down before picking up the sections. If used immediately, the sections will spread out on the drop. This is useful in removing the compression artifact produced during the sectioning procedure *(5)*. Too much spreading, however, can damage the morphology of lightly fixed material. The surface tension on the sucrose droplet may pull apart tissue sections or cell pellets. Alternatively, drops of 2% methyl cellulose alone, or 2% methyl cellulose containing 0.3% uranyl acetate can be used *(20)*. The methyl cellulose drops have a lower surface tension and will not spread out the sections when they are picked up. This results in improved morphology of the sample.

19. Blocking: Any "sticky" protein can be used to cover nonspecific binding sites on the thawed sections. Alternative blocking agents include ovalbumin, bovine serum

albumin, bovine gelatin (2%) or cold water fish skin gelatin. Be aware that the chosen protein must not react with any of the affinity reagents to be used (e.g., rabbit serum cannot be used with protein A gold; bovine serum (FBS) cannot be used as a blocking agent if antibodies to serum proteins being studied).

20. Immunolabeling: There are many causes for poor results in immunocytochemistry and below are a list of common problems found when using cryosections for immunolabeling reactions, and some possible solutions. (*see* **refs.** *6* and *26* for detailed discussions of immunolabeling) (*see* **Table 3** for annotated labeling protocols).

    a. No Label Detected.

       Possible Causes:

         i. Antigen destruction, masking or extraction: The antigen may not have been fixed enough, allowing it to be washed away during processing. Demonstrate antibody labeling by light microscopy and use the same preparative conditions for electron microscopic labeling. Some antigens are sensitive to glutaraldehyde, but not formaldehyde. Other antigens may be masked by other proteins leaving the antigens inaccessible to the primary antibody, protein A-gold or both, unless the cells are extracted with detergents. Occasionally, antigens will not be immobilized by the chemical fixative and will either wash out of the section or be redistributed during the immunolabeling procedure. For such antigens, it may be best to immunolabel sections through resin-embedded material *(27)*.

       ii. Primary antibody does not work: Causes of this problem may be the result of a wrong dilution of the antibody (do serial dilutions). Make up the working dilutions fresh for each experiment and do not store diluted antibody, rather, only make up the amounts needed for each experiment. Inadequate storage procedures producing repeated freeze-thawing, or growth of contaminants may also result in a deactivation of the antibody. Store frozen in small aliquots or in 50% glycerol containing 0.01% sodium azide to prevent contamination. If antibodies must be stored at 4°C, then include sodium azide or chloroform to prevent microbial contamination. Prepare positive controls using other methods to demonstrate specific labeling. Change the antibody if necessary.

      iii. Low amount of antigen: One of the advantages of using protein A-gold to detect antibody binding is that it can be quantitative. In some systems, the amounts of antigen present will be very low. Most often an estimation of antigen amounts can be obtained from biochemical experiments. If these amounts are low, then the actual amount of labeling detected will also be low. Increasing the signal will be difficult in these cases unless antigen masking is occurring. Increasing antigen accessibility by detergent treatment may help.

      iv. The magnification of the image is too small to detect the gold particle: Colloidal gold probes can be made in many sizes. Some multiple labeling experiments require the use of 5 nm or 4 nm gold particles. These

Table 3
Labeling Protocols Using Specific Antibodies and Protein A-Gold

A. Antibodies that will bind protein A-gold.

1. 10% FBS in PBS	5 min.
2. Primary antibody	20 min.
3. Wash in PBS	5 × 2 min.
4. Protein A-gold	20 min.
5. Wash in PBS	5 × 2 min.
6. Wash in water	4 × 1 min.

B. Antibodies that do not bind protein A-gold. Using a second, bridging antibody that will bind to the primary antibody and will bind protein A-gold.

1. 10% FBS in PBS	5 min.
2. Primary antibody	20 min.
3. Wash in PBS	5 × 2 min.
4. Bridging antibody	20 min.
5. Wash in PBS	5 × 2 min.
6. Protein A-gold	20 min.
5. Wash in PBS	5 × 2 min.
6. Wash in water	4 × 1 min.

C. Double label using two primary antibodies produced in rabbits. Two antibodies that will bind to protein A-gold *(12,21,22)*.

1. 10% FBS in PBS	5 min.
2. 1st. Primary antibody	20 min.
3. Wash in PBS	5 × 2 min.
4. Protein A-gold (1st. size)	20 min.
5. Wash in PBS	5 × 2 min.
6. 1% glutaraldehyde	10 min.
7. 0.12% glycine in PBS	3 × 2 min.
8. 10% FBS in PBS	5 min.
9. 2nd. Primary antibody	20 min.
10. Wash in PBS	5 × 2 min.
11. Protein A-gold (2nd. size)	20 min.
12. Wash in PBS	5 × 2 min.
13. Wash in water	4 × 1 min.

are very small and require that the screen magnification in the electron microscope be approx x18,000 to x25,000. Gold particles less than 3 nm in particle size are extremely difficult to visualize in the transmission electron microscope without specialized detection methods.

   v. The protein A-gold does not bind to the primary antibody: Many antibodies, especially mouse monoclonals, do not have binding sites for protein A. If this is so, then include a bridging antibody (one that will recognize the primary antibody and bind protein A) between the primary antibody incubation and the protein A-gold step, or use secondary antibody coupled to gold particles.

vi. The image contrast is too strong: Small gold particles cannot be easily seen on images of resin sections that have been contrasted with large amounts of heavy metal stains. To improve the visibility of the smaller gold particles, reduce the specimen contrast as needed.

vii. The diluted gold preparation has been centrifuged: Unlike the primary antibodies, the gold suspensions should not be centrifuged prior to use. Centrifugation will remove the gold probe from the buffer resulting in a dilution of the reagent. If overcentrifuged, or if the gold particles are large, then most of the reactive probe will be removed from the buffer. This will result in lower levels of labeling than expected, the amount of label varying between experiments. In extreme cases, no labeling will be detected.

b. Background (more gold particles than expected): Background labeling is that signal produced by the primary antibody or visualization probe which is judged to be nonspecific labeling.

Possible Causes:

i. Antibody preparation: Many background problems arise from the primary antibody. Suitable negative controls will identify this.

Control experiments;

i)   incubate in the absence of primary antibody only.

ii) incubate with antibodies from the same species as the primary antibody but which do not react with the cells or tissue under study.

If the antibodies are producing background labeling then it may be sufficient to treat the sections with blocking agents dissolved in PBS to prevent nonspecific binding. These include 1% gelatin (calf skin or fish skin), 2% bovine serum albumin, 5–10% fetal bovine serum or 1% ovalbumin. Additionally, specimens that have been fixed with aldehydes can be treated with primary amines which will quench any free aldehyde groups which could crosslink antibodies. These can be dissolved in PBS, and include glycine (0.15%) or ammonium chloride (50 mM). A 10 min incubation is sufficient.

If background persists then try diluting the antibody a little more. The best antibody dilution is that which produces as strong a specific signal as possible with a low background signal. Often this means using the primary antibody as concentrated as possible and accepting some background labeling.

Background labeling can also be caused by using sera instead of purified immunoglobulin fractions. The primary antibodies that usually produce the best results are affinity purified IgG fractions.

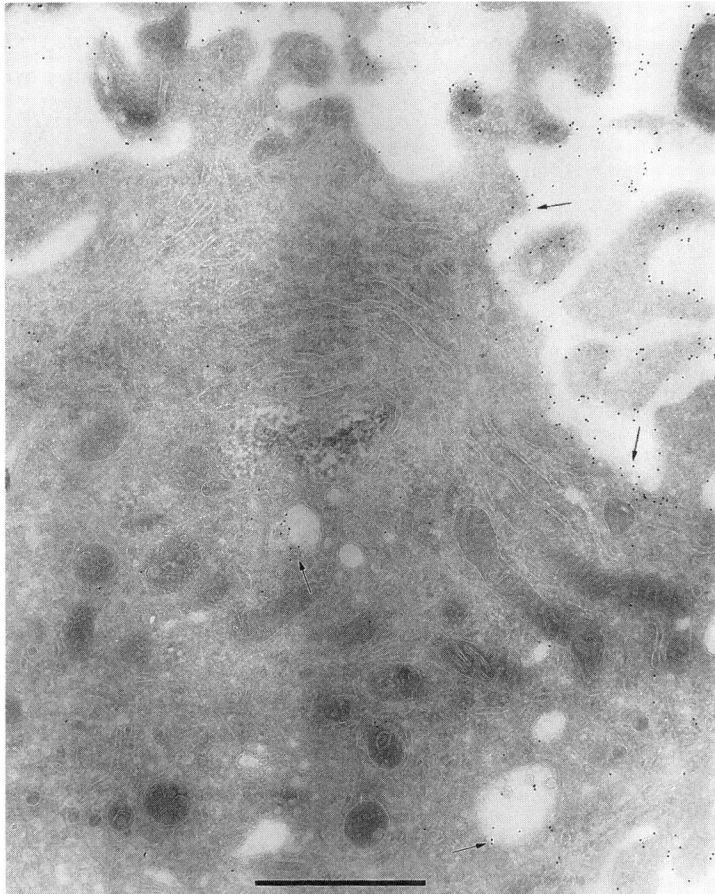

Fig. 6. Micrograph of a cryosection through a bone marrow dendritic cell. Prior to fixation, the cell was biotinylated, on ice, for 10 min and then warmed to 37°C for 30 min. After fixation with 4% formaldehyde buffered in 100 mM phosphate buffer and sectioning as described in this chapter, the section was sequentially labeled with antibodies to biotin and with 10 nm protein A-gold. Morphology is preserved in this fragile cell preparation using the cryosectioning method described. Surface and internalized biotin is easily visualized with the colloidal gold probe (arrows). Bar = 1 μm.

If the background labeling cannot be removed, it is possible that it is a specific signal. Check the results with other labeling procedures to confirm this. Specific signal in polyclonal antibodies can often be removed using affinity adsorption techniques.

ii. Background caused by blocking agents: Sometimes nonspecific binding of protein A-gold can be caused by using protein A-binding molecules as blocking agents. Rabbit or pig serum, if used as a blocking

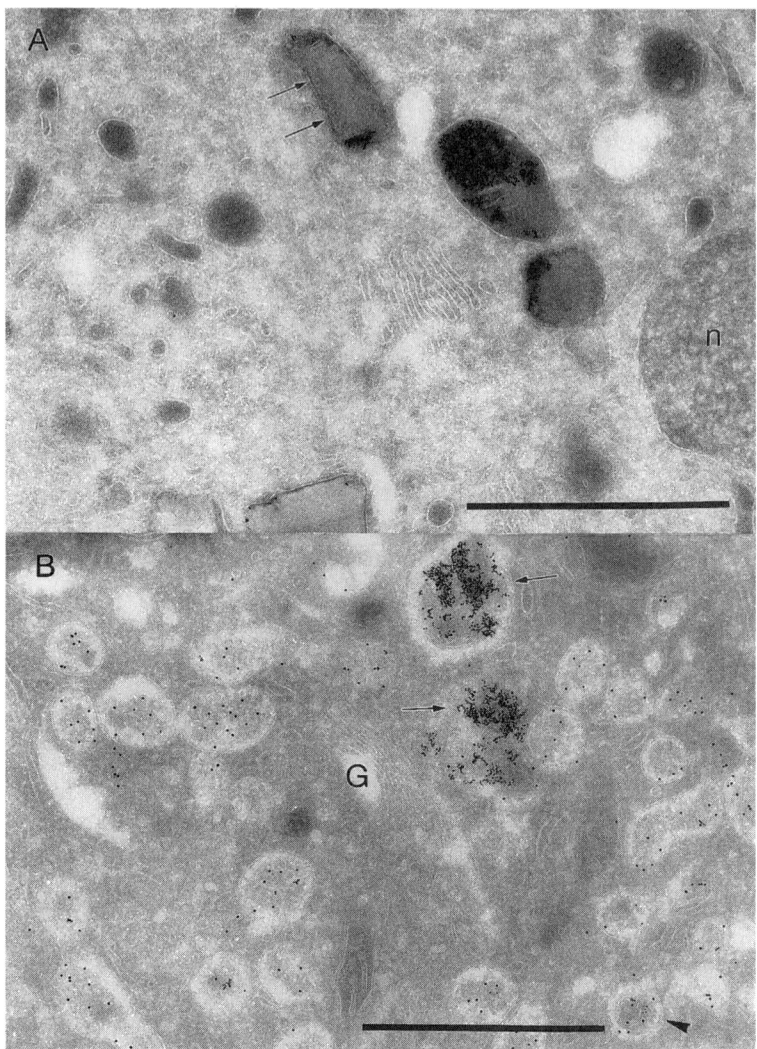

Fig. 7. Micrographs of cryosections through bone marrow dendritic cells obtained using the described cryosectioning procedure. The cells were incubated with BSA-gold particles for 30 min at 37°C, washed and then incubated for a further 3 h in the absence of the gold probe prior to fixation. This protocol, designed to load lysosomes with endocytosed gold, results in the small BSA-gold particles being found in many intracellular structures. **(A)** The BSA-gold is found in structures close to the nucleus (n) which contain crystalline structural arrays (arrows) that are not visualized in resin-embedded material. Bar = 1 μm. **(B)** The small BSA-gold particles aggregate within large multivesicular structures (arrows). This section was labeled with polyclonal antibodies to MHC class II molecules (visualized using 10 nm particle size protein

agent on the sections, will bind protein A-gold and produce a nonspecific signal.

iii. Antigen migration: If the antigen has not been properly immobilized by the fixation step it may become redistributed throughout the cell prior to its being washed away *(27)*. Try a different fixation protocol.

iv. Protein A-gold: Background labeling can be caused by using too high a concentration of protein A-gold. Diluting the protein A-gold will remove this background. Some preparations of protein A-gold will produce background if they are not diluted in PBS containing blocking agents. Addition of a suitable (i.e., nonprotein A-binding) blocking agent will remove this background.

v. Background is specific signal: If all attempts to remove background labeling have failed then it is possible that the signal is specific. Reexamine all the scientific data.

c. Contamination of the specimen: Contamination can be easily avoided by using filtered solutions and being scrupulously clean during all the handling procedures. However, even if the most careful precautions are taken, it is still possible to produce fine precipitates on sections that make evaluation of the immunolabeling impossible.

Possible Causes:

i. Buffers reacting with specimen grids: If copper specimen grids are being used then neither the antibodies nor the gold probes can be diluted in Tris-HCL buffer. The HCl will react with the copper. If this is happening then the antibody droplets will turn blue after the incubation. Change either the buffer in which the antibodies are stored or diluted or use nickel or gold grids on which to mount the specimens. Copper grids will react with PBS and produce small amounts of a fine precipitate over the specimen if the support film is not intact. For this, either the source of the grids can be changed or grids made from other metals substituted for the copper ones. Alternatively, grids can be covered with plastic, by dipping them in formvar solution prior to coating.

ii. Uranyl acetate precipitation: Salt solutions are used for antibody dilutions and washes. If these salts are present during contrasting, then the uranyl acetate will precipitate out of solution. Wash sections with water before incubation with uranyl acetate.

21. Morphology: There is a misconception that cryosections produce poor morphology in biological material. However, if care is taken during sample preparation and sectioning and section retrieval, there will be no damage to the subcellular morphology. Material frozen in the presence of sucrose can be thawed out and

---

A-gold), which bind over the BSA-gold containing organelles, as well as other structures in close proximity. Endocytosed gold particles is found in only a few of the MHC class II labeled organelles (arrowhead). Bar = 1 μm.

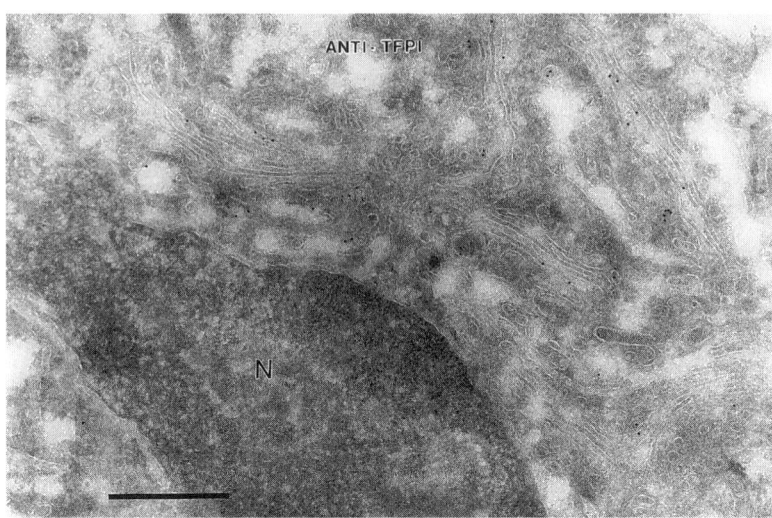

Fig. 8. Cryosection through human umbilical vein endothelial cell labeled with antibodies to tissue factor pathway inhibitor (TFPI). These cells, maintained in primary culture for up to 8 d prior to fixation and sectioning, have to be carefully handled, otherwise the morphology in cryosections is not well preserved. These cells are a primary site of TFPI production as can be seen from the specific anti-TFPI label over the Golgi complex, which is in close proximity to the cell nucleus (n). For more details of this work *see* Hansen, J. B., Olsen, R., and Webster, P. 1997. *Blood* **90,** 3568–3578). Micrograph courtesy of R. Olsen, University of Tromsø, Norway. Bar = 0.5 μm.

embedded in epoxy resin, providing a simple way for evaluating morphology of frozen material. Alternatively, the cryoprotected, frozen material can be embedded in resin after being dehydrated by freeze substitution *(27)*. Below are a few of the most common causes of poor morphology in cryosections:

a. Trypsinized cells: Cells growing on flat substrates are routinely passaged by detaching them from their substrate. This is achieved by treating them with a trypsin solution. When placed into fresh culture dishes the cells will rapidly reattach to the substrate and will look normal by light microscopy within a few hours. For electron microscopy, however, the cells may require at least 2 d before returning to normal. Sometimes good morphology is obtained only after the cells have been grown for many days in the presence of excess nutrients.

b. Centrifugation: If cells are to be pelleted by centrifugation, it is best to do this after chemical fixation. If this is not possible, then only a gentle centrifugation should be used otherwise the morphology may be adversely affected by the rough treatment given to the cells by centrifugation.

c. Support film: Cryosections are usually dried after immunolabeling so that they can be examined in the electron microscope. To protect them from surface tension forces, which damage the morphology, the sections are embedded in a plastic film to protect the sections during drying. Drying artifact, caused by too thin a support film being applied over the sections, is often mistaken for poor fixation. If the morphology is better near the grid bar where the support film is thicker, the support film may be too thin over the rest of the grid.

d. Blunt knife: Poor morphology can result from sectioning using a blunt knife. The knife will tear the sections, pulling them apart at their weakest points. Biological materials that have only been fixed in low concentrations of form- aldehyde are especially susceptible to being ripped apart in this way. In addi- tion to producing knife marks and other obvious cutting artifacts, the damage to subcellular morphology may also be subtle and not immediately be obvi- ous that it is the knife at fault.

e. Surface tension effects: Sections of lightly fixed material are also susceptible to being ripped apart when they are first picked up from the knife surface, onto the sucrose droplet. The surface tension effects, used to spread the sec- tions, may be too great and pull the section apart. Using a mixture of methyl cellulose and sucrose to pick up sections, instead of sucrose alone, will avoid this artifact *(20)*.

f. Mishandling trauma: Well-fixed organs and tissues from perfused animals can still show poor morphology if they have been dissected out carelessly. Squeezing or pulling will cause ultrastructural damage.

g. Sections too thin: Very thin sections of lightly fixed material may appear to have poor morphology. This may be due to poor crosslinking of cellular con- stituents resulting in their loss from the section. However, careful examina- tion of the sections will reveal the remaining structures to be well fixed. Very thin sections of Lowicryl embedded material may also exhibit similar poor morphology.

22. No sections obtained: There are many reasons why cryosections cannot be obtained, but the most common problems are as follows.

a. Loose sample. Make sure the sample is securely clamped into the specimen arm. Check also to see that the thermal bridge attaching the sample holder to the specimen arm is secure.

b. Loose knife. Check to *see* that the knife is securely fixed into its holder before use.

c. Blunt knife. It is important that the knife be sharp enough to section through the sample and produce thin sections. Good quality diamond or glass knives are required.

d. Sucrose crystals in the sample. These crystals are produced if the sample is left too long drying before being frozen. The result is to produce uneven sec- tioning qualities when thin sectioning. The unevenness is often localized to a few areas of the block.

Examples of the preservation, morphology, and labeling of cells processed using this method are given in **Figs. 6–8** (which appear on pages 69-72).

Acknowledgments

This simple set of instructions incorporates the efforts of many years of work by many people. Fernandez-Moran, Bernard, Leduc and Christensen did most of the pioneer work (*see* **ref. 4**). The innovative developments of Professor K. Tokuyasu made cryosectioning possible and justifiably gave the technique his name. The modern innovations generated by Jan W. Slot and Gareth Griffiths, their continuous application of the technique, and their tireless teaching, has made the technique widely accessible. The author thanks Professors Tokuyasu, Slot, and Griffiths for their many years of teaching. Thanks also to the EMBO course organizers (Professors Griffiths, Raska, Roos, and Slot) who invited the author to their courses to teach the technique to their students. Thanks to all the EMBO course participants who accepted the author as an expert, and to Linda Chicoine and Lisa Hartnell, two people who endured the author's instruction and who helped provide the figures in this chapter. Figure 8 was a gift from R. Olsen, a cryosectioning expert at the University of Tromsø, Norway. A special thanks goes to Janice Griffith from Utrecht, who is probably the one person who has been cryosectioning for the longest continuous time, she is a patient teacher who has passed on technical innovations and new ideas for improvements as they become available. The development and application of this method owes much to her patience. Much of this chapter is based on her technical explanations to me. The author takes sole credit for any mistakes that may have been made in this manuscript.

References

1. Tokuyasu, K. T. (1973) A technique for ultracryotomy of cell suspensions and tissues. *J. Cell. Biol.* **57,** 551–565.
2. Tokuyasu, K. T. (1976) Membranes as observed in frozen sections. *J. Ultrastruct. Res.* **55,** 281–287.
3. Tokuyasu, K. T. (1986) Application of cryoultramicrotomy to immunocytochemistry. *J. Microsc.* **143,** 139–149.
4. Tokuyasu, K. T. and Singer, J. S. (1976) Improved procedures for immunoferritin labelling of ultrathin frozen sections. *J. Cell Biol.* **71,** 894–906.
5. Griffiths, G. (1984) Selective contrast for electron microscopy using thawed frozen sections and immunocytochemistry, in *Proc. Congr. Specimen Preparation*, Traverse City, Michigan (Revel, J. P., Barnard, T., and Harris, G. H., eds.), SEM Inc., Chicago, IL, pp. 153–159.
6. Griffiths, G. (1993) Fine structure immunocytochemistry. Springer Verlag, Heidelberg, Germany.
7. Griffiths, G., Simons, K., Warren, G., and Tokuyasu, K. T. (1983) Immunoelectron microscopy using thin frozen sections: Applications to studies of the intracellular transport of Semliki Forest virus spike proteins. *Methods Enzymol.* **96,** 466–484.

8. Ericsson, M., Sodeik, B., Krijnse-Locker, J., and Griffiths, G. (1997) In vitro reconstitution of an intermediate assembly stage of vaccinia virus. *Virology* **235**, 218–227.

9. Griffiths, G. Hoflack, B., Simons, K., Mellman, I., and Kornfeld, S. (1988) The mannose 6-phosphate receptor and the biogenesis of lysosomes. *Cell* **52**, 329–341.

10. Liou, W., Geuze, H. J., Geelen, M. J., and Slot, J. W. (1997) The autophagic and endocytic pathways converge at the nascent autophagic vacuoles. *J. Cell Biol.* **136**, 61–70.

11. Sodeik, B., Ebersold, M. W., and Helenius, A. (1997) Microtubule-mediated transport of incoming herpes simplex virus 1 capsids to the nucleus. *J. Cell Biol.* **136**, 1007-1021.

12. Slot, J. W., Geuze, H. J., Gigengack, S., Lienhard, G. E., and James, D. E. (1991) Immuno-localization of the insulin regulatable glucose transporter in brown adipose tissue of the rat. *J. Cell Biol.* **113**, 123–135.

13. Webster, P. and Grab, D. J. (1988) Intracellular colocalization of variant surface glycoprotein and transferrin-gold in *Trypanosoma brucei*. *J. Cell Biol.* **106**, 279–288.

14. Webster, P., Vanacore, L., Nairn, A., and Marino, C. (1994) Endosomes are a predominant site of CFTR localization in salivary gland duct cells. *Amer. J. Pathol.* **267(Cell Physiol. 36)**, C340–334.

15. Geuze, H. J., Slot, J. W., van der Ley, P. A., and Scheffer, R. C. (1981) Use of colloidal gold particles in double-labeling immunoelectron microscopy of ultrathin frozen tissue sections. *J. Cell Biol.* **89**, 653–665.

16. Köhler, S., Delwiche, C. F., Denny, P. W., Tilney, L. G., Webster, P., Wilson, R. J. M., et al. (1997) A plastid of probable green algal origin in Apicomplexan parasites. *Science* **275**, 1485–1488.

17. Pollock, J. A., Ellisman, M. H., and Benzer, S. (1990) Subcellular localization of transcripts in Drosophila photoreceptor neurons: Chapoptic mutants have an aberrant distribution. *Genes Dev.* **4**, 806–821.

18. Tong, Y., Zhao, H. F., Simard, J., Labrie, F., and Pelletier, G. (1989) Electron microscopic autoradiographic localization of prolactin mRNA in rat pituitary. *J. Histochem. Cytochem.* **37**, 567–571.

19. Griffiths, G. and Hoppeler, H. (1986) Quantitation in immunocytochemistry, correlation of immunogold labelling to absolute numbers of membrane antigens. *J. Histochem. Cytochem.* **34**, 1389–1398.

20. Liou, W., Geuze, H. J., and Slot, J. W. (1996) Improving structural integrity of cryosections for immunogold labeling. *Histochem. Cell Biol.* **106**, 41–58.

21. Slot, J. W. and Geuze, H. J. (1981) Sizing of protein A-colloidal gold probes for immunoelectron microscopy. *J. Cell Biol.* **90**, 533–536.

22. Slot, J. W. and Geuze, H. J. (1985) A new method of preparing gold probes for multiple-labeling cytochemistry. *Eur. J. Cell Biol.* **38**, 87–93.

23. Griffith, J., and Postuma, G. Long term storage of cryosections dried in a layer of methyl celluose and sucrose, in preparation.

24. Webster, P. (1992) A simple modification to the LKB knifemaker which allows glass knives to be made using a balanced break. *J. Microsc.* **169,** 85–88.
25. Stang, E. (1988) Modification of the LKB 7800 series knifemaker for symetrical breaking of 'cryo' knives. *J. Microsc.* **149,** 77–79.
26. Larsson, L. I. (1988) *Immunocytochemistry: Theory and Practice*, CRC, Boca Raton, FL.
27. van Genderen, I. L., van Meer, G., Slot, J. W., Geuze, H. J., and Voorhout, W. F. (1991) Subcellular localization of Forssman glycolipid in epithelial MDCK cells by immuno-electronmicroscopy after freeze-substitution. *J. Cell Biol.* **115,** 1009–1019.

5

High-Pressure Freezing for Preservation of High Resolution Fine Structure and Antigenicity for Immunolabeling

Kent McDonald

1. Introduction

What is high-pressure freezing (HPF)? HPF is a method of specimen preparation for electron microscopy (EM) that freezes noncryoprotected samples up to 0.5 mm in thickness without significant ice crystal damage. Other freezing methods are limited to 1/100 (plunge and impact freezing) to 1/10 (propane jet freezing) that sample thickness. With HPF you are working in the realm of whole cells and tissues instead of small cells or cell cortical regions. For more details on the theory of how HPF works, and some examples of its application, there are several reviews available *(1–4)*.

What's so special about freezing? When cells are ultrarapidly frozen, all the cell contents are physically immobilized in millseconds. By contrast, cells that are fixed by chemicals at room temperature take orders of magnitude longer to fix because it takes that long for fixatives to diffuse through the tissues. If there are significant diffusion barriers such as cell walls, or embryonic layers the fixation may take minutes to hours. Furthermore, only cell components that can react with the fixative, usually glutaraldehyde, are crosslinked during this primary fixation. Postfixation with osmium and compounds such as tannic acid or uranyl acetate adds to the preservation of different components, but between these different fixative steps, cell components may be lost or rearranged. For a detailed discussion of fixation chemistry, *see* **refs. *5–7***, for a demonstration of the differences between frozen and chemically-fixed material, *see* **refs. *8–13***, and for general reading about freezing methods, *see* **refs. *14–17***.

What do you do with your cells after they are frozen? There is a wide range of options here depending on the goal of the research. For minimal pro-

From: *Methods in Molecular Biology*, vol. 117: *Electron Microscopy Methods and Protocols*
Edited by: N. Hajibagheri © Humana Press Inc., Totowa, NJ

cessing of the sample, you could simply fracture the cells and coat them in a freeze fracture apparatus and look at them frozen in a cryo field emission scanning EM *(18,19)*, or look at the replica in a conventional Transmission EM *(3,4)*. Frozen tissues can be sectioned and observed in a cryoelectron microscope *(20)*, or they can be freeze dried and used for X-ray microanalytical work *(17)*. More typically, the cells would be freeze-substituted, embedded in resin, sectioned, and imaged in a conventional transmission EM. Freeze substitution "substitutes" organic solvent such as acetone or methanol for the cell water at temperatures from –78 to –90°C. At this temperature, it is possible to remove the water molecules and minimize the inevitable collapse and shrinkage that occurs during room temperature dehydration. By putting fixatives such as osmium or glutaraldehyde in the freeze substitution medium, you also avoid the diffusion gradient problem of conventional processing because the fixative diffuses into the cells while they are immobilized by low temperature and when they warm up to a temperature that permits chemical crosslinking, the fixatives are already dispersed throughout the cells and fixation occurs quickly. For a good overview of freeze substitution, see the article by Hippe-Sanwald *(21)*.

How can HPF serve the needs of the contemporary biologist? For purposes of this article in the context of this book, we consider the biologist to be someone who uses molecular, genetic, and cell biological techniques to attack their research question. HPF is a tool used primarily by electron microscopists, and EM is a tool to localize or characterize the structure of molecules and macromolecular assemblies in vivo. If you have an antibody to a gene product that you have localized by light microscopy, there is a chance that you need more specific information about its subcellular location, or its interaction with specific organelles in vivo. By using HPF and EM immunocytochemistry *(7,22–24)*, it is possible to get that high resolution information in very well preserved tissue. In the text that follows, we will explain how to do HPF for those who have never used the method.

2. Materials

2.1. General Equipment

1. High-pressure freezer (*see* **Note 1**).
2. Specimen carriers for HPF (*see* **Note 2** and **Figs. 1** and **2**)
3. Dissecting microscope with fiber-optic light source.
4. Dissecting microscope mounted on a pole stand plus fiber-optic light source.
5. Specimen loading station (*see* **Note 3**).
6. 2 pr. angled fine-tipped tweezers (Ted Pella, Inc., 4595 Mountain Lakes Blvd., Redding, CA 96003 USA, Catalog. no. 9, Cat. no. 5624).
7. 4 L dewar for LN2.
8. LN2 workstation (*see* **Note 4**).

9. 1 L widemouth dewar for temporary storage or transfer of samples.
10. LN2 refrigerator or storage dewar.
11. 2 mL Nalgene Cryotubes (*see* **Note 5**).
12. Nalgene cryotube rack with locking bottoms to match cryotubes (*see* **Note 5**).
13. Device for splitting specimen carriers under LN2 (*see* **Note 6**).
14. Long tweezers (Ted Pella, Inc., Cat. no. 5306) with foam or Velcro insulation (*see* **Note 7**).
15. Halogen or fiber optic light source to illuminate HPF console working area.
16. Soft (number 2 or softer) pencil for writing on cryotubes.
17. Small container for receiving used specimen carriers.
18. Gloves for handling cold samples (*see* **Note 8**).
19. Liquid nitrogen (LN2) source.
20. Hair dryer, or similar hot air heat gun (*see* **Note 9**).
21. Spare O-rings for specimen carrier holder.
22. Silicon grease to lubricate O-rings.
23. Large forceps for handling cooled cryotubes.

2.2. Cell Suspensions

1. Microfuge, preferably with horizontal rotor.
2. Clinical centrifuge with swinging rotor.
3. Microcentrifuge tubes—2 mL.
4. 15 mL conical plastic tubes.
5. 15 mL vacuum filtration apparatus.
6. 0.4 μm polycarbonate filters, 25 mm circles.
7. No. 542 Whatman filter paper discs, 24 mm circles.
8. Toothpicks or small, round wooden sticks.
9. Dog toenail clippers (*see* **Note 10**).
10. Fine-tip transfer pipets (Fisher Scientific, Cat. no. 13-711-26).
11. Paper points (Ted Pella, Inc., Cat. no. 115-18).
12. Micropippetor (1–10 μL) plus tips.
13. 2% agar in 60 mm plastic Petri dishes.

2.3. Plant Leaves, Cartilage, and Other Firm Samples

1. Custom-made punches—1.6, 1.8, and 2.0 mm diameter (*see* **Note 11**).
2. Razor blades or scalpels.
3. 1-hexedecene (Sigma, P.O. Box 14508, St. Louis, MO 63178, USA, Cat. no. H2131).
4. Cutting surface, such as dental wax, or silicone rubber mat.
5. Vacuum dessicator.

2.4. Soft Animal Tissues

1. Biopsy needle apparatus (*see* **ref.** *26*).
2. Razor blades or scalpels.
3. Cutting surface, such as dental wax or silicone rubber mat.

   4. Ringer's solution.
   5. 1-hexedecene.

2.5. Drosophila and Similar Embryos

   1. 2% agar in 60 mm Petri dishes.
   2. No. 0 red sable paint brush.
   3. 10% methanol in water.
   4. Baker's yeast.
   5. Microfuge tubes and rack.

3. Methods

3.1. Starting up the High-Pressure Freezer (see Note 12)

   1. Check the room ventilation. If there is a fan, turn it on. if not, open a window. Make sure the door(s) to the room are open. When liquid nitrogen changes to a gas, there is about a 700-fold increase in volume. During the inital cooling of the HPF, large volumes of LN2 are spilled onto the floor and if ventilation is inadequate, there is a danger of asphyxiation. As a precaution you can obtain and wear an oxygen monitor.
   2. Make sure the LN2 dewar (usually a 160 L size) attached to the HPF is at least half full.
   3. Empty alcohol overflow bottle (inside left door panel) if it is more than half full.
   4. Fill alcohol reservoir with ethanol or isopropyl alcohol. Make sure nitrogen gas cylinder is turned off, then open reservoir. On later models, a nut is removed through a hole in the top of the machine console. Fill until level indicator on the front console panel shows full, then replace nut and tighten with torque wrench set to 15 lb.ft.
   5. Check that hydraulic oil pressure relief valve is closed.
   6. Check oil level in oil tank.
   7. Check level of water in recirculator. If lower than 2 cm from top, refill.
   8. Turn main switch on.
   9. Open valve on dry nitrogen tank all the way. Pressure should be set to 80 psi, the alcohol pressure gage should read 4–5 bar.
  10. Turn on the liquid nitrogen valve all the way.
  11. Press the START button, wait until the PISTON IN POSITION light comes on.
  12. Press NITROGEN button and the LN2 will begin to flow and the NITROGEN LEVEL light will come on. Hold a glove firmly over the dewar vent in the back of the machine for 1 min. This will flush the system with dry nitrogen gas.
  13. When the NITROGEN LEVEL light goes off, press DRIVE IN.
  14. When DRIVE IN light goes off, press AUTO. Oil pressure should rise to about 315 bar.
  15. Check Object Head Temperature readout on console. It should read between 30 and 37°C. If outside that range, adjust knob on water recirculator to bring temperature within range.

6. With test specimen holder and locking pin in place, check to see that Ready light comes on.
7. Do 1–2 test runs with test specimen holder to make sure readings are within normal limits. On the latest models, Pressure Maintenance should be about 0.5 s, Cooling Time should be about 7 s, and Temperature Maintenance should be 3–5 s.
8. Replace and grease O-ring on specimen carrier holder if more than 25 shots were carried out last session, or if more than 25 shots are likely to be carried out in the present session.

3.2. Loading Samples into Specimen Carriers

This is without doubt the most important step in HPF. What is done at this step will determine whether the sample is well frozen or not. The HPF machine is a product of robust engineering and works reliably most of the time. In those rare times that it malfunctions, there are warning indicators that let the operator know that something is wrong. So the key variable during a HPF run is the condition of the sample in the holder, not the HPF machine.

There are several guidelines about loading specimens that apply to all types of samples:

1. Work quickly. This is primarily to prevent the samples from drying out, but it is also good to limit the exposure of the sample to air or external cryoprotectant to the shortest time possible. Remember that even the largest wells hold less than one microliter of sample.
2. Use the smallest possible volume that will accomodate your sample. At present, the smallest chamber that is readily available is a 100-μm deep well with a diameter of 2 mm. The largest is two 300 μm deep wells placed fact to face to make a 600-μm deep well (**Fig. 1**). The ideal specimen carrier would just fit the geometry of the sample. If the project is important enough, consider having a machine shop make custom carriers. At present, not having a wide range of specimen carrier geometries to choose from is the single most important barrier to progress in getting consistently good results from HPF.
3. Fill the specimen carrier completely, but not too full (**Fig. 2**). The ideal is to fill the volume entirely with your sample. But if that is not possible, you need to add something to fill in the spaces (*see* **Note 13**).

3.2.1. Yeast and Bacteria

1. Arrange ahead of time to have the cells at an optimum growth density. For most cell types in our experience, early log phase is best. Cells in stationary phase will not look good, even if they freeze well.
2. If the cells require aeration and/or higher-than-normal temperatures, arrange to have a shaker and/or water bath or incubator set up next to the HPF. Cells should remain in optimal growth conditions throughout the freezing run and aliquots taken as needed.

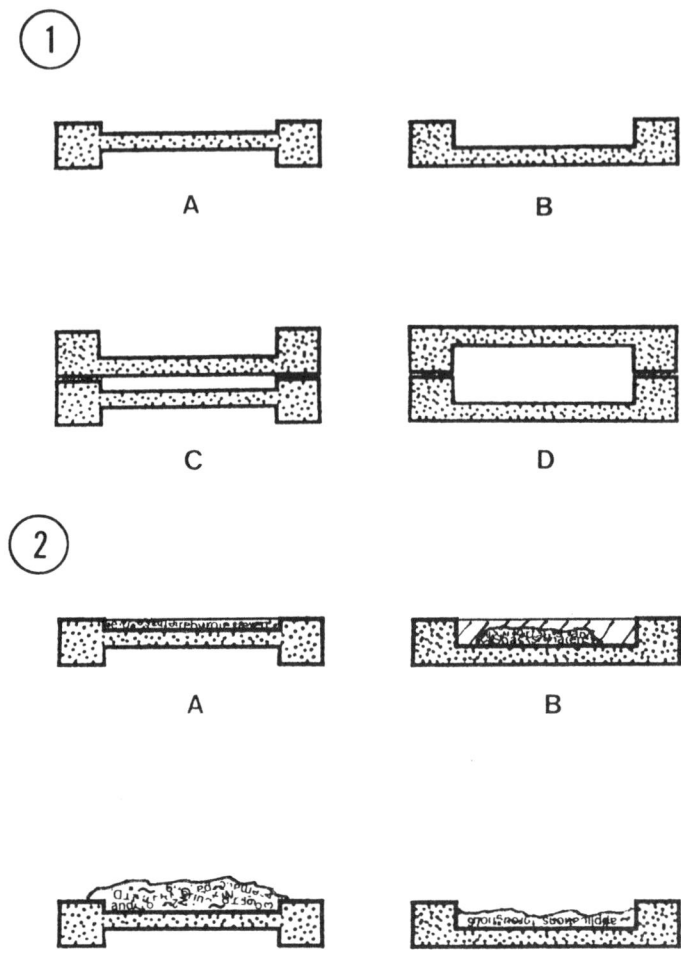

Fig. 1. Cut-away side view of two types of high-pressure freezing specimen carriers and two of the six ways that they can be arranged. (**A**) In Type A, there are wells on either side, one that is 100 μm deep (shown here on the top side) and the other 200 μm deep (shown here on the bottom). (**B**) In Type B, there is one flat side, and a well 300 μm deep on the other side. (**C**) Configuration of specimen carriers to give the shallowest (100 μm) freezing well. (**D**) Configuration of specimen carriers to give the deepest (600 μm) freezing well.

Fig. 2. Properly (**A,B**) and improperly (**C,D**) loaded specimen carriers as seen in cut-away side view. (**A**) The ideally loaded specimen carrier is completely filled with the sample of interest. (**B**) Sample plus external cryoprotectant completely fills the specimen carrier. (**C**) Sample carrier with too much material will not close completely or if forced, the sample may be crushed prior to freezing. (**D**) Too little material in the specimen carrier will lead to collapse of the specimen carrier wall.

3. Make sure you have enough cells to do multiple HPF runs. It will take about 10 mL of cells in log phase for each freezing run. We typically try to have a minimum of 100 mL cells to work with.

4. Set up a vacuum filtration apparatus near the workspace for loading specimens. We use a 15-mL column and Osmonic Poretics Products 0.4 μm polycarbonate filters (Cat. no. 11028) on top of a Whatman No. 542 filter paper circle.

5. Sharpen a toothpick or small stick so that the end is flat and can scrape cells off the filter. Use a different stick for each separate culture type.

6. Load about 10 mL of cell suspension from the flask on the shaker/water bath into the filtration apparatus and apply vacuum. When the liquid surface just touches the filter, shut off the vacuum.

7. With forceps, remove the polycarbonate filter and place on the agar plate. Observe through a dissecting scope. Use the flattened stick to scrape off cells from the surface. If the cells seem too "soupy," put a piece of dry Whatman 542 filter paper under the polycarbonate filter to wick off excess medium. If too dry, discard and try again, being careful not to let the vacuum stay on too long.

8. Transfer cells from the tip of the stick to the well of a 100-μm deep specimen carrier. Adjust volume so that the cells just fill the well. One way to do this is to use the side of a paper point as a scraper and wipe off excess cells in a sweeping motion across the top of the well. You will need to hold the specimen holder down with the angled forceps as you do this to prevent the whole assembly from just sticking to the paper point.

9. Place the 100 μm well in the tip of the specimen carrier holder cells side up and cover with the flat side of a Type B carrier.

10. Close hinge of specimen carrier holder, tighten down and proceed to HPF immediately to freeze. The elapsed time of the above steps from placing the filtered cells on the agar plate to freezing should take less that 1 min.

3.2.2. Tissue Culture and Other "Delicate" Cell Types

Cells without cell walls may not survive the vacuum filtration and scraping described above. For these cells, concentration by centrifugation works better. For extremely delicate cells, such as plant protoplasts, concentration by gravitational settling may have to be used.

1. Concentrate cells as much as possible by whichever method gives the most cells and the least damage.

2. If microfuge tubes were used, remove excess liquid above the pellet with a fine-tipped plastic transfer pipet, then use dog toenail clippers (*see* **Note 10**) to cutoff the tip of the tube with the cells in it. Use a combination of pipetting and wicking with paper points to remove all excess fluid above the concentrated cells. Use a micropipettor to pull up a microliter or so from the bottom of the tube and transfer the cells to a specimen carrier on a silicone pad.

3. If low speed centrifugation or gravitational settling was used, use a transfer pipet or micropipettor to transfer cells to a specimen carrier.

4. If necessary, use two paper points, one on each side, to wick off excess fluid from the specimen carrier and further concentrate cells. With practice, this can be done quickly and without wicking off too many cells. Add more cells if the level goes below the top of the specimen carrier. Repeat as necessary to get the densest possible concentration of cells (*see* **Note 14**).
5. Transfer specimen carrier to specimen carrier holder, add top, close tip and proceed to HPF.

3.2.3. Use of Dialysis Tubing for Cell Suspensions

There is a published article by Hohenberg et al. *(25)* describing a method of concentrating cells using dialysis tubing. We have not used this technique, but can summarize the basic steps here. For more detail, *see* the original article.

1. Obtain dialysis tubing. In the United States, they are available from Spectrum, 23022 La Cadena Dr., Suite 100, Laguna Hills, CA 92653, USA. Ask for the Spectra/Por Hollow Fiber Bundles, 200 µm inner diameter, Cat. no. 132 290.
2. Using capilliary action, or perhaps suction by fitting the tubing into an appropriately sized micropipet tip, draw concentrated cells into the tubing.
3. Submerge in 1-hexedecene in a small plastic Petri dish.
4. Using a scalpel blade, cut the region of the tubing containing cells into lengths that will fit into the 2 mm wells of the specimen carriers. The ends should be crimped so that the cells will not leak out.
5. Transfer from the 1-hexedecene dish to the well of a specimen carrier.
6. Fill the specimen carrier with 1-hexedecene, add a top and freeze.

3.2.4. Plant Tissues and Cartilage

1. Select a leaf, root, etc., or cut a thin (100–200 µm preferably) layer of cartilage.
2. Working on a cutting surface such as a silicone rubber mat, use one of the custom punches to cut a disk of sample from the plant or cartilage. Remember to keep the cartilage moist with appropriate fluid.
3. Working with leaves it may be necessary to remove air from the tissues. Transfer the leaf disk to a small well with 1-hexedecene and *see* if it sinks. If not, you may have to apply a gentle vacuum to remove air from the tissue. When the leaf disk sinks, transfer to the well of a specimen carrier.
4. The cartilage disk can be transferred quickly to the specimen carrier of appropriate depth.
5. For both tissues, add 1-hexedecene to fill space, if necessary.
6. Add a top specimen carrier, close tip of specimen carrier holder and freeze.

3.2.5. Soft Animal Tissues

These are some of the more difficult samples for HPF because it is hard to dissect out the tissue without subjecting it to physical stresses and damage. Again, Hohenberg et al. *(26)* have addressed this issue very sensibly by using a biopsy needle to sample the tissue. This may not be applicable for every type of soft animal tissue, but where it is, it is an elegant solution. For details the reader is referred to the original article.

1. The alternative to the biopsy needle is to use dissecting tools, scalpels, or razor blades to remove and cut up pieces of tissue into suitably small pieces that will fit into a specimen carrier.
2. Fill in empty spaces in the specimen carrier with 1-hexedecene.
3. Add a top specimen carrier, close tip of the holder and freeze.

3.2.6. Drosophila *Embryos and Other "Dry" Tissues*

These materials are easy to work with because they do not have to be constantly immersed in some fluid. There are embryonic tissue layers that keep the cells hydrated, and in the case of adult insects, there is a cuticle layer. These layers act as diffusion barriers and are the reason that such tissues are very difficult to fix by conventional diffusion chemistry.

1. Harvest *Drosophila* embryos by whatever methods usually used.
2. Make up a mixture of Baker's yeast and 10% methanol to the consistency of a stiff paste. This is an extremely useful nonpenetrating cryoprotectant and space filler (*see* **Note 15**).
3. Using the Size 0 paint brush, pick up about 50-100 embryos and transfer them to the surface of an agar dish.
4. Place several Type A and several Type B specimen carriers in the same dish.
5. Finally, place a small amount of yeast paste in the dish using a small wooden stick.
6. Dry the paint brush, then pick up a tiny amount (roughly half the amount to fill the 200-µm well of a Type A specimen carrier) of yeast paste on the end. Use this to pick up about an equal volume of embryos (approx 20), and mix together on the agar surface. Transfer the mixture to the 200 µm well, adding or removing yeast to fill the well just full (**Fig. 2**).
7. Transfer to the specimen carrier holder, add a Type B specimen carrier (flat side down), close the holder tip and freeze.

3.3. Freezing Samples in the HPF

For all but the easiest samples, e.g., yeast, it is useful to have one person load the samples into the specimen carrier holder, and another person freezing the samples, distributing the specimen carriers to the proper storage or fixative vials, and drying off the specimen carrier holder.

1. Fill LN2 workstation to proper level, and prechill forceps (with foam- or Velcro-insulated handles).
2. Turn on light above the LN2 workstation.
3. Place loaded specimen carrier holder into the machine, insert the safety locking bolt, and check to make sure the Ready light is lit.

4. With left hand on safety bolt, press Jet with right hand and immediately grip the top of the specimen carrier holder.
5. When Pressure Over light comes on, or when the exhaust has clearly finished, pull out locking bolt with the left hand, and as quickly and smoothly as possible, transfer the tip of the specimen carrier holder to the slot in the aluminum block in the LN2 workstation.
6. Rotate the specimen carrier holder four turns counterclockwise, then lift the tip out of the slot and into the well in the aluminum block. Make sure the level ot the LN2 is sufficient to cover the specien carrier, but not so high that it is over the O-ring.
7. Using the precooled tweezers, swing out the hinged tip of the specimen carrier holder to about 90°, and push out the specimen carrier into the well in the aluminum block.
8. Dry off and warm up the tip of the specimen carrier holder with a hair dryer or heat gun and return to the specimen loading area (*see* **Note 9**).
9. Use specimen carrier splitter (*see* **Note 6**) to separate the top and bottom specimen carriers.
10. Transfer the carrier with material in it to either a storage vial or to a vial of prefrozen fixative. Even if you are not planning to process the sample immediately, it is sometimes convenient to store samples in the fixative vials. Otherwise, you have to take samples and fixatives out of storage, set up the LN2 workstation and transfer from storage to fixative vial. If you know how you want to fix the samples, it saves time and effort to load them into fixative as you freeze them.
11. Repeat steps 3–12 as needed.
12. Lightly lubricate the O-ring on the specimen carrier holder about every five shots.
13. During lengthy freezing sessions (30–40 shots, or more) check both the alcohol level and Nitrogen Level indicators to make sure you have not run out of either (*see* **Note 16**).
14. At the end of the session, transfer samples to storage or to the next step in processing, e.g., freeze fracture or freeze substitution.

3.4. Shutting Off the HPF

1. Wait for the Piston in Position light to come on.
2. Press the Stop button.
3. Turn off the dry nitrogen.
4. Turn off the liquid nitrogen.
5. Open the Oil Pressure Relief valve until the pressure drops to zero, then retighten the knob by turning clockwise until it stops.
6. Leave the main switch on so the machine warms up and bakes out overnight. Turn off the main switch the next day.
7. Turn off the oxygen monitor if you have been using one.
8. Record number of shots in the log book.

3.5. Freeze Substitution

Because freeze substitution is probably the most common method of processing samples following HPF, we include here the instructions for how to make up freeze substitution fixatives and how to carry out a simple freeze substitution with readily available lab supplies.

3.5.1. Materials for Making Fixatives

1. About 35 Nalgene cryovials.
2. No. 2 pencil.
3. Nalgene cryogenic vial rack.
4. Liquid nitrogen in 4 L dewar.
5. Styrafoam box large enough to contain Nalgene cryogenic vial rack.
6. 100% methanol.
7. 100% acetone.
8. Crystalline osmium tetroxide.
9. 10% glutaraldehyde in acetone (Electron Microscopy Sciences, Cat. no. 16530).
10. Uranyl acetate crystals.
11. Aluminum foil.
12. Vial rocker.
13. 50-mL conical tube.
14. Repeater pipettor to dispense up to 2 mL aliquots (*see* **Note 17**).
15. Sonicator.

3.5.2. Procedure for Making 50 mL of Fixative (Glutaraldehyde or Osmium Tetroxide) in Acetone

1. *Note! You must wear gloves and work in the hood. Osmium in acetone is especially volatile and therefore, especially dangerous.*
2. Label with pencil about 35 Nalgene 2.0 mL cryovials "2.1 - Os-UA," which stands for 2% osmium tetroxide plus 0.1% uranyl acetate. If you are making glutaraldehyde fixative, label the vial ".2.1 G-UA," which stands for 0.2% glutaraldehyde plus 0.1% uranyl acetate. If you are using a freeze substitution solvent other than acetone, indicate that also.
3. Put vials in vial rack with tops off.
4. Make up stock solution of 5% uranyl acetate in methanol by adding 0.1 g of uranyl acetate (*see* **Note 18**) to 2 mL pure methanol, cover with foil and put on rocker to dissolve (about 10 min).
5. Sonicate 1 g. OsO_4 vial for about 10 s to loosen crystals from glass.
6. Fill 50 mL conical tube with 45 mL pure acetone.
7. Add 1 g OsO_4 crystals (or 1.0 mL of a 10% solution of glutaraldehyde in acetone) to the 45 mL of acetone, cover and mix until completely dissolved.
8. Add 1 mL 5% uranyl acetate stock to OsO_4-acetone, add pure acetone to make 50 mL and mix well.
9. Use repeater pipettor to dispense 1.5 mL into each cryovial. Work quickly, but not carelessly. Cap cryovials as soon as they are filled.
10. Add LN2 to Styrafoam box to a level that will cover cryovials in rack.
11. Slowly lower rack into LN2. Vials must remain upright so fixative will remain in the bottom of the vial.
12. When completely frozen, transfer vials to LN2 storage device until ready to use.

3.5.3. Materials for Freeze Substitution

1. Freeze substitution apparatus such as the Leica AFS freeze substitution system or use dry ice.
2. Styrafoam box.
3. Aluminum block with 13 mm holes (Fisher Cat. no. 11-718-14).
4. Digital thermometer (Cole-Parmer Instrument Co., 625 East Bunker Court, Vernon Hills, IL 60061-1844, USA, Model no. E-08528-20).
5. Type T thermocouple probe (Omega Engineering, Inc., One Omega Drive, Box 4047, Stamford, CT 06907, USA, Cat. no. 5SC-TT-T-30-36).
6. Rotary shaker.
7. Elastic cord.
8. Cryotube with hole in the cap to admit the thermocouple probe.
9. Acetone.

3.5.4. Freeze Substitution Procedure

1. If you have a commercial freeze substitution device, follow the instructions of the manufacturer.
2. If not, load a Styrafoam box 2/3 full with dry ice.
3. Put the aluminum block on the ice to cool.
4. Into one of the holes in the block place a cryotube filled with acetone and into which has been inserted a thermocouple probe through the vial cap.
5. When the acetone has cooled to about –78°C (in about an hour), transfer HPF-frozen samples in fixative from liquid nitrogen storage into the holes of the block.
6. Fill the remainder of the box with dry ice, covering the top of the aluminum block and the vials with something such as aluminum foil.
7. Put the lid on the box and put the whole assembly onto the rotary shaker. You must have a shaker that will take this much weight.
8. Use an elastic cord to tie down the Styrafoam box on the shaker.
9. Start the shaker and rotate at least 100 rotations/min.
10. Continuously shake until the ice has sublimed and the temperature of the vials has risen to the appropriate temperature for the kind of resin you intend to use for infiltration and embedding.
11. For epoxy embedding such as Epon or Epon-Araldite, warm to room temperature, rinse and embed as usual.
12. For low-temperature embedding, follow the manufacturers instructions for resin:solvent exchange and polymerization (*see* **ref. 24**).

3.6. Examples

Figures 3–10 are illustrations of biological material prepared by HPF and freeze substitution, and in some cases, labeled with colloidal gold. Because this type of material is less extracted, and therefore more dense than conventionally processed samples, it is useful to cut thinner-than-normal sections in order to get a crisp image in the microscope. We cut most samples at 60 nm thickness and adjust up or down depending on the cell density or magnification

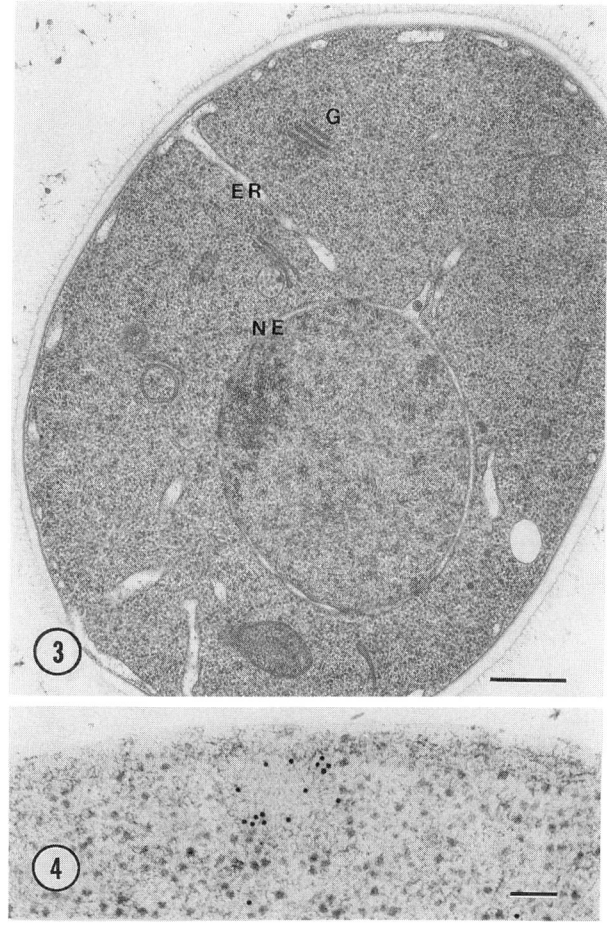

Fig. 3. Low magnification view of high-pressure frozen *Saccharomyces cerevisiae* yeast freeze-substituted in 2% OsO4 plus 0.1% uranyl acetate in acetone for 48 h at –78°C, then warmed to –20°C over 6–8 h and held at –20°C for 12 h. Following a gradual (4 h) warm-up to room temperature, the cells were infiltrated with Epon-Araldite resin and embedded and polymerized as usual. The extra period in OsO4 at –20 enhances the membrane contrast. Further contrast is achieved by cutting 40 nm sections and poststaining with methanolic uranyl acetate instead of aqueous. Sections were also poststained with Reynold's lead citrate. Endomembranes such as the nuclear envelope (ne), Golgi (g) and endoplasmic reticulum (er) are easily visualized. Bar = 0.5 µm.

Fig. 4. *S. cerevisiae* high-pressure frozen and freeze substituted in 0.2% glutaraldehyde plus 0.1% uranyl acetate in acetone and embedded in Lowicryl HM20 polymerized by UV at –35°C. An actin patch in the cortex is revealed by labeling with 10 nm gold. Actin antibody provided by Dave Drubin, Department of Molecular and Cellular Biology, University of California, Berkeley, CA). Bar = 0.1 µm.

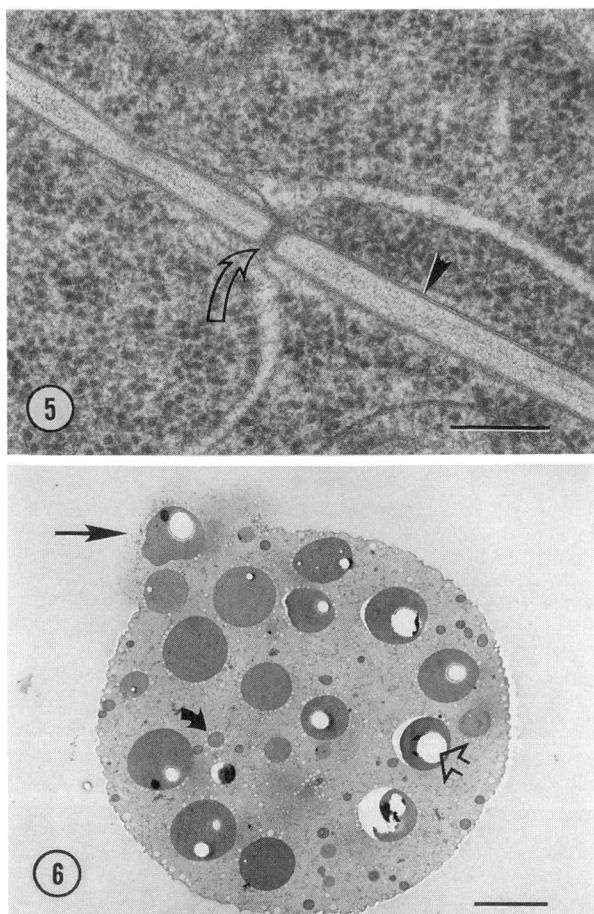

Fig. 5. High-pressure frozen *Arabidopsis* root tip cells showing a plasmodesmatum (curved arrow) and ER. Notice the straight, smooth plasmamembrane (arrowhead). Bar = 0.2 μm. (*Arabidopsis* supplied by Regina McClinton, Dept. of Plant Biology, University of California, Berkeley).

Fig. 6. Low magnification view of a barley aleurone protoplast (supplied by Jennifer Lonsdale and Russell Jones, Department of Plant Biology, University of California, Berkeley, also for **Figs. 7** and **8**) prepared by HPF and embedded in Epon-Araldite following freeze substitution in 2% OsO4 plus 0.1% uranyl acetate in acetone. The arrow indicates an "artefact" of high-pressure freezing that is sometimes seen, i.e., an eruption of the cell contents. More commonly, one sees this with nuclei and certain types of vacuoles or storage granules in other cell types. The curved arrow indicates the portion of the cell shown at higher magnification in **Fig. 7**. The open areas in the granules (open arrow) are the result of incomplete infiltration and polymerization due in part to the presence of phytic acid crystals. Bar = 2 μm.

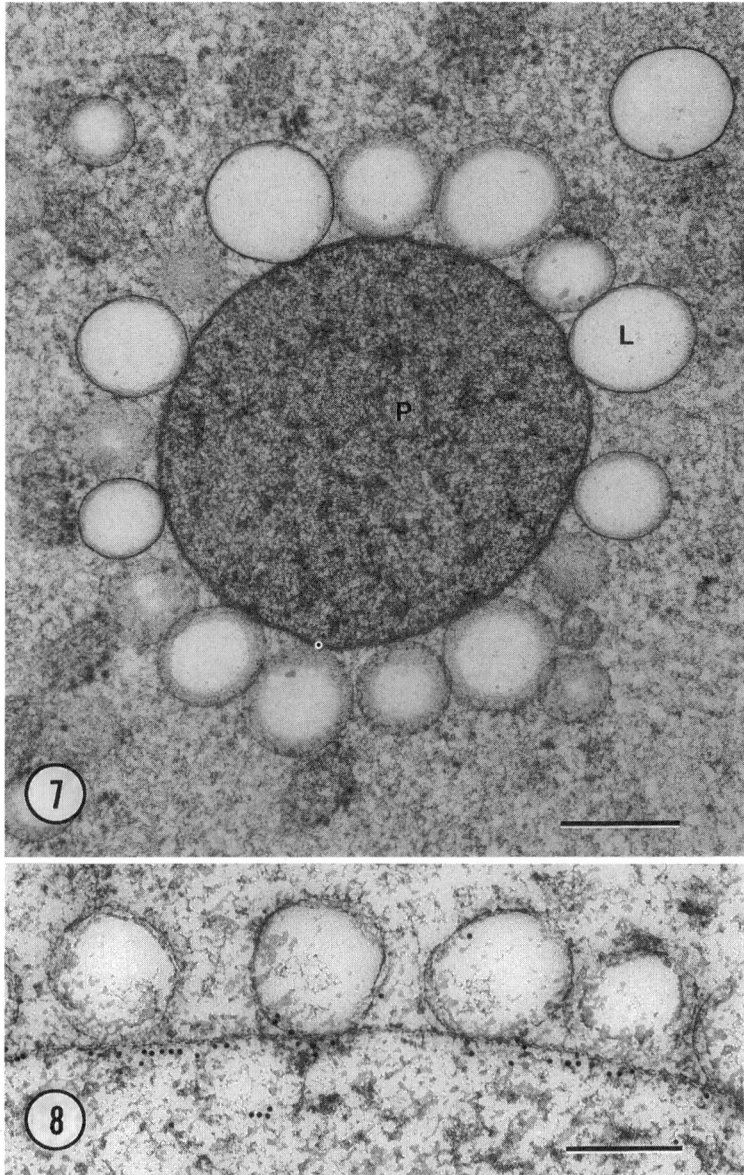

Fig. 7. Higher magnification view of barley aleurone protoplast showing the protein storage granule (p) and associated lipid bodies (l). Bar = 0.2 μm.

Fig. 8. Immunolocalization of alpha-tonoplast intrinsic protein with 10 nm gold. Cells were high-pressure frozen, freeze substituted in 0.2% glutaraldehyde plus 0.1% uranyl acetate in acetone and embedded in Lowicryl HM20 polymerized by UV at –35°C. Bar = 0.2 μm.

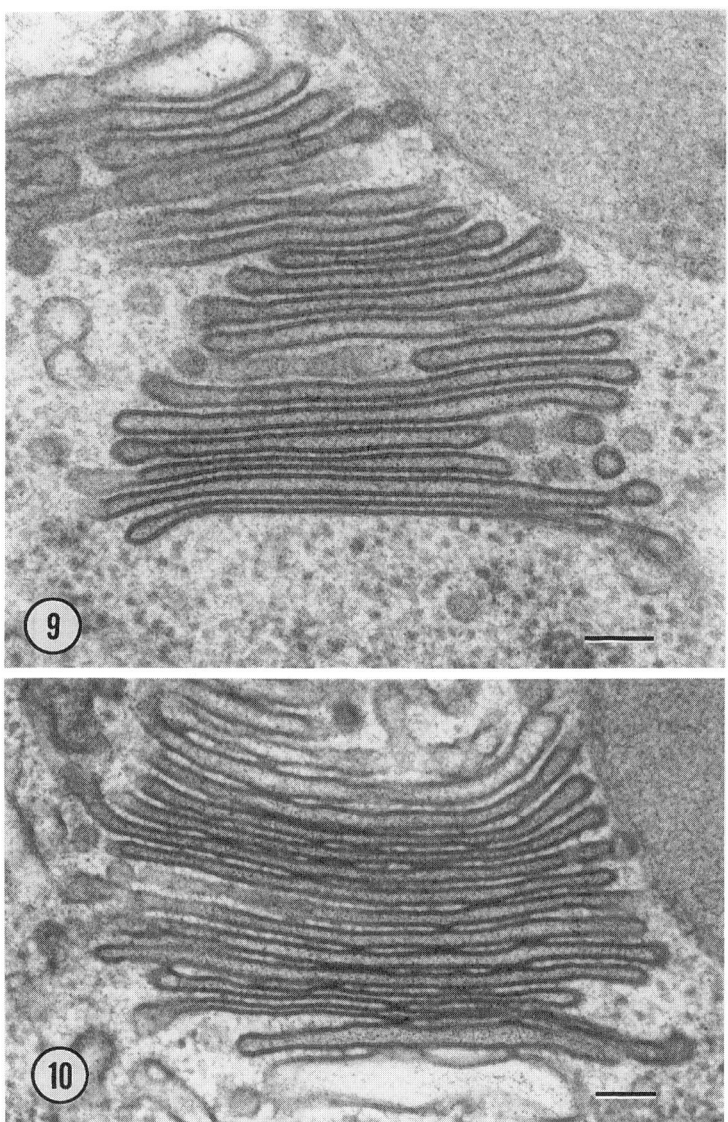

Fig. 9. High magnification view of a Golgi apparatus in the cytoplasm of a starfish oocyte following HPF and freeze substitution in 2% OsO4 plus 0.1% uranyl acetate in acetone. Cells were embedded in Epon-Araldite. Material provided by Steve Stricker, Department of Zoology, University of New Mexico, Albuquerque, NM, USA. Bar = 0.1 μm.

Fig. 10. As for **Fig. 9**, except that this Golgi has suffered damage from small ice crystals which distort the high resolution fine structure. To detect this level of ice damage, it is usually necessary to observe the cells at a high final magnification, i.e., close to x100,000. Bar = 0.1 μm.

we are working at. For yeast and bacteria, we usually cut 40-nm thick sections. Each material is different, and cutting a range of thicknesses from 30 to 80 nm at the beginning of a study is an easy way to determine the optimum combination of section thickness and staining time. For extremely thin sections, it will probably be necessary to use uranyl acetate dissolved in alcohol for poststaining.

For other examples of HPF material, consult review articles *(1–4)* or use electronic literature searching methods. Also check the website (www.em.biol.ethz.ch/) at the Laboratory for Electron Microscopy I at the Swiss Institute of Technology, Zurich. Click on the <"High Pressure Freezing" Literature Compilation> listing for a comprehensive survey of high-pressure freezing papers. The head of this lab, Dr. Martin Müller, and his staff, students, and postdoctorals are responsible for most of the developments in high-pressure freezing technique in recent years. One can learn a lot about this method and its application at this website.

4. Notes

1. There are two vendors who sell and service HPF. There is Bal-Tec Ltd. (Lichtenstein) and Leica (Vienna, Austria). Bal-Tec began selling HPFs in about 1985 and most of the machines currently in use are theirs. Leica entered the HPF business about 4 yr ago. All the work illustrated in this chapter was done either on the HPF at Boulder, Colorado, USA (in the laboratory of Prof. Andrew L. Staehelin), the first U.S. machine, or the machine in the laboratory at Berkeley, currently the last machine installed in the U.S. Note that the instructions for how to operate the HPF described in this chapter are for a Bal-Tec machine.

2. A large supply of specimen carriers is essential to productive work with the HPF. Unfortunately, Bal-Tec only supplies about 20 pairs of specimen carriers with the machine. One can buy more from them, but they are expensive. Now, however, one can buy specimen carriers in lots of 50 from Ted Pella, Inc. (Cat. no. 39200 for Type A and Cat. no. 39201 for Type B) for considerably less. Before Pella began offering them, we used to custom order our specimen carriers from Swiss Precission, Inc., 908 Industrial Ave., Palo Alto, CA 94303, USA (Ph. 415-493-0440, attention Mr. Rick Chan). Swiss Precision also makes the Craig et al. *(27)* holders, which are useful for freeze-fracture work. These specimen carriers are designed to fit into the specimen carrier holder tip No. BU 012 117-T in the Bal-Tec catalogue. Be sure that you have this tip if you use these holders. There is another tip (Bal-Tec Cat. no. BU 012 110-T) that is larger and if you use that with the Pella specimen carriers, you will lose most of your sample during freezing.

3. It is important to set up a dedicated workspace next to the HPF to prepare and load samples. HPF vendors sell a Loading Station, however, we have never used one and cannot evaluate its effectiveness. Mainly, one needs something to hold the specimen carrier holder while loading, and we find that a small Styrafoam box (ours formerly held a micropipettor) works fine. One should also have a working surface in the same box to load the specimen carriers. It should be at the same

height as the tip of the holder and in close proximity to it. We use a silicone rubber pad (Pelco Cat. no. 10521) for most loadings because the specimen carriers are more easily loaded and controlled on a soft surface. Using forceps that are angled to 45° also provides a good grip when picking up the sample from the flat surface and transferring it to the specimen carrier holder tip. We load under a dissecting microscope mounted on a pole stand so we can have a large, flat area to work on instead of the small stage of most dissecting microscopes. A fiberoptic light is a good source of illumination.

4. As with the specimen loading station, one can buy an LN2 worksation from the vendor, or make one yourself. We use a Styrafoam box approx 16 cm wide by 40 cm long by 8 cm deep. In this box, we place a piece of aluminium 6.5 cm square by 1.5 cm thick into which has been cut a groove 3.5 mm wide and 8 mm deep. In the center of the block, a well has been cut with a radius of 25 mm, and within that, another well of radius 12 mm. The depth of the first well is 4 mm and the second 8 mm. After freezing, the specimen carrier holder tip is placed in the groove and turned about four turns. Using a pair of insulated forceps, the hinged tip is swung open and the specimen carrier pushed out into the center well.

5. To store samples or hold fixative for freeze substitution, we use Nalgene cryovials (Nalge Co., Rochester, N.Y., 14602-0365, USA, Tel 716-586-8800, Cat. no. 5000-0020). These are supposedly not for use in LN2, but we did not find this out until we had been using them for several years without incident. If this concerns you, use LN2-certified vials. To open and close these vials in LN2, it is very convenient to use a rack specifically designed to hold the bottom of the cryovial fast while you are unscrewing or screwing on the top. We buy Nalgene racks (Cat. no. 5030-0510) and cut them down to fit into the LN2 workstation.

6. When transferred to the aluminium well in the LN2 workstation specimen carriers will sometimes fall into their two halves, but more frequently they will stick together. We like to split our specimen carriers before we put them in storage or into freeze substitution solutions. That way, we can recycle at least half the specimen carriers. To split the carriers, we had a shop build a splitting device based on the one shown in the article by Craig et al. *(27)*. For the Pella specimen carriers, it is necessary to make the well a little deeper than the one in the article.

7. An easy way to insulate tweezers is to buy self-adhesive Velcro strips at the hardware store and cut pieces that will fit on the sides of the tweezers.

8. Neoprene gloves sold at Scuba diving stores work well for handling cold samples. Get the ones with textured palms and fingers for better dexterity and Velcro closures at the wrists so LN2 cannot get inside the gloves.

9. It is essential that the tip of the specimen carrier holder be warmed up and dried off between freezing runs. After unloading the specimen carrier in the LN2 workstation, we dry the specimen carrier holder tip with a heat gun. An easy way to do this is to work on a cart or tabletop that is heat resistant. Lay the heat gun down on one side of the table and the specimen holder a distance away such that it will not get too hot, then turn on the heat gun. At this point, you can go back to the

HPF and take care of the specimen you just froze. Just do not leave the heat gun on so long that the tip of the specimen holder gets too hot.

10. When trying to remove a pellet from the bottom of a microfuge tube, it is much easier to work with just the tip of the tube than the whole thing. Dog toenail clippers work very well for cutting off the tips of microfuge tubes. They are available at most pet supply stores.

11. We have found that a quick and easy way to prepare certain samples for loading into specimen carriers is to punch out a disc that is just the diameter of the carrier. If the tissue is very stiff, then you can use a 2.0-mm diameter punch (Bal-Tec Cat. no. B8010 111 25) and the disc will entirely fill the cavity of the specimen carrier. If the sample is only moderately stiff, it will sometimes "relax" a little after you cut out a disk. In this instance, you can use a smaller diameter (e.g., 1.8 mm) punch to compensate for the expansion. If you cannot find punches of this diameter, a machine shop can make them.

12. This section and **Subheadings 3.3.** and **3.4.** are modifications of a checklist prepared by Tom Giddings for the Balzers HPM 010 HPF in Andrew Staehelin's lab at the Dept. of Molecular, Cellular, and Developmental Biology, University of Colorado, Boulder, CO 90309, USA. The instructions that come with the Bal-Tec machine are much more detailed that this, and this abbreviated version works better for us as a day-to-day operating document. When using any cryopreparation equipment be sure to observe all safety precautions *(28)*.

13. When using the Types A and B specimen carriers, you can tell if you had an air space in the carrier if it is collapsed inward (toward the sample cavity) when you remove it from the specimen holder tip. We generally discard these samples, though we do not actually know if the samples are poorly frozen.

14. Paper points are small wicking tools used by dentists in root canal surgery. They are also very useful for controlled wicking of small volumes such as specimen carriers. With a little practice, this is a quick and easy method for wicking away excess fluid from delicate specimens. We like to use one on either side of the specimen carrier because this reduces the tendency of cells to crawl up onto the paper point.

15. The yeast need to be inactivated by heating or by using very old yeast. Otherwise, they begin to grow and produce gases that create small, collapseable spaces. With their small size, they are excellent at space-filling, and if not fully hydrated, they act as little microsponges to absorb excess fluids transferred into the specimen carrier with the sample. You can also use a fluid that is compatible with your sample to make up the paste. Finally, prior to embedding and polymerization, it is easy to separate most samples from the yeast paste making it much easier to orient your sample for sectioning, if so desired.

16. If the Nitrogen Level light comes on during a run, confirm that the tank is empty and shut off the valve on the tank. Switch quickly to a new tank and open the valve all the way. When the Nitrogen Level light goes off, proceed as before. You should do a test run or two in case gaseous nitrogen has entered the line. If the alcohol level drops below the minimum level indicator (or the red light on the

console of older HPF models indicates you are out of alcohol), shut off the nitrogen gas cylinder, open up the alcohol fill plug, fill with alcohol, replace the plug and turn on the nitrogen gas. Proceed as before.

17. You can do this by using plastic transfer pipets, or a bulb pipettor, but it is much easier, quicker, and more reliable to use a repeater pipettor that will dispense 1.5 mL aliquots.

18. The solubility properties of uranyl acetate in water and other solvents seems to vary considerably. In general, more recent vintages seem more soluble than some, but not all, of the older reagents. If your uranyl acetate does not dissolve easily in methanol, try different bottles, and, if necessary, buy a new bottle.

Acknowledgments

The author would like to thank all those who contributed material used in this chapter. Their names and affiliations are listed with the appropriate figures. The author also wants to thank Colleen Lavin of the Integrated Microscopy Resource in Madison, WI, USA for giving me the address for ordering dialysis tubing in the U.S. Thanks also to Mary Morphew, Dick McIntosh, Tom Giddings, and Andrew Staehelin of the Department of Molecular, Cellular and Developmental Biology, Boulder, CO, USA, for their help with high-pressure freezing and microscopy when the author was working in that department.

References

1. Moor, H. (1987) Theory and practice of high pressure freezing, in *Cryotechniques in Biological Electron Microscopy* (Steinbrecht, R. A. and Zierold, K., eds.), Springer-Verlag, Berlin, pp. 175–191.
2. Dahl, R. and Staehelin, L. A. (1989) High-pressure freezing for the preservation of biological structure: Theory and practice. *J. Electron Microsc. Tech.* **13**, 165–174.
3. Studer, D., Michel, M., and Müller, M. (1989) High pressure freezing comes of age. *Scanning Microscopy* **S3**, 253–269.
4. Kiss, J. Z. and Staehelin, L. A. (1995) High pressure freezing, in *Rapid Freezing, Freeze Fracture, and Deep Etching* (Severs, N. J. and Shotton, D.M., eds.), Wiley-Liss, New York, pp. 89–104.
5. Hayat, M. A. (1981) *Fixation for Electron Microscopy*, Academic, New York.
6. Hayat, M. A. (1989) *Principles and Techniques of Electron Microscopy*, 3rd ed., CRC Press, Boca Raton, FL.
7. Griffiths, G. (1993) *Fine Structure Immunocytochemistry*, Springer-Verlag, Berlin.
8. Steinbrecht, R. A. and Müller, M. (1987) Freeze substitution and freeze drying, in *Cryotechniques in Biological Electron Microscopy* (Steinbrecht, R. A. and Zierold, K., eds.), Springer-Verlag, Berlin, pp. 149–172.
9. Studer, D., Hennecke, H., and Müller, M. (1992) High pressure freezing of soybean nodules leads to an improved preservation of ultrastructure. *Planta* **188**, 155-163.

10. Kellenberger, E., Johansen, R., Maeder, M., Bohrmann, B., Stauffer, E., and Villiger, W. (1992) Artefacts and morphological changes during chemical fixation. *J. Microscopy* (Oxford) **168,** 181–201.

11. Steinbrecht, R. A. (1993) Freeze-substitution for morphological and immunocytochemical studies in insects. *Microsc. Res. Tech.* **24,** 488–504.

12. McDonald, K. (1994) Electron microscopy and EM immunocytochemistry. *Meth. Cell Biol.* **44,** 411–444.

13. Keene, D. R. and McDonald, K. (1993) The ultrastructure of the connective tissue matrix of skin and cartilage after high pressure freezing and freeze substitution. *J. Histochem. Cytochem.* **41,** 1141–1153.

14. Robards, A. W. and Sleytr, U. B. (1985) Low temperature methods in biological electron microscopy, in *Practical Methods in Electron Microscopy*, vol. 10 (Glauert, A. M., ed.), Elsevier, Amsterdam.

15. Gilkey, J. C. and Staehelin, L. A. (1986) Advances in ultrarapid freezing for the preservation of cellular ultrastructure. *J. Electron Microsc. Tech.* **3,** 177–210.

16. Kellenberger, E. (1987) The response of biological macromolecules and supramolecular structures to the physics of specimen cryopreparation, in *Cryotechniques in Biological Electron Microscopy* (Steinbrecht, R. A. and Zierold, K., eds.), Springer-Verlag, Berlin, pp. 35–63.

17. Echlin, P. (1992) *Low-Temperature Microscopy and Analysis*, Plenum, New York.

18. Walther, P., Chen, Y., Pech, L. L., and Pawley, J. B. (1992) High-resolution scanning electron microscopy of frozen-hydrated cells. *J. Microsc.* (Oxford) **168,** 169–180.

19. Walther, P. and Müller, M. (1997) Double-layer coating for field-emission cryo-scanning electron microscopy—present state and applications. *Scanning* **19,** 343–348.

20. Michel, M., Hillmann, T., and Müller, M. (1991) Cryosectioning of plant material frozen at high pressure. *J. Microsc.* (Oxford) **163,** 3–18.

21. Hippe-Sanwald, S. (1993) The impact of freeze substitution on biological electron microscopy. *Microsc. Res. Tech.* **24,** 400–422.

22. Nicolas, M.-T. and J.-M. Bassot. (1993) Freeze substitution after fast-freeze fixation in preparation for immunocytochemistry. *Micros. Res. Tech.* **24,** 474–487.

23. Villiger, W. (1991) Lowicryl resins, in *Colloidal Gold: Principles, Methods and Applicatons*, vol. 3 (Hayat, M. A., ed.), Academic Press, San Diego, CA, 59–71.

24. Newman, G.R. and Hobot, J.A. (1993) *Resin Microscopy and On-Section Immunocytochemistry*, Springer-Verlag, Berlin.

25. Hohenberg, H., Mannweiler, K., and Sleytr, U. B. (1994) High pressure freezing of cell suspensions in cellulose capillary tubes. *J. Microsc.* (Oxford) **175,** 34–43.

26. Hohenberg, H., Tobler, M., and Müller, M. (1996). High-pressure freezing of tissue obtained by fine-needle biopsy. *J. Microsc.* (Oxford) **183,** 133–139.

27. Craig, S., Gilkey, J. C., and Staehelin, L. A. (1987) Improved specimen support cups and auxiliary devices for the Balzers high pressure freezing appartus. *J. Microsc.* (Oxford) **48,** 103–106.

28. Sitte, H., Neumann, K., and Edelmann, L. (1987) Safety rules for cryopreparation, in *Cryotechniques in Biological Electron Microscopy* (Steinbrecht, R. A. and Zierold, K., eds.), Springer-Verlag, Berlin, pp. 285–289.

6

The Application of LR Gold Resin for Immunogold Labeling

J. R. Thorpe

1. Introduction

The routine preparation of biological samples for transmission electron microscopy (TEM) usually involves a double-fixation in, first, glutaraldehyde and subsequently in osmium tetroxide (OsO_4). The specimens are then dehydrated and embedded in a (heat-polymerized) epoxy resin. While this procedure may produce excellent ultrastructural preservation (*see* Chapter 1), the antigenicity of most tissue proteins is severely affected. Subsequent successful immunolabeling of such sections with a specific antibody, visualized with either protein A (pA)- or secondary antibody (IgG)-bound gold probe, is thus rare. This loss of antigenicity may reflect denaturation (through fixation, exposure to alcohol, or to the heat of resin polymerization) or inaccessibility (because of high-density crosslinkage of proteins with the fixatives and resin). Protein integrity may be maintained to a degree by adopting a more minimal fixation regime (e.g., omission of OsO_4 [*see* **Note 1**], decreased concentration of aldehyde and lowered temperature). Additionally, certain antigens in epoxy resin-embedded specimens may also be "unmasked" by "etching" of the sections (e.g., in saturated sodium metaperiodate). However, the remaining component of heat denaturation may still preclude (or hinder; *see* **Note 1**) successful immunolocalizations.

Recently, a number of different (UV light-polymerized) low-temperature resins have been devised (*see* Chapter 7). The use of these resins has dramatically increased the range of proteins that may be successfully immunolocalized. However, their use in the recommended procedures often necessitates the use of relatively expensive cryopreparative equipment. This chapter describes a

From: *Methods in Molecular Biology*, vol. 117: *Electron Microscopy Methods and Protocols*
Edited by: N. Hajibagheri © Humana Press Inc., Totowa, NJ

Table 1
Antigen Localizations Utilizing LR Gold Resin

Specimen	Antigen(s)	Ref(s)
Sheep pituitary tissue	Prolactin, growth hormone	*(1)*
Sheep pituitary cells	Prolactin, growth hormone	*(2)*
Rat/mouse islet of Langerhans	Insulin, glucagon, somatostatin, opioid peptides	*(3,4)*
Locust neuron	Vasopressin-like reactive peptide	*(5)*
Snail auricle	Neuropeptides	[a]
Shrimp hepatopancreas	α-amylase, chymotrypsin and trypsin	[a]
Yeast	Horseradish peroxidase	[a]
Yeast mitochondria (isolated)	Heat-shock protein	*(6)*
Wheat and pea leaves	Glycine decarboxylase	*(7)*
Pea leaf mitochondria (isolated)	Glycine decarboxylase	*(8)*

[a]Unpublished.

very simple procedure utilizing the aromatic, crosslinked, acrylic resin "LR Gold" (London Resin Co., Reading, U.K.), which requires no such equipment.

The described methodology has been used to localize a wide variety of plant and animal antigens in the author's laboratory over the last decade or so (*see* **Table 1**). Some of these antigens were not localizable at all following embedding in standard epoxy resins. With others, the enhanced retention of protein antigenicity with LR Gold embedding enabled the use of antibodies at dilutions some 100-fold greater, such that the specificity of immunolabeling was markedly improved. To summarize, the technique involves a minimal fixation (usually in a mixture of formaldehyde and glutaraldehyde), a thorough buffer rinse, followed by dehydration in ethanol and embedding in LR Gold resin. The whole preparative procedure, including the (near UV) light-polymerization of the resin, is carried out at 4°C.

2. Materials

1. Glutaraldehyde: This is widely commercially available. A 25% solution is most common for use in fixation. An "EM" grade or equivalent should be used. For certain sensitive antigens a purer form of glutaraldehyde (e.g., monomeric, vacuum-distilled) may be required. (Storage: At 4°C; Stability: Several years; Hazards: Harmful by ingestion and inhalation. Extremely irritating to the eyes. Prolonged skin contact can cause dermatitis and sensitization. Toxicity data: LD50 134 mg/kg oral, rat. *Should be handled in a fume hood with gloves.*)

2. Formaldehyde: This should be prepared from (powdered) paraformaldehyde as a 10% aqueous solution. This is heated to 60°C with stirring, and then 1 *M* NaOH is added drop by drop until the solution clears. (Storage: At 4°C; Stability: Best to make up freshly for each fixation but may be stored for a few months; Hazards: Paraformaldehyde is harmful by ingestion and irritating to the skin, eyes and respiratory system. Can cause sensitization by skin contact. Toxicity data: LD50 800 mg/kg oral, rat. Formaldehyde solution is toxic by ingestion and inhalation. Prolonged exposure to vapour can cause conjunctivitis, laryngitis, bronchitis, or bronchial pneumonia. Causes burns to eyes and skin, with cracking and ulceration, particularly around fingernails. Possible risk of cancer caused by chronic inhalation. Toxicity data: LD50 800 mg/kg oral, rat; LC50 0.9 mg/L inh., rat. *Should be handled in a fume hood with gloves.*)

3. Buffer: Standard EM buffers should be used. Phosphate-buffered saline (PBS) or sodium cacodylate/HCl (50–200 m*M*) is suitable for animal tissues; the latter or sodium phosphate (50–200 m*M*) buffers for plant tissue. With a new tissue it is always advisable to try a range of buffer molarities in an initial check for the best ultrastructural preservation. All buffers should be at pH 7.0. (Storage: At +4°C.; Stability: A few weeks; Hazards: Sodium cacodylate is toxic by inhalation, ingestion and skin contact. Symptoms include nervousness, thirst, vomiting, diarrhoea, cyanosis, and collapse. Irritating to skin, eyes, and mucous membranes. Inhalation of small concentrations over a long period will cause poisoning. Danger of cumulative effects. Should be treated as a suspected carcinogen and may cause adverse mutagenic or teratogenic effects. Toxicity data: LD50 2600 mg/kg oral, rat.)

4. Ethanol: This is used for specimen dehydration. It is advisable to dry the absolute ethanol by adding molecular sieve pellets (type 3A).

5. LR Gold resin and benzil: The resin and catalyst for light-polymerization may be purchased from most EM suppliers for embedding. (Storage: At 4°C; Stability: Several years; Hazards: LR Gold resin and benzil are irritants.)

6. Polymerization: The requirements are as follows:
   a. a transformer (to provide a 100 W, 12V supply);
   b. a projector lamp (Philips Type 6834);
   c. a retort stand and two clamps;
   d. a glass plate;
   e. a 96-well microtiter plate;
   f. gelatin capsules.

7. Sectioning: Standard copper or nickel grids may be used; the high transmission type are generally preferable.

8. Immunolabeling: Requirements are as follows:
   a. fine forceps;
   b. glass Petri dishes;
   c. Parafilm;
   d. a 96-well microtiter plate;
   e. a humid box (small plastic box and lid with damp tissue paper).

9. Blocking incubation: Requirements are nonimmune sera from the donor species for the IgG-gold probe if used or a modified phosphate-buffered saline (PBS+) if protein A(pA)- or protein G(pG)-gold probes are used. PBS+ is made up as follows *(1)*:

Bovine serum albumin	10.0 g/L
Na$_2$HPO$_4$ (anhydrous)	0.524 g/L
KH$_2$PO$_4$ (anhydrous)	0.092 g/L
NaCl	8.76 g/L
NaEDTA	0.372 g/L
NaN$_3$	0.2 g/L
Tween-20	500 μL/L

This buffer is thoroughly mixed and the pH adjusted to 8.2. It is best to filter immediately before use (0.22 μm). This buffer is subsequently referred to as PBS+ and is used as diluent for all antisera and gold probes as well as for all the buffer rinse steps. (Storage: At 4°C; Stability: a few months; Hazards: NaN$_3$ is very toxic by inhalation and ingestion, causing somnolence, headache, irritability, and gastric disturbances. Irritating to skin and eyes. Evidence of mutagenic effects. EDTA is harmful if ingested in quantity. May be irritating to the eyes. Evidence of reproductive effects. Toxicity data: NaN$_3$ LD50 27 mg/kg oral, rat; EDTA no data.)

10. Specific antibody incubation: The absolute prerequisite is a good specific antibody (*see* **Note 2**).

11. Gold Probes: Either pA-, pG-, or IgG-bound gold probes may be used, dependent on the antibody used (*see* **Note 3**).

12. Section Poststaining: A 0.5% aqueous solution of uranyl acetate is made up and then filtered (Whatman No. 1). (Storage: At +4°C in dark; Stability: several months; Hazards: Very toxic by ingestion or if inhaled as dust. Toxic by skin contact if skin is cut or abraded. Danger of cumulative effects. Radioactivity 10,400 Bq/g. Should be treated as a suspected carcinogen. May cause adverse mutagenic and teratogenic effects.)

Lead citrate: This is made up from three stock solutions as follows:

Stock solutions:	A	trisodium citrate	37.7 g/100 mL water
	B	lead nitrate	33.1 g/100 mL water
	C	1 *M* NaOH	4.0 g/100 mL water

To make up, dispense 0.64 mL distilled water into an Eppendorf tube, then add 0.12 mL A, 0.08 mL B, and finally 0.16 mL C (whirlimix after each addition). Spin down for 10 min in a standard bench centrifuge. (Storage: At 4°C; Stability: several months; Hazards: lead nitrate is harmful by ingestion and if inhaled as dust causes severe internal injury, with vomiting, diarrhea, and collapse. Major hazard is because of the cumulative effects of lead. Symptoms include digestive disturbances, pallor, anemia, blue line on gums, and irritation of the eyes. Evidence of reproductive effects. Trisodium citrate is harmful by ingestion of large amounts and may cause diarrhoea, nausea, vomiting, hypernoea, and convulsions. May irritate eyes and respiratory system if inhaled as dust. NaOH is corrosive and irritant and harmful as dust. If ingested causes severe internal irritation and damage. Toxicity data: no data.)

3. Methods

Steps 1–4 are carried out at 4°C.

1. Fixation: Specimens are sliced directly into fixative; individual pieces should not be larger than 1 mm³, and preferably smaller still. The fixative should be buffered in PBS or sodium cacodylate/HCl (0.05–0.2 M) for animal tissue or the latter or sodium phosphate for plant tissue. (N. B. the pH of any buffered fixative solution should always be checked after preparation). With any new specimen/antibody combination it is best to carry out an initial range of fixative combinations to optimize the compromise between ultrastructural preservation and the antigenicity of the particular protein under study (*see* **Note 1**). This would typically be: 1. 2.5% glutaraldehyde, 2. 4% formaldehyde/1% glutaraldehyde, 3. 4% formaldehyde/ 0.25% glutaraldehyde, and 4. 4% formaldehyde.

2. Dehydration: After 3 h the specimens are thoroughly rinsed in buffer (minimum of 4 changes in 2 h, but better to leave overnight) and then dehydrated through an ethanol series. The latter comprises 50%, 75% and ×3 absolute ethanol (all for 20 min each, with the last few changes in dried absolute alcohol [*see* **Subheading 2.4.**]).

3. Embedding: The final (dried) absolute ethanol is removed and replaced with LR Gold resin. The following day the resin is replaced with fresh resin to which 0.1% (w/v) benzil has been added. The duration required for complete resin infiltration varies with specimen size and type (*see* **Note 6**). As a rough rule of thumb, very small tissue pieces may be polymerized after 3–4 d whereas larger samples—and especially plant tissues—should be left for a week. The LR Gold resin + (0.1% w/v) benzil is changed every other day or so during this infiltration period.

4. Polymerization: The projector lamp is held in place by one clamp at the base of the retort so that it shines upwards. The other clamp is used to hold the glass plate directly above the lamp at a distance of c. 20 cm. (Hazards: the projector lamp should be surrounded by light-proofing to avoid near UV damage to the eyes.) Specimens are pipetted into gelatin capsules (the resin must fill the capsule completely as the polymerization process is anaerobic) and the capsule top replaced. The capsules are then placed in a 96-well microtiter plate and placed on the glass plate of the polymerization apparatus directly above the lamp. Polymerization is complete after 24–48 h.

5. Sectioning: Ultrathin sections are cut as for standard (epoxy) EM resins. Properly embedded LR Gold blocks have good sectioning quality and the acquisition of ultrathin sections should be easily achievable. These should be allowed to dry down naturally. It is usually best to cut sections fresh for any immunolabeling experiment.

6. Immunolabeling: For all incubations grids are placed on 20 µL droplets on Parafilm within a Petri dish. Buffer rinses are carried out by floating grids in the (filled to convexity) wells of a (96-well) microtiter plate. When transferring grids from one solution to another, great care should be taken not to wet the opposite face to that being labeled; an exaggerated "lifting" action should be used.

7. Blocking Step: For pA- or pG-gold immunolabeling this should be in PBS+ alone (contains 1% BSA as blocking agent). If an IgG-gold probe is used then a 1:10 dilution (in PBS+) of nonimmune serum from the (gold probe) IgG donor species should be used. Both incubations should be 20–30 min at room temperature. After blocking, the grids are transferred directly to the specific antibody.

8. Incubation in Antibody: A first immunolabeling with any new specimen/ polyclonal antibody combination should include a titration series of the specific antibody to establish the optimal dilution (*see* **Note 2**). This would routinely be 1:100, 1:1000 and 1:10,000. Concurrent incubations should be carried out in nonimmune serum (at similar dilutions) from the primary antibody donor species. This titration series is unnecessary for a monoclonal antibody (as there are no unwanted antibodies to be diluted out); dilutions of 1:5 to 1:20 are commonly used. The incubation may be either 2 h at room temperature or overnight at 4°C (in a humid box). The latter is preferable as it appears to produce lower levels of nonspecific binding (Thorpe, unpublished data). The grids are subsequently rinsed in PBS+ (3×2 min) before transferring to the gold probe.

9. Gold Labeling: Gold probes are diluted 1:10 in PBS+ immediately before use. Incubation for 1 h at room temperature is usually optimal. (For the applicabilities of different gold probe types and sizes, *see* **Note 3**.) The grids are then rinsed thoroughly in buffer (3×10 min) and then double glass-distilled water (4×3 min).

10. Section Poststaining: LR Gold resin is hydrophilic and takes up the uranyl and lead stains considerably more rapidly than the standard epoxy resins. For 5–10 min in 0.5% aqueous uranyl acetate and 1–2 min in lead citrate is usually sufficient to give good contrast (*see* **Note 3**). Lead citrate should be dispensed from the top of the solution. Staining should be carried out in a covered Petri dish containing a small dish of NaOH pellets. Do not breathe on the lead citrate at any stage as insoluble lead carbonate will form which produces dense spots over the sections.

The methodology described herein is a simple way to improve the retention of protein antigenicity in samples prepared for EM immunogold labeling (*see* **Fig. 1**). Polymerized blocks are ready in 6–9 d, dependent on the type of specimen being processed. **Table 1** (*see* **Subheading 1.**) summarizes the range of plant and animal antigens that have been successfully immunolocalized in the author's laboratory.

4. Notes

1. Initial Preparation: The reader should note that any new antibody/antigen investigation will entail an initial range of preparative/immunological procedural variations as described herein. **Table 2** shows quantified data from such an initial investigation that illustrates the potential advantages of the described methodology. In this particular case, for example, the omission of osmium significantly enhanced the immunolabeling; this reflects the common occurrence of epitope susceptibility to osmication. However, the most optimal immunolabeling was achieved after

embedding in LR Gold resin in place of Araldite (with the rest of the preparation being identical). This particular antibody could then be used at a 10-fold greater dilution while specific immunolabeling was still doubled. The improvement in specific: nonspecific binding ("signal-to-noise") was from 141 to 925:1.

2. Antisera: The primary requirement for successful immunolabeling is a good antiserum. Polyclonal antisera have the advantage that they are generally multivalent and thus produce stronger labeling. They should preferably be of high titer so that they may be used at high dilution; this reduces background nonspecific labeling to a minimum. Monoclonal antibodies might intrinsically appear to be preferable, as they should contain no unwanted antibodies and nonspecific labeling is thus usually minimal. However, the particular epitope against which the monoclonal is directed may be present in related (or other) proteins so that cross reaction may give rise to misleading immunolabeling. Also, this epitope may be destroyed by the preparative process. Whichever antibody is used, its specificity must be verified immunocytochemically as well as immunologically. A minimum approach would be i) to immunolabel a known positive tissue and ii) to preabsorb the antibody with the homologous and related antigens before use (to check for the absence and presence of labeling, respectively). For an excellent overview of immunological and methodological problems associated with immunocytochemical work refer to Van Leeuwen *(9)*.

   Some thought should be given to the preparation (or purchase) of antibodies if double-immunolabelings are to be pursued. If the primary antibodies are raised in different donor species then the most simple methodology may be used: visualization with two secondary antibody-gold probes (of different sizes) against the two primary donor species (e.g., goat antirabbit 10 nm and goat antimouse 5 nm gold probes). Double (pA- or IgG-gold) immunolabelings for two antisera raised in the same animal are possible, but the methodologies are more complicated and prone to immunological and method artefacts (e.g., *see* **ref.** *10*).

3. Choice of Gold Probe: The two major considerations in the choice of gold probe are type (protein A[pA]-, protein G [pG]- or IgG-bound) and particle size. *Staphylococcal* pA binds to the Fc portion of IgG molecules of many animal species: the strongest binding is to human, rabbit, guinea pig, swine, monkey, and dog; it has less affinity for cow, goat, and mouse and considerably weaker affinity for sheep, horse, and rat *(11–13)*. *Streptococcal* pG binds strongly to all human IgG subclasses and rabbit, cow, horse, goat, sheep, and swine IgG; it also has moderate affinity to mouse, rat, guinea pig, and dog IgG *(14)*. IgG-gold probes will generally give a higher immunolabeling density as there are, potentially, multiple binding sites on the primary antibody. Protein A-gold probes may be preferred where relative quantitative assessments of immunolabeling density are required (*see* Chapter 11).

   The usual range of gold probe size for TEM is 5–20 nm. A smaller (5 nm) probe size has the advantage of improved spatial resolution over the tissue (may be important with membrane-associated antigens, for example). Immunolabel density will also be higher for a given OD of gold probe incubation dilution (there is an eight-

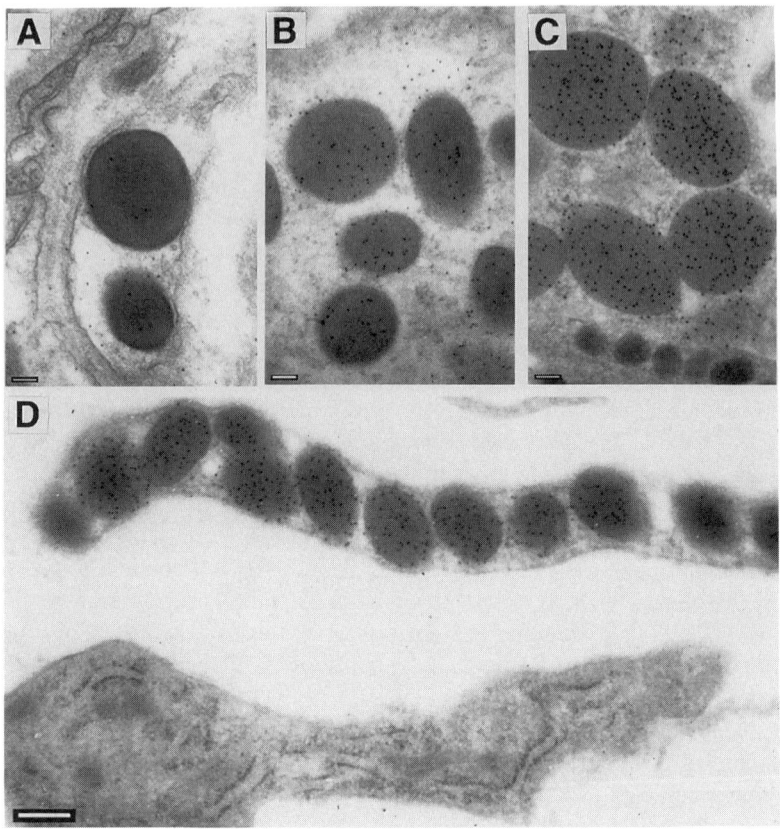

Fig. 1. Comparative Immunolabeling Densities with Different Preparative Procedures. Images (digital TEM) of sections from (snail auricle) tissue prepared by three different procedures (from which the data in Table 2 was derived). In all cases, the tissue was fixed initially in a mixture of 1% formaldehyde and 2.5% glutaraldehyde for 4 h at 4°C. The LR Gold sample was processed as described herein. Araldite samples were resin-infiltrated at room temperature and then heat-polymerized. All sections were immunolabeled with the same (polyclonal rabbit antineuropeptide) antibody (araldite samples at a dilution of 1:100 and the LR Gold sample at 1:1000). (Sections from araldite samples were etched for 1 hour in saturated aqueous sodium metaperiodate before immunolabeling). The 10 nm protein A-gold probe was used throughout. **(A)**: Araldite-embedded tissue additionally postfixed in 1% osmium tetroxide for 30 min at 4°C. Note very low level of specific labeling (over neuropeptide-containing granule). (Original magnification ×50,000; bar = 100 nm). **(B)** Araldite-embedded tissue (no postfixation): Note much lower immunolabeling density (at 1:100 antibody dilution) compared with the LR Gold samples (at 1:1000 antibody dilution) and rather high nonspecific background over the cytoplasm. (Original magnification ×50,000; bar = 100 nm). **(C, D)** LR Gold-embedded tissue: Note very dense specific immunolabeling and virtually complete absence of background. **(C)** original magnification ×50,000; bar = 100 nm; **(D)** original magnification ×20,000; bar = 250 nm).

**Table 2
Comparisons of Immunolabeling Densities with Different Preparative
Procedures**

Fixation	Post-fixation	Embedding	Section pre-treatment	Antibody dilution	Immunolabeling density specific	non-specific
4%F/2.5%G	None	LR Gold	None	1:1000	54.37 ± 4.29	0.06 ± 0.03
4%F/2.5%G	None	Araldite	Etched[a]	1:100	26.18 ± 2.45	0.19 ± 0.03
4%F/2.5%G	0.5% Os	Araldite	Etched[a]	1:100	4.47 ± 0.86	1.67 ± 0.30

[a]Sections etched for 1 h in saturated aqueous sodium meta-periodate.
Abbreviations: F = formaldehyde; G = glutaraldehyde; Os = osmium tetroxide. 10 nm pA-gold probe used throughout.
Results are from snail auricle tissue immunolabeled for a neuropeptide (unpublished data). All primary fixations were for 4 h at 4°C; postfixation 30 min at 4°C. LR Gold samples were processed as described herein; araldite samples were resin-infiltrated at room temperature and then heat-polymerized. Immunolabeling densities are expressed as number of gold particles per mean (specific peptide-containing) granule profile area (specific = labeling over granules; nonspecific = labeling over neighboring muscle fibres; n = 20 ± s.e.m.).

fold reduction in number with a doubling in size). With the larger gold probes effects of steric hindrance may also decrease the observed immunolabel density.

Some consideration should be given to the gold probe size used and the electron-density of the expected location of labeling when section poststaining times are chosen. For example, 5 nm gold particles may be hard to visualize over densely stained granules.

4. High Nonspecific Labeling: Unacceptably high levels of background may occur for a number of reasons. If gold particles are visible over areas of resin away from tissue then the problem is methodological (e.g., insufficient rinsing or gold probe contamination of, and drying down onto, the reverse section face). High tissue-specific background may occur because a) the antibody (if polyclonal) needs further dilution to minimize the proportion of unwanted antibody populations, b) the blocking step was incomplete or incorrect (check and remake solutions if necessary), or c) there is cross-reaction of the antibody with a ubiquitous epitope within the tissue. The latter problem, unfortunately, is often insoluble unless the cross-reactive antigen is known and can be preabsorbed with the antibody before immunolabeling.

5. Negative Results: Although the described methodology has been used to successfully immunolocalize a wide variety of antigens, there may be occasions when there is no positive specific labeling visible. If your antibody is known to be good, and a range of specimen fixations and dilutions of antibody, as described herein, have been performed, then the antigenicity of the protein may have been lost. If a monoclonal antibody has been used then it is worth trying a polyclonal antiserum (if available); this will verify that it is not just that one particular

epitope, against which the monoclonal antibody is directed, which has been affected. Negative results have been noted by this author especially with membrane-bound antigens (e.g., some proteins on mitochondrial and yeast cell membranes). It should be noted that this may sometimes reflect a rarity of epitope exposure (the chance of sectioning a "glancing face" of the membrane) rather than loss of antigenicity. In this instance, it may be that an alternative method of preembedding immunolabeling may be of choice if the antigen is on the outer membrane of an isolated organelle or cell preparation (*see* Chapter 12).

Having taken all the above into account, it may be that a negative result reflects the failure of the preparative procedure to preserve antigenic integrity. A first step would be to lower the temperature of dehydration and embedding into LR Gold (to –25°C as originally recommended *[15]*) if the required low-temperature facilities are available (*see* Alternative Method 1 below). If this is still unsuccessful then unfixed tissue processing may be attempted (*see* Alternative Method 2 below). Of course, ultrastructural preservation will be compromised. If neither of the latter produce positive labeling then the aliphatic, crosslinked acrylic resin Lowicryl K4M may be used (dependent on the low temperature equipment being available); this may be cured at –35°C (*see* Chapter 7). Alternatively, cryosubstitution may be attempted (*see* Chapter 9).

Alternative Method 1: After fixation as in Method 1 above, samples are prepared as follows:

50% methanol	4°C	30 min
70% methanol	–25°C	60 min
90% methanol	–25°C	60 min
100% methanol	–25°C	60 min
LR Gold resin + (0.1% w/v) benzil	–25°C	60 min
LR Gold resin + benzil	–25°C	overnight
Polymerization	–25°C	

Alternative Method 2: (for unfixed tissue): The procedure is as follows (PVP = polyvinyl pyrollidone; mol wt 44,000):

50% methanol + 20% PVP	0°C	15 min
70% methanol + 20% PVP	–25°C	45 min
90% methanol + 20% PVP	–25°C	45 min
50% LR Gold resin/50% methanol + 10% PVP	–25°C	30 min
70% LR Gold resin/30% methanol + 10% PVP	–25°C	60 min
LR Gold resin	–25°C	60 min
LR Gold resin + [0.1% w/v] benzil	–25°C	60 min
LR Gold resin + benzil	–25°C	overnight
LR Gold + benzil	–25°C	c. 24 h.
Polymerization	–25°C	

6. Poor Embedding: If embedding is found to be incomplete—evidenced by poor sectioning quality and the presence of holes—then this may reflect insufficient dehydration or duration of resin infiltration. The specimen size may be too large; if this is unavoidable (because of sectioning considerations) then the

length of infiltration should be increased; otherwise, of course, smaller pieces should be excised. Insufficient dehydration can be remedied by increasing the number of changes in absolute ethanol (perhaps to 5 or 6 over 1–2 h). In the author's experience plants in general, and leaf tissue and yeast cells in particular, are prone to these problems. Indeed yeast is notoriously difficult to embed in any EM resin and the production of spheroplasts (by enzymic digestion of the cell wall) will greatly help here.

References

1. Thorpe, J. R., Ray, K. P., and Wallis, M. (1990) Occurrence of rare somatomammotrophs in ovine anterior pituitary tissue studied by immunogold labelling and electron microscopy. *J. Endocrinol.* **124,** 67–73.
2. Thorpe, J. R. and Wallis, M. (1991) Immunocytochemical and morphometric investigations of mammotrophs, somatotrophs and somatomammotrophs in sheep pituitary cell cultures. *J. Endocrinol.* **129,** 417–422.
3. Khawaja, X. Z., Green, I. C., Thorpe, J. R., and Bailey, C. J. (1990) Increased sensitivity to insulin-releasing and glucoregulatory effects of dynorphin A (1-13) and U50,488h on insulin release and glucose homeostasis in genetically obese (ob/ob) mice. *Diabetes* **39,** 1289–1297.
4. Khawaja, X. Z., Green, I. C., Thorpe, J. R., and Titheradge, M. A. (1990) The occurrence and receptor specificity of endogenous opioid peptides within the pancreas and liver of the rat - comparison with brain. *Biochem. J.* **267,** 233–240.
5. Thompson, K. S. J., Baines, R. A., Rayne, R. C., Thorpe, J. R., Alef, A., and Bacon, J. P. (1993) Locust VPLI neurons contain three vasopressin-related peptides; VPLI activity reduces cyclic AMP levels in the CNS. *Soc. Neurosci. Abstr.* **19,** 1274.
6. Moore, A. L., Walters, A. J., Thorpe, J. R. Fricaud, A.-C., and Watts, F. Z. (1992) Schizosaccharomyces pombe mitochondria: Morphological, respiratory and protein import characteristics. *Yeast* **8,** 923–933.
7. Tobin, A. K., Thorpe, J. R., Hylton, C. M., and Rawsthorne, S. (1989) Spatial and temporal influences on the cell-specific localization of glycine decarboxylase in leaves of wheat (Triticum aestivum) and pea (Pisum sativum). *Plant Physiol.* **91,** 1219–1225.
8. Tobin, A. K., Thorpe, J. R., Day, D. A., Wiskitch, J. T., and Moore, A. L. (1990) Immunogold localization of glycine decarboxylase in isolated pea leaf mitochondria (7th. Congress of the Federation of European Societies of Plant Physiology). *Physiologia Plantarum* **79,** Abstract 414.
9. Van Leeuwen, F. (1982) Specific immunocytochemical localizations of neuropeptides: A Utopian goal?, in *Techniques in Immunocytochemistry*, vol. 1 (Bullock, G. R. and Petrusz, P., eds.), pp. 283–299.
10. Thorpe, J. R. (1992) A novel methodology for double protein A-gold immunolabeling utilizing the monovalent fragment of protein A. *J. Histochem. Cytochem.* **40,** 435–441.
11. Biberfeld, B., Ghetie, V., and Sjoquist, J. (1975) Demonstration and assaying of IgG antibodies in tissues and on cells by labelled Staphylococcal protein A. *J. Immunol. Meth.* **6,** 249–259.

12. Goudswaard, J., van der Donk, J. A., Noordzig, A., van Dam, R. H., and Vaerman, J. P. (1978) Protein A reactivity of various mammalian immunoglobulins. *Scand. J. Immunol.* **8,** 21–28.

13. Goding, J. W. (1978) Use of Staphylococcal protein A as an immunological reagent. *J. Immunol. Meth.* **20,** 241–253.

14. Bjorck, L. and Kronvall, G. (1984) Purification and some properties of streptococcal protein G, a novel IgG-binding reagent. *J. Immunol.* **133,** 969–974.

15. McPhail, G. D., Finn, T., and Isaacson, P. G. (1987) A useful low temperature method for post-embedding electron immunocytochemistry in routine histopathology. *J. Pathol.* **151,** 231–238.

7

Low-Temperature Embedding in Acrylic Resins

Pierre Gounon

1. Introduction
1.1. Rationale of Low-Temperature Embedding

It has been demonstrated that enzymes *(5,6)* can maintain their structure and their activity at very low temperature in concentrated organic solvent. Therefore, in order to minimize molecular thermal vibration, which can have adverse effects on specimens weakly fixed with paraformaldehyde, one can dehydrate samples partially or totally at low temperature. Carlemalm et al., 1982 *(3)* introduced the PLT technique (progressive lowering of temperature) that combines increasing organic solvent concentration with decreasing temperature, after which infiltration and polymerization are carried out. The results obtained with Lowicryls clearly show the advantages of this approach to obtain good structural preservation of cellular contents and ultrastructure. Furthermore, the PLT method employs low temperature to reduce protein denaturation and to maintain a degree of hydration, which may be important in preserving protein structural conformation. Specimens suffer most during dehydration by organic solvents, mainly ethanol, whereas final infiltration by resin monomers and polymerization seems to be less critical.

1.2. How to Manage Low-Temperature Embedding

1. It is best to have specialized equipment (BalTec FSU10 or Leica AFS; *see* **Fig. 1**, for example) with adequate temperature control and easy handling of specimens.
2. At a temperature of −20°C a freezer is sufficient, however, great care must be exercised to prevent water condensation and frost. Infiltration must be carried out in small glass vials with screw caps (*see* **Fig. 2**).
3. Ensure good protection from vapor and liquid components of acrylic resins, which are allergenic by accumulation and probably harmful. Gloves, masks and a fume hood should be used.

From: *Methods in Molecular Biology*, vol. 117: *Electron Microscopy Methods and Protocols*
Edited by: N. Hajibagheri © Humana Press Inc., Totowa, NJ

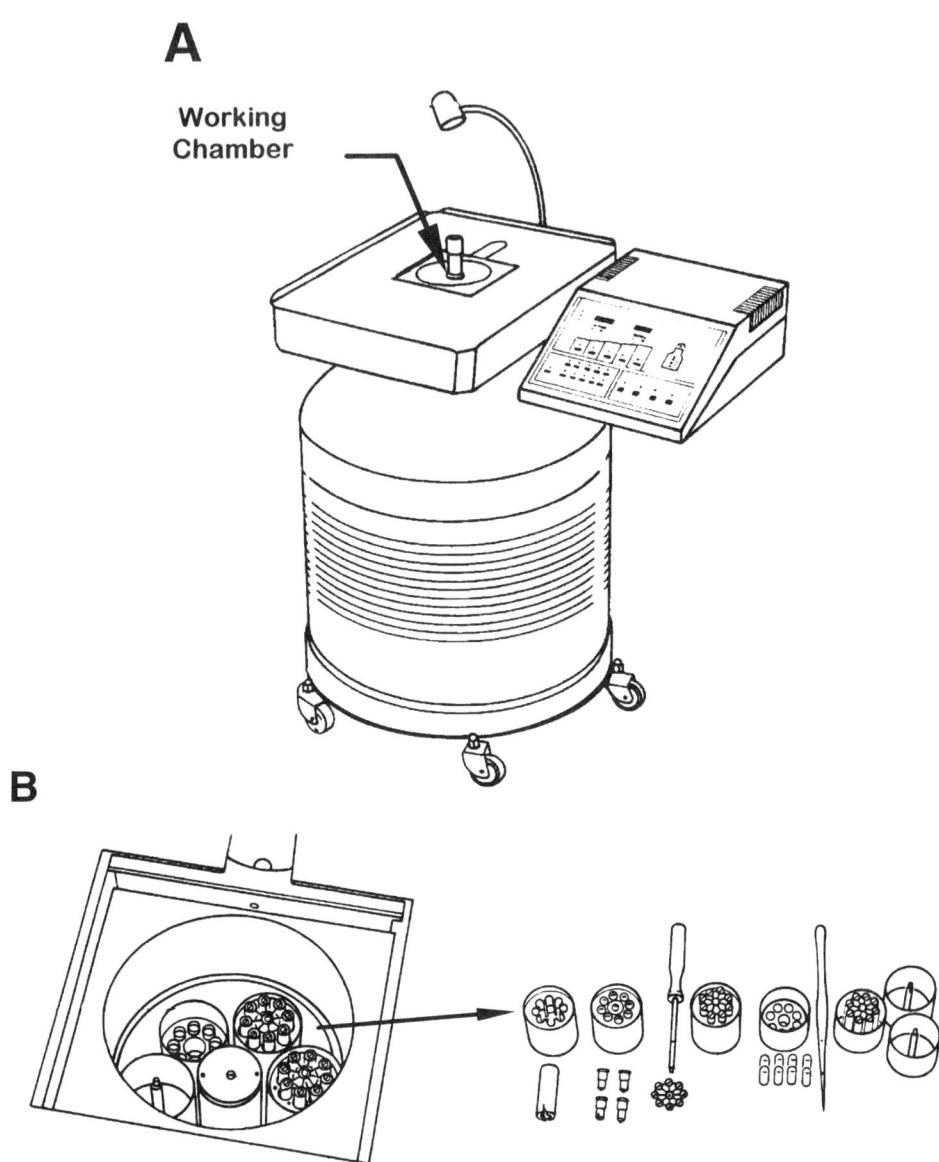

Fig. 1. (A) Leica (Reichert) AFS can be used for automatic freeze substitution, but also for controlled progressive lowering temperature (PLT). Temperature and slopes between steps are easily programmed. A warning can be emitted at different steps of the program. This system consists of a cylinder fitted into the neck of a liquid nitrogen dewar. The temperature is controlled both by heaters and nitrogen evaporation. The samples remain under nitrogen vapor, thus avoiding water condensation. (B) In the working chamber on the left (closed with a lid, not shown) many different embedding tools can be placed (right). All manipulations are performed with forceps, pipets, and screwdrivers to avoid contact with acrylic resins (drawing from Leica with permission).

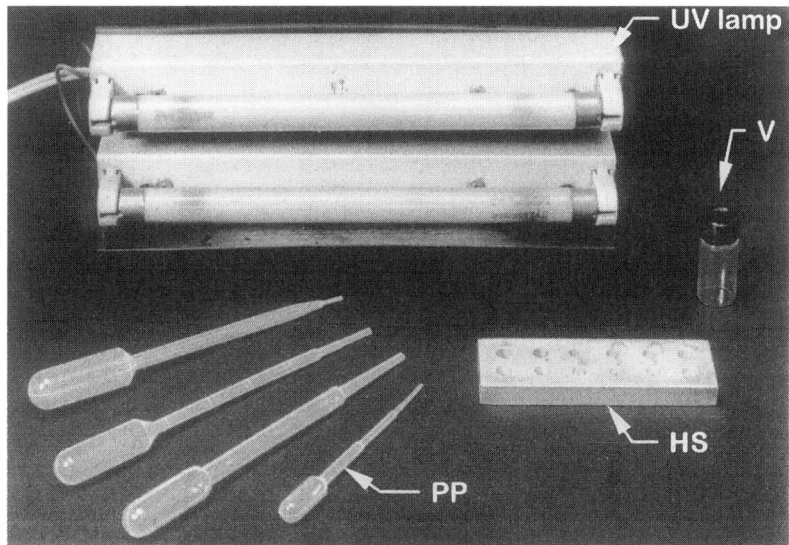

Fig. 2. Some tools for acrylic embedding and UV polymerization. For embedding in a chest freezer, a heat sink could be helpful. It is simply made of an aluminium plate with drilled holes to the exact size of gelatine capsules gage "O" (HS). A set of polyethylene pipets (PP) and small vials with screw cap (v) are also very useful. UV lamp made up with 2×6 W strips tubes (Sylvania F6W/BL350) fitted on plastic plates (UV lamp).

4. Disposal of resins and solvents is achieved after polymerization at room temperature with ultraviolet (UV) light (*see* **Fig. 2**).

1.3. How to Choose a Protocol

Hundreds of protocols are available throughout the literature for low-temperature embedding and the spectrum of resins is so large that selection is not easy. An extensive study on effect of tissue processing has been reported by Bendayan *(18)* giving interesting comparisons. Nevertheless, these simple rules will give good guidance to the beginner.

1. Lowicryl K4M is more popular than Lowicryl HM20. This may be due to the classification of HM20 as "hydrophobic" or "apolar" similar to epoxide or polyester families. Certainly the miscibility with water is less than Lowicryl K4M, but HM20 cannot be considered as a true apolar resin.
2. Lowicryl HM20 polymerizes easily in two days at –50°C under UV light. Thin sectioning is easier since the block surface does not have a tendency to wet, as with Lowicryl K4M, thus the water level in the knife trough does not have to be very low.
3. Immunocytochemical background observed with Lowicryl HM20 is equivalent to that observed with Lowicryl K4M.

4. Unicryl is easy to manipulate as it is very fluid at –20°C and even at –35°C. Although Unicryl polymerizes slowly, this does not affect either staining properties, or immunocytochemical sensitivity.
5. Unicryl is useful for observation of tissue in semithin sections.
6. Generally, Lowicryl K4M, Lowicryl HM20, and Unicryl give similar results as far as the ultrastructural details and immunocytochemical sensitivity are concerned. However the author prefers Lowicryl HM20 for its high performance at very low temperature and Unicryl is good for staining properties and fast infiltration of specimens at –20°C.

1.4. Choice of Dehydrating Solvents

Various dehydrating agents can be used for PLT protocols with Lowicryl. Ethanol, methanol, and acetone are miscible with Lowicryl. Ethylene glycol and dimethyl formamide, which are proposed in the literature (19), give excellent results in terms of structural preservation but are less employed owing to their limitation. Dimethyl formamide (DMF) and ethylene glycol are miscible with Lowicryl K4M, but do not mix with Lowicryl HM20.

Acetone can be used but care should be exercised during PLT because 70% acetone freezes around –25°C (**Fig. 3**). Moreover, acetone prevents acrylic resin polymerization. Thus, if any amount of acetone is left in the specimen, polymerization can be partial and blocks are difficult to cut into thin sections. Normally, excess solvent must be eliminated by infiltration with pure resin.

Ethanol is most commonly used with Lowicryl at any temperature, although methanol can also be used. It offers excellent preservation of nuclear ultrastructure and organization of DNP and RNP fibrils (15).

Acrylic resins tolerate a small amount of water. It is, therefore, not necessary to dry the final solvent over a molecular sieve. Nevertheless, polymerization of acrylic resin with some water, which can preserve the hydration shell of proteins, leads to softer blocks that are sometimes very difficult to section.

Figure 3 shows curves of approximate freezing points for some solvent/water mixtures that can be used as a guide for developing new dehydration procedures. At the same concentration, the various solvents used for dehydration during the PLT method have very different freezing points. For example, at 50% concentration, acetone will freeze near –25°C and ethanol near –35°C.

2. Materials

2.1. Major Equipment and Supplies

1. Low-temperature embedding system (for example, Leica-Reichert AFS or BalTec FSU10).
2. Small gelatine capsules ("0" gage). The volume is approx 0.4 mL.
3. Polyethylene gloves (do not use Latex gloves).

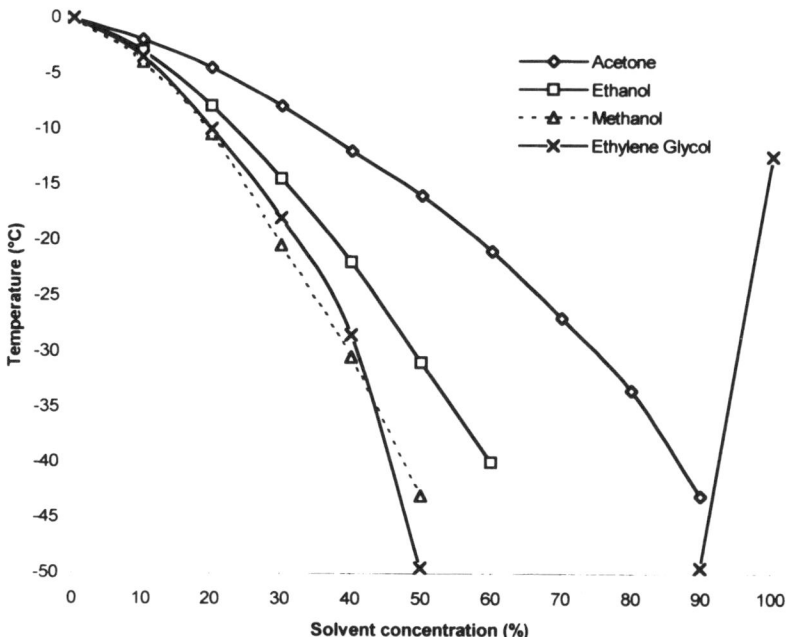

Fig. 3. Diagram showing the freezing points for some solvent-water mixtures. Note that the freezing point for ethylene glycol-water (EGOH) is higher at high solvent concentrations.

4. Plastic pipets (polyethylene).
5. Ethanol or methanol.
6. Agarose (for pre embedding culture cells and bacteria) type VII or IX (Sigma n° A4018 or n° A5030).
7. UV light (two 6-W UV tubes, for polymerization of waste, *see* **Fig. 2**).
8. Lowicryl Resin K4M or HM20 (K11M or HM23 have similar properties to K4M and HM20, but can be used at lower temperatures).

2.2. Fixation

Fixation may vary and strongly depends on the tissues or cells under study. However, the general rule is to use formaldehyde (freshly made from paraformaldehyde) in the range of 1–4% in a 0.1 *M* phosphate buffer (*see* **Note 1**) adjusted to pH 7.4 (*see* **Note 2**).

2.3. Preparation of the Lowicryl Resins

The Lowicryl resins are manufactured and sold by Poly Sciences Ltd. Lowicryl kits contain a rage of monomers, crosslinks and initiators. The recipes for mixing the various Lowicryl resins (*see* **Note 3**) are as follow.

K4M		HM 20	
Crosslinker (A)	2.7 g	Crosslinker (D)	3 g
Monomer (B)	17.3 g	Monomer (E)	17 g
Initiator (C)	0.1 g	Initiator (C)	0.1 g

Slowly mix the three components in a light protected, well-sealed glass flask (brown glass). Avoid introducing oxygen from air as it interferes with polymerization. Mixing the components is simply achieved by gently tilting and rolling the bottle (no mechanical agitation needed).

2.9. Counterstaining of Sections

Before observation, it is necessary to stain or counterstain thin sections of Lowicryl or Unicryl. This is not difficult, however, it requires some attention. Two methods are recommended: 1) Double staining with uranyl acetate and Millonig's lead tartrate stain and 2) Double staining with ethanolic uranyl acetate and ethanolic phosphotungstate. The second method seems to be the best as it prevents lead from precipitating. However, it cannot be employed with Unicryl embedded material since alcohol swells polymerized Unicryl.

2.9.1. First Staining Method

AQUEOUS URANYL ACETATE 4% *(13)*

Type of resin	*Staining time*
LR White/Unicryl	5 min.
K4M	5–15 min.
HM20	35 min.

Wash with distilled water. Dry carefully.

LEAD TARTRATE *(9)*

Staining time for all types of resins is 1–3 min.

PREPARATION OF LEAD TARTRATE *(9)*:

Stock solution

A)	NaOH	20 g
	K-Na Tartrate	1 g
	H$_2$O	50 mL (final volume)
B)	Lead acetate	20% (W/v)

Preparation: 1 mL of solution (A), 5 mL of solution (B). Slowly add B to A. This forms a cloudy solution which needs to be diluted 5–10 times with distilled water before use. The solution is cleared by filtration. The clear solution is stable for a few days and should remain clear (discard if solution turns cloudy).

2.9.2. Second Staining Method

This method has been proposed by Horowitz and Woodcock *(8)* and is strongly recommended for ultrastructural *in situ* hybridization *(4)* .

Ethanol-Uranyl acetate (EUA) and ethanol-phosphotungstic acid (EPTA) are freshly prepared by mixing a 2% aqueous uranyl acetate or phosphostungstic acid solution with an equal volume of absolute ethanol.

Sections are stained "on drops" with EUA (3–4 min), are washed first 30 s with distilled water, then with EPTA 1–2 min, washed again with water and are finally air dried.

3. Methods

3.1. Typical Dehydration Schedule for Lowicryl K4M

Times and temperatures given below are minimal values. Times have to be adjusted according to the specimen size as well as the type of tissue (*see* **Note 2**).

1. Immerse the specimen in 30% (v/v) ethanol at 0°C for 30 min.
2. 50% (v/v) ethanol at –20°C for 60 min;
3. 70% (v/v) ethanol at –35°C for 60 min;
4. 95% (v/v) ethanol at –35°C for 60 min.
5. 100% (v/v) ethanol at –35°C for 60 min.
6. 100% (v/v) ethanol at –35°C for 60 min.
7. Ethanol/resin mix (1:1) at –35°C for 60 min.
8. Ethanol/resin mix (1:2) at –35°C for 60 min.
9. Immerse in pure resin at –35°C for 60 min.
10. Immerse in pure resin at –35°C overnight.
11. Polymerize for 2 d at low temperature with ultraviolet (UV) light at –35°C.
12. Samples are then slowly brought back to room temperature under UV light (this step takes 2–6 h).

3.2. Embedding in Unicryl Resin

Unicryl is a fairly new resin introduced by Scala et al. *(11)* and initially known as Bioacryl *(12)*. This resin is used in histology where it gives good staining on semithin sections and in immunocytochemistry where it offers high sensitivity and good structural preservation. In 1996 Bogers et al. *(2)* compared Unicryl and Lowicryl K4M. They concluded that embedding in Unicryl gives more sensitive results in tissues where the expected concentration of antigen is low. A comparison between LR White and Unicryl also demonstrated the superiority of Unicryl *(7)*. The author proposes this protocol, which is suitable for many tissues.

3.3. Fixation for Unicryl Resin

Tissues are fixed with a combination of paraformaldehyde and glutaraldehyde, typically 2–4% paraformaldehyde and 0.05–0.2% glutaraldehyde in

0.1 *M* HEPES buffer (*see* **Note 1** and Griffiths *[16]*). Specimens are then washed with the fixation buffer, and optionally incubated in 50 m*M* NH$_4$Cl in PBS (for 10 min) in order to block free aldehyde groups (*see* **Note 9**).

3.4. Dehydration and Infiltration for Unicryl

30% ethanol	10 min	4°C
50% ethanol	10 min	4°C
75% ethanol	15 min	–20°C
90% ethanol	15 min	–20°C
100% ethanol	15 min	–20°C
Unicryl ethanol 1:2	30 min	–20°C
Unicryl ethanol 2:1	30 min	–20°C
Unicryl (pure)	60 min	–20°C
Unicryl (pure)	overnight infiltration	–20°C

Optionally, add fresh resin (60 min, -20°C) before final embedding of the samples in capsules.

3.5. Embedding and Polymerisation for Unicryl

The embedding and polymerization are performed at low temperature (–20°C). The actual polymerization is carried out for 4 d in a polymerization chamber fitted with UV light using 2 (6 W) tubes (type Sylvania blacklite 350. F6W/BL 350), and then followed by final curing for 7 d at 4°C (*see* **Note 8**). It is possible to shorten the time needed for curing either by reducing the distance between the samples and the UV strip lights or by increasing the curing temperature (*see* **Table 1**).

An another modification for fast curing has been proposed by Gounon *(19)*. Benzoin ethyl ether at 0.1% or 0.05% is added to the neat resin during infiltration and for final embedding. In this case, polymerization can occur in less than 2 d between –30°C and –35°C.

3.6. Immunogold Labeling

Sections on grids can either be incubated on drops (placed on a parafilm sheet in a moist chamber) or immersed in the different reagents. Both methods give good results, however, in the second case, care must be exercised to wash carefully both sides of the grid.

Usually, Lowicryl sections are fragile. They have to be deposited on fine grid mesh (300 or 400 mesh) thus reducing visible areas. If larger areas are desirable one can use larger mesh grid (100–200 mesh) covered with a formvar film. In this case, only one side of the grids can be stained which reduces the overall intensity of the immunolabeling.

Table 1
Minimal Temperature and UV-Irradiation Time for Unicryl Resin Polymerization

Temperature	Specimen-lamp distance (cm) for 2 × 6 W lamps[a]			
	5 cm	10 cm	15 cm	20 cm
−20°C	Not advisable[b]	Not advisable[b]	1–2 d	2–3 d
+4°C	Not advisable[b]	1–2 d	2–3 d	3–4 d
−10°C	2 d	2–3 d	3–4 d	4–5 d

[a]Times given are approximate and depend on the quantity of resin, opacity of the vials, and efficiency of the lamps at different temperatures.
[b]These combinations of distance-temperature are not advisable as they give brittle blocks induced by overheating during fast polymerization).

Immunolabeling on sections is usually performed according to the following schedule (**Figs. 4** and **5**):

1. PBS (or TBS), 50 mM NH₄Cl — 5–10 min.
2. PBS, 1% BSA + 1% NGS — 5–10 min.
3. Specific antibody diluted in PBS, BSA 1%, NGS 1% — 30 min–many hours.
4. PBS BSA 0.1% — 3 × 5 min
5. PBS gelatine 0.01% — 5 min.
6. Colloidal gold antibodies diluted in PBS 0.01% gelatine — 30 min.
7. PBS — 3 × 5 min.
8. 1% Glutaraldehyde in PBS — 5 min.
9. PBS — 5 min.
10. Distilled water — 3 × 5 min
11. Dry Sections
12. Counterstaining (*see* **Subheading 3.3.**).

4. Notes

1. The buffer can either be phosphate or cacodylate buffer or even preferably, PIPES or HEPES buffers for culture cells. As mentioned by Griffiths *(16)*, fixation is a very complex step. Chemical composition of the buffer, osmolarity, ionic strength of the buffer vehicle, and finally, pH has implications in the final result. PIPES, HEPES, and MOPS are actually the best candidates. Mixtures of glutaraldehyde with larger concentrations of formaldehyde (for example monomeric 0.1%–0.2% glutaraldehyde with 2% formaldehyde) are now very popular and have been shown to preserve both tissue reactivity and ultrastructure. Blocks of tissue should be of minimal size (1 mm³) and fixed for 3 h. Prolonged buffer washes must be avoided whenever possible as fixation reversal will result. Cells are fixed *in situ* in the Petri dish for 30 min to 1 h. After a buffer wash cells are carefully scraped with a rubber policeman and preembedded in melted 2% Agar before being pelleted in an Eppendorf tube. After gelling the agar on crushed ice, the tip of the tube is removed and small cubes are prepared with a chilled scalpel blade.

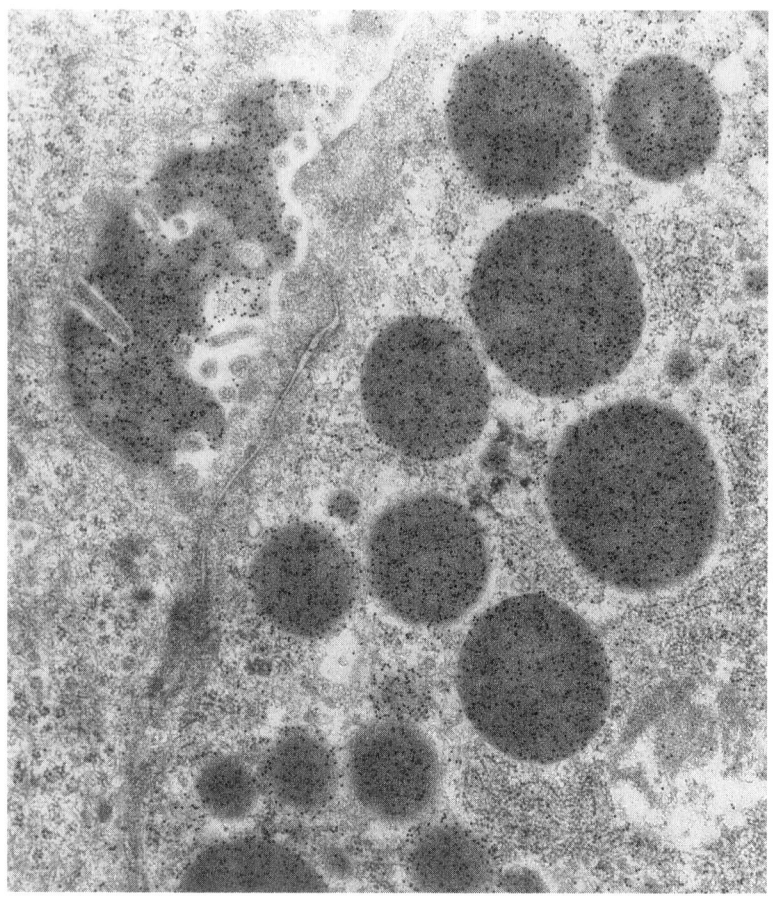

Fig. 4. Example of immunogold labeling on resin section. Mouse pancreas was fixed with 1% glutaraldehyde and embedded in Unicryl +0.05% BEE at –35°C by the PLT method. Zymogen granules were stained with antiamylase followed with antirabbit gold conjugate (10 nm size) × 60,000.

2. It is advisable to add a step of postfixation with uranyl acetate 0.5% in water for 30 min. at 4°C. This step proposed by Benichou et al. *(1)* improves sectioning by a mechanism of increased cohesiveness between cells and surrounding resin. This step does not affect immunogold labeling of a variety of antigens.
3. UV polymerization is performed at temperatures ranging between –50°C and +30°C; chemical polymerization takes place between +50°C to +60°C.
4. Some tissues or cells are very permeable to organic solvents and neat resin, this is also strongly dependent on fixation as highly crosslinked specimens are poorly permeable. Great care should be taken decreasing temperature and infiltrating with organic solvents (ethanol or methanol).

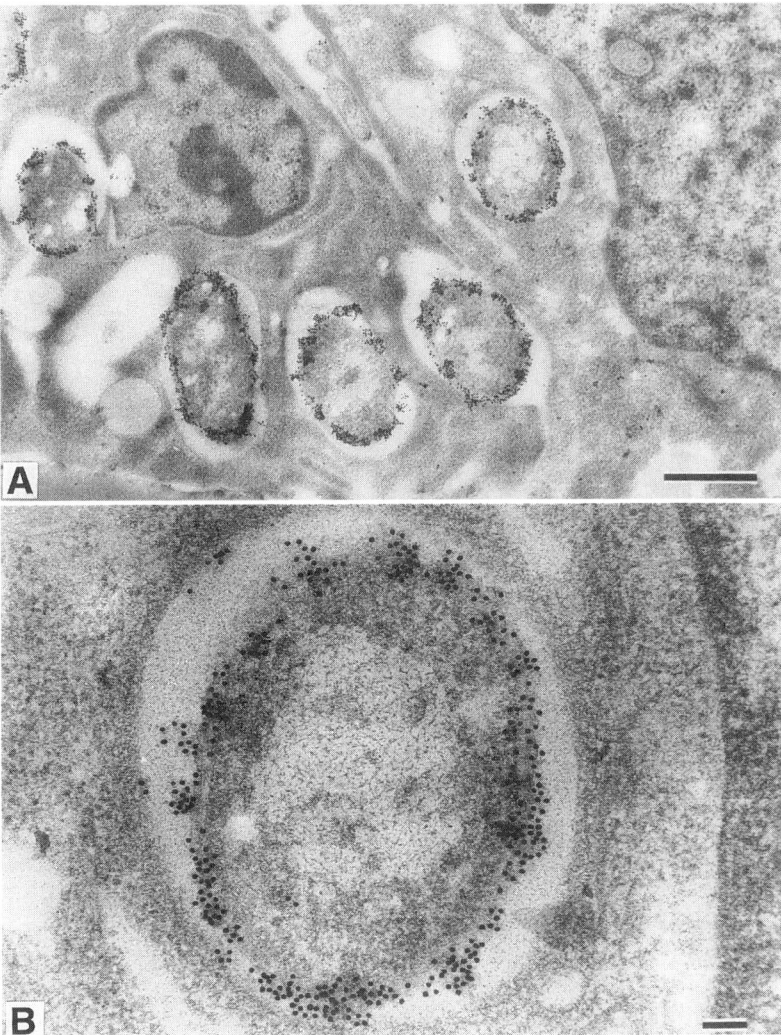

Fig. 5. Immunolocalization of lipopolysaccharide type 5a of invasive enterobacteria Shigella flexneri. Cells (T84 cell line) infected with S.flexneri were fixed with 2% formaldehyde and 0.2% glutaraldehyde in 0.1 *M* HEPES for 1 h. Cells were embedded in Lowicryl HM20 at –50°C. Immungold labeling was performed exactly as described in **Subheading 3.1.** with specific LPS rabbit antiserum and antirabbit gold conjugate (15 nm size). Bars = 0.1 μm.

5. This step can be omitted. In this case, samples must be maintained in 50% ethanol at 4°C for at least 20 min before storage in the freezer at –20°C in order to avoid freezing samples insufficiently impregnated with ethanol.
6. Methanol can be used instead of ethanol.

7. A small amount of poorly polymerized and sticky resin may remain at the end of the block. This is because the inhibition of polymerization of acrylic resin in the presence of oxygen trapped in the small bubble at the top of the capsule. If the block seems to be insufficiently cured (softness of the block during cutting, or instability under the electron beam), further polymerization can be achieved by leaving the capsules under UV light at room temperature. However opened capsules cannot be further polymerized. Closed capsules are generally stored in a dry area at + 4°C–8°C.

8. Rapid polymerization results in brittle blocks often impossible to trim correctly and to thin section. This problem affects all acrylic resins. Slow polymerization does not seem to have an adverse effect on the final properties of the specimens. However insufficient or partial polymerization leads to "gummy" specimen.

9. The blocking of free aldehydes can be achieved in various ways. PBS-50 mM NH$_4$Cl seems to be the most efficient method. This step reduces immunocytochemical background.

10. Trouble shooting in immunolabeling. Many pitfalls arise during immunogold labeling. In addition to the problems of weak antibodies, a low amount of available antigens or too strong and inappropriate fixation, possible sources of background are very high. However, many solutions have been proposed.

Possible Source of Background	**Solution**
A. Nonspecific attraction of antibodies to cellular components through fixative residues and tissue aldehydes.	Block the surface of the sections with BSA (Bovine Serum Albumin) before and during incubations.
B. Highly positively charged components (collagen, elastin, lysine residues in histones) attracting negatively charged gold particles.	Block the surface of the sections with BSA. Incubate with buffer at a higher pH (i.e., 7.5 to 8.0). Increase the salt content of the incubation medium (usually 0.15 M NaCl, can be increased to 0.2 M NaCl or higher).
C. Hydrophobic tissue components (tryptophan residues) attracting hydrophobic gold particles.	Add surfactant to the incubation buffer. The best is Tween 20 (0.1% to 1%). Wash the sections thoroughly after this incubation step as Tween 20 is not recommended with colloidal gold.
D. Receptors in tissue attracting non-specifically, secondary labeled antibodies.	Block the surface of the section with normal goat serum before incubation with specific antibodies.

Add normal serum (1%) in all incubation buffers.
Protein A, Protein G or Protein AG Conjugates instead of IgG gold conjugates.

E. Sulphur containing components (in the tissue, cystein or epoxy resins) attracting gold particles.

Increase BSA (5%) in the gold incubation medium. Do not add gelatine since it will increase the attraction of gold to the surface of the section.

References

1. Benichou, J. C., Frehel, C., and Ryter, A. (1990) Improved sectioning and ultrastructure of bacteria and animal cells embedded in Lowicryl. *J. Electron Microsc. Tech.* **14,** 289-97.
2. Bogers, J. J., Nibbeling, H. A., Deelder, A. M., and Van Marck, E. A. (1996) Quantitative and morphological aspects of Unicryl versus Lowicryl K4M embedding in immunoelectron microscopic studies. *J. Histochem. Cytochem.* **44,** 43–48.
3. Carlemalm, E., Garavito, R. M., and Villiger, W. (1982) Resin development for electron microscopy and an analysis of embedding at low temperature. *J. Microsc.* (Oxford) **126,** 123–143.
4. Chevalier, J., Yi, J., Michel, O., and Tang, X. M. (1997) Biotin and digoxigenin as labels for light and electron microscopy in situ hybridization probes: Where do we stand? *J. Histochem. Cytochem.* **45,** 481–491.
5. Douzou, P. (1977) Enzymology at sub-zero temperatures. *Adv. Enzymol.* **45,** 157–272.
6. Fink, A. L. and Ahmed., I. A. (1976) Formation of stable crystalline enzyme-substrate intermediates at sub-zero temperatures. *Nature* **263 (5575),** 294–297.
7. Goping, G., Kuijpers, G. A. J., Vinet, R., and Pollard, H. B. (1996) Comparison of LR White and Unicryl as embedding media for light and electron immunomicroscopy of chromaffin cells. *J. Histochem. Cytochem.* **44,** 289–295.
8. Horowitz, R. A. and Woodcock, C. L. (1992) Alternative staining methods for Lowicryl sections. *J. Histochem. Cytochem.* **40,** 123–133.
9. Millonig, G. (1961) A modified procedure for lead staining of thin sections. *J. Biophys. Biochem. Cytol.* **11,** 736–739.
10. Newman, G. R. and Hobot, J. A. (1993) *Resin Microscopy and On-Section Immunocytochemistry* (Springer Lab., ed.) Springer-Verlag, Berlin, Germany.
11. Scala, C., Cenacchi, G., Ferrari, C., Pasqualini, G., Preda, P., and Manara, G. C. (1992) A new acrylic resin formulation: A useful tool for histological, ultrastructural, and immunocytochemical investigations. *J. Histochem. Cytochem.* **40,** 1799–1804.
12. Scala, C., Preda, P., Cenacchi, G., Martinelli, G. N., Manara, G. C., and Pasquinelli, G. (1993) A new polychrome stain and simultaneous methods of histological, his-

tochemical and immunohistochemical stainings performed on semithin sections of Bioacryl-embedded human tissues. *Histochem. J.* **25,** 670–677.

13. Villiger, W. (1991) Lowicryl resins, in *Colloidal Gold: Principles, Methods, and Applications*, vol. 3 (Hayat, M. A., ed.), Academic, San Diego, pp. 59–71.

14. Weibull, C., Villiger, W., and Carlemalm, E. (1984) Extraction of lipids during freeze-substitution of Acholeplasma laidlawii cells for electron microscopy. *J. Microsc.* **134,** 213–216.

15. Puvion-Dutilleul, F. (1993) Protocol of electron microscope in situ nucleic acid hybridization for the exclusive detection of double-stranded DNA sequences in cells containing large amounts of homologous single-stranded DNA and RNA sequences: Application to adenovirus type 5 infected HeLa cells. *Microsc. Res. Tech.* **25,** 2–11.

16. Griffiths G. (1993) *Fine Structure Immunocytochemistry*, Springer-Verlag, Berlin, Germany.

17. Roth, J., Bendayan, M., Carlemalm, E., Villiger, W., and Garavito, R. M. (1981) Enhancement of structural preservation and immunocytochemical staining in low temperature embedded pancreatic tissue. *J. Histochem. Cytochem.* **29,** 663–671.

18. Bendayan, M. (1995) Colloidal gold post-embedding immunocytochemistry. *Prog. Histochem. Cytochem.* **29,** 1–159.

19. Gounon, P. and Rolland, J.-P. (1998) Modification of Unicryl composition for rapid polymerization at low temperature without alteration of immunocytochemical sensitivity. *Micron* **29(4),** 293–296.

8

Quantitative Aspects of Immunogold Labeling in Embedded and Nonembedded Sections

Catherine Rabouille

1. Introduction

A new challenge facing electron microscopists in the early 1980s was to combine their experience, taste, and knowledge of the cell ultrastructure with the newly developed antibodies *(1)* in order to expand morphology into biochemistry and *"to bridge the gulf between the so called sciences of matter and science of form" (2)*. The preparation of samples for conventional electron microscopy is harsh. The infiltration of the samples with Epon or Araldite resins after the aldehyde fixation, the osmium postfixation, and the dehydration with ethanol and propylene oxide do not allow antigenicity to be preserved (except in some rare cases). New, softer methods for preserving some antigenicity along with a reasonable morphology were developed, allowing visualization of antigens (mainly proteins) on sections.

First, new dehydration techniques known as "cryosubstitution" and "progressive lowering temperature" were designed and new resins (LR white, LR gold, or Lowicryl HM20, *see* **refs. 3,4**) were introduced that could, in a similar manner as above, infiltrate the sample and polymerize, allowing sectioning at room temperature. Second, a new technique called immunolabeling of cryosections or cryoimmunolabeling or cryoimmuno EM was developed. The samples are also fixed by aldehydes, but there is no resin embedding. (Only in cases of cells in culture or suspension is the cell pellet embedded in gelatine.) The hardness and plasticity of the blocks required for proper sectioning is conferred by freezing the blocks (instead of resin polymerization) after infiltration in sucrose. Sectioning itself is performed at about −130°C in an ultramicrotome. Tokayasu in San Diego, CA, Geuze and Slot in Utrecht, the Netherlands, and Griffiths in Heidelberg, Germany *(5–8)* have been pioneers in the field and have developed

From: *Methods in Molecular Biology*, vol. 117: *Electron Microscopy Methods and Protocols*
Edited by: N. Hajibagheri © Humana Press Inc., Totowa, NJ

(and still develop, *9*) this powerful technique over a period of 20 years. Many scientists use it now routinely.

Both techniques consists in obtaining ultrathin sections from mildly fixed (low percentage aldehydes) biological material that can be labeled with specific antibodies against proteins of interest. These antibodies are subsequently visualized using electron dense reagents such as colloidal gold or ferritin coupled to IgG or protein A. This is a postlabeling embedding method (first embedding and then labeling) that fulfills three fundamental requirements: preservation and display of ultrastructural details, maintenance of immunoreactivity, and fixation of antigen molecules at their natural location

How to get such ultrathin cryosections which could be suitable for immunolabeling is detailed in Chapter 4, and the resins techniques are described in Chapters 5–7. Proper fixing, freezing, and sectioning of the specimen (which can be tissue, cells, organelle fractions, virus) are the three crucial steps in order to preserve the morphology of the sample in the best possible way. The author would like to stress this point because morphology is the first leg of the bridge. Without it, no quantitation is meaningful. The second leg of the bridge is the labeling and the quantitation of this labeling on relevant structures. The author will now review it.

2. Labeling Cryosections

2.1. Why are We Interested in Labeling Proteins on Sections

CryoimmunoEM is particularly powerful to study the location of proteins at a subcellular level. The resolution of fractionation is limited and has made difficult the assessment of fine localization. A very good example is the different compartments of the Golgi apparatus that cannot be resolved biochemically. Indirect immunofluorescence has provided enormous amount of information on localization, but again the fine resolution of compartments localized very close to one another has been compromised. The power of the confocal microscopes associated with the newly developed use of the Green Fluoresescent Protein (GFP) has led to unprecedented resolution *(10)*. However, in fluorescence microscopy, only the fluorophore is visualized, not the membranes independently of the marker. Clearly, immunoelectron microscopy still has an important role to play in localizing proteins to compartments while visualizing their boundaries.

2.2. What Does it Take to Label a Protein?

Observing labeling on a cell section means that the labeling density, LD (number of gold particles/μm^2 of section; *see* also **Subheading 4.3.1.**) has reached a minimal value. LD is a function of two major parameters difficult to control and influence: the labeling efficiency (LE) and the copy number of the antigen of interest.

2.2.1. The Labeling Efficiency (LE)

The labeling efficiency is the ratio between the number of gold particles observed, Ng, and the number of antigen present, Na (Ng/Na). It is asking how many gold particles are present per antigen. It can be affected by fixation, configuration of the protein (oligomerization, sugar modification), accessibility, and penetration (dense versus less dense organelle, section thickness, *see* also **Subheading 5.2.**) and the antibody used for the labeling. The more affinity an antibody has for the antigen and the less sensitive to configuration the antigen, the higher the LE. In general, the labeling efficiency cannot be measured, but is estimated to be between 1–5 gold particles for 100 antigenic molecules.

2.2.2. The Copy Number

The copy number is the measure of how abundant the antigen is. Taken into account the LE, a protein of low abundance will be less labeled than a protein of higher abundance (regardless of other parameters such as antibody, organelle density, etc.). Consequently, when an organelle is scored positive for gold particles, it means that the number of antigen present on the fraction of this organelle visible on the section is high enough to be detected. For instance, the plasma membrane contains "*n*" number of antigen homogeneously dispersed. However, only a very small fraction of the plasma membrane is visible on the section. So the same small fraction of "*n*" can be visualized. Even with excellent antibodies of high specificity, that could be too little to be detected. Let us imagine that the same "*n*" antigen are now on the Golgi apparatus (whereas being synthesized, for instance). Because the density of membrane in the Golgi apparatus is higher, a length of Golgi membranes "*x*" times higher than the plasma membrane length is present in each section. This "*x*" increase could mean that some of the antigens could now be detected even if "*n*" was the same in both cases. In other words, for large compartments to be scored positive, the number of antigen has to be higher than for denser organelles.

3. Observing Gold Particles

The first step in assessing localization and performing quantitation is to recognise the subcellular organelles. That is why the morphology should be preserved as best as possible.

3.1. How to Recognize an Organelle

The text books and the old issues of *Journal of Cell Biology* are a valuable source of information concerning the 2-D morphology of cell ultrastructure. In addition to mere pattern recognition, performing double labeling *(11)* of the antigen and a marker known to localize on the organelle of interest is required in order to unambiguously identify it. There are numerous antigens that have been localized precisely and can be used as reference. However, multiple labeling is not without problems and should be performed with caution (*see* **Subheading 5.1.**).

3.2. Sampling (12)

Sampling is one of the most serious matters in quantitation. At least 20 profiles should be observed and quantified to provide converging and significant information. If the labeling is low, more profiles should be counted in order to reach a critical number of gold particles. The profiles should be dispatched on the grid, taken from different grids, sectioned different days from different blocks, from different animals or cells. In principle, random pictures should be taken or random fields counted. The nucleus, ER, plasma membrane, and cytoplasm occupy most of the space. Mitochondria, endosomes, and lysosomes are rather randomly distributed around the plasma membrane. The problem occurs with organelles such as the Golgi apparatus whose distribution and relatively small dimensions would underscore it greatly if random sampling is used. Together with the fact that cryosectioning is a difficult technique making every section a precious source of information, the author would suggest using every section as material for assessing gold labeling.

3.3. Five Different Possible Patterns for Gold Labeling (Fig. 1).

It seemed necessary to take some examples in order to face the problems inherent to the interpretation of the labeling pattern. Since the author has long worked on the Golgi apparatus, the author took the example of a protein that is suspected to be localized to the Golgi apparatus from immunofluorescence data.

Here the author describes five patterns that can occur after labeling this protein.

1. This situation is the best one, where the labeling is clearly localized on a particular organelle, here the Golgi stack.
2. Here, we found reproducibly one gold particle on the Golgi apparatus and another one somewhere else, which is depicted on the nucleus. The situation is potentially good, but more work should be performed to improve the signal, such as a lower dilution of the antibody, an enhancement method or a new antibody.
3. The observation reveals that each organelle contain 1 gold particle. This situation is not promising because the labeling does not seem specific. Either the dilution could be too high or the antigenicity has been lost during the fixation.
4. The background is very high which makes information difficult to retrieve. The situation is potentially interesting, but the background needs to be reduced maybe

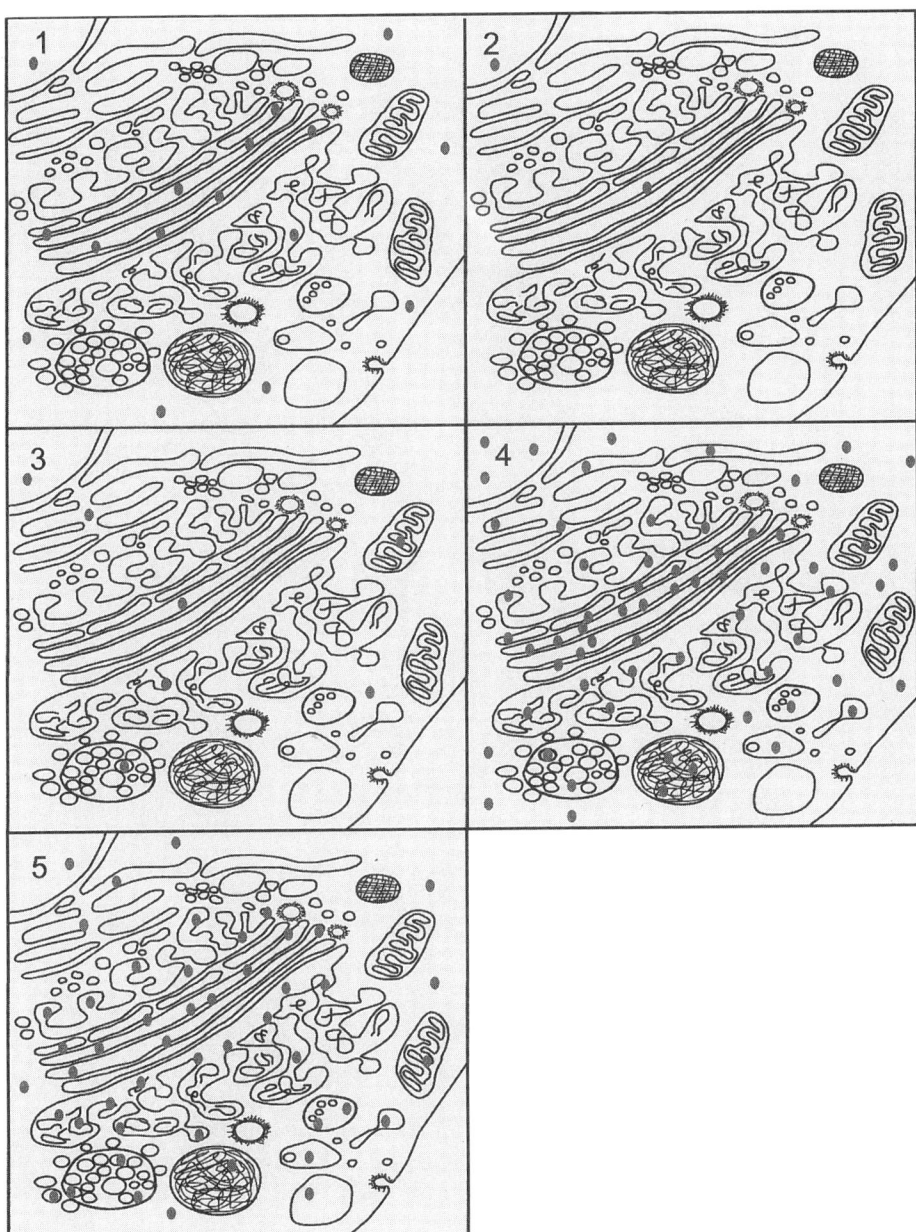

Fig. 1. Five different patterns observed after immunogold labeling of a cell section. The intracellular membrane compartments are depicted in a cartoon-like fashion. The gray dots represent the gold particles, which are distributed differently in each pattern.

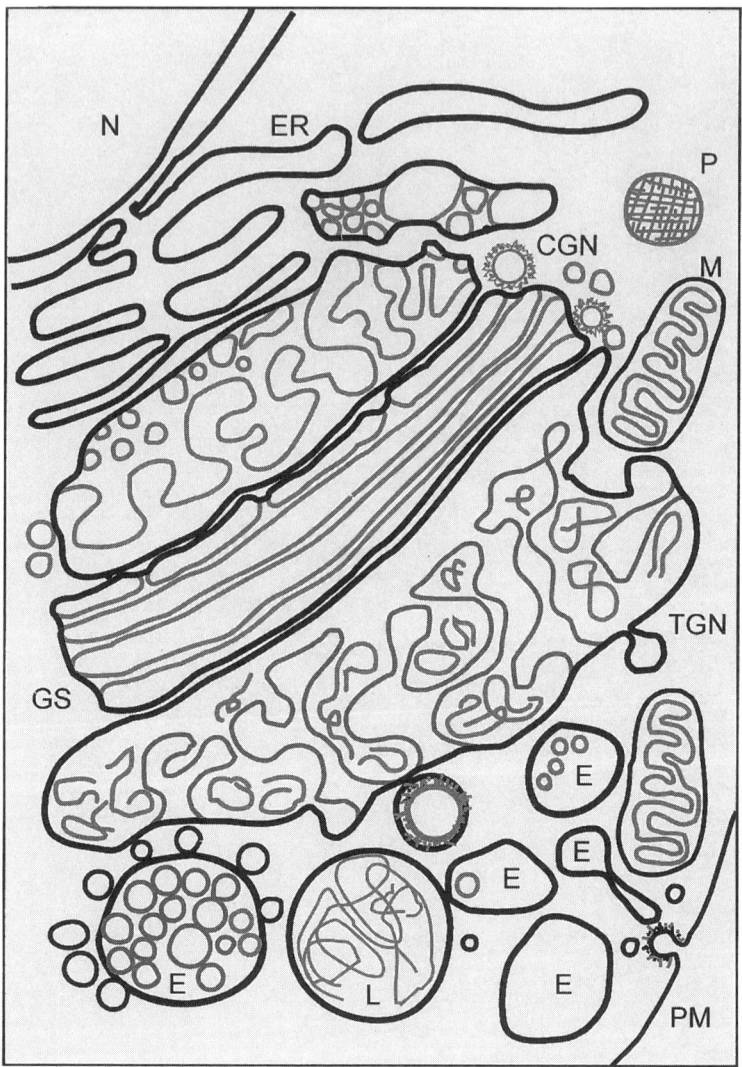

Fig. 2. The boundaries of 7 different compartments have been drawn in black around the membranes (as depicted in **Fig. 1**) and labeled as follows; N is the nucleus, ER is the endoplasmic reticulum, CGN is the Cis Golgi Network, GS is the Golgi stack, TGN is Trans Golgi Network, E is endosome, L is lysome, M is mitochondria, P is peroxisome and PM is plasma membrane. Note that the boundaries of the Golgi stack, the CGN and the TGN do not eactly follow the membranes (in gray). The boundaries of the Golgi stack were drawn by following the outmost membranes of the cis and trans cisternae. The boundaries of the CGN and TGN were drawn by following the interface between the outmost membranes and the more amorphous cytoplasm *(23)*. (*See* Color Plate 4, following p. 180.)

Table 1
Number of Gold Particles Falling Within the Boundaries of Each Organelle Ng(org) as Depicted in Figs. 1 and 2

Patterns	nucl	ER	mito	Golgi Stack	CGN	TGN	endo/ lyso	cyto
1	1	0	0	4	0	1	1	6
2	1	0	0	1	0	0	0	0
3	1	1	1	1	0	1	1	1
4	3	4	1	20	5	10	8	13
5	1	2	1	10	7	12	10	5

nucl, Nucleus; ER, Endoplasmic Reticulum; mito, Mitochondria; CGN, Cis Golgi Network; TGN, Trans Golgi Network; endo/lyso, the endosomal/lysosomal compartment; cyto, Cytoplasm.

by using a higher dilution of the antibody or using a blocker in a higher concentration.

5. The gold particles are definitely labeling the Golgi stack as well as possibly the CGN, the TGN, and the endosomal/lysosomal compartment.

4. Quantitating Specimens or How to Make the Pictures Reveal their Secrets

So far, there has not been satisfactory methods to relate the number of gold particles to the number of antigens (absolute quantitation, *see* **Subheading 4.3.**). Therefore, the method that follow deals with the quantitation of the gold particles and is therefore semiquantitative.

4.1. Counting Gold Particles

After taking pictures and recognizing the organelle, the author suggests to draw the boundaries of each organelle (such as in **Fig. 2**) and counting the number of gold particles falling within the boundaries of each organelle. It can occur that the labeling appears a clusters of three or more gold particles. It is suggested to count every gold particles (because it is difficult to distinguish between a cluster and the actual labeling of two very close antigens). The counting for the examples of **Fig. 1** are recorded in **Table 1** for each of the five patterns. The counting is expressed in number of gold particles per organelle. Labeling is seldom perfect, so there are gold particles outside the Golgi apparatus for each pattern. Are those gold particles background? Are the gold particles falling within the boundaries of the Golgi apparatus significant, that is to say significantly higher than background? Is the labeled protein also localized in another organelle than the Golgi apparatus?

4.2. Estimation of the Background Labeling Density (Background LD)

4.2.1. How?

The critical aspect of quantitation is to extract the signal out of the background. In the examples given, the antigen is supposedly located in the Golgi apparatus and is unlikely to be in the nucleus. So in this case, the background could be estimated from the nucleus and will represent the background on the whole cell, the whole section and possibly the whole grid. If the antigen was nuclear, another part of the cell, possibly the organelles could be used to determine the background. It is not advisable to use mitochondria in determining background because they can be sticky for some antibodies/gold conjugates.

The background estimation implies the determination of two parameters related to the organelle where no antigen is expected, here the nucleus.

1. The number of gold particles falling on the nucleus, Ng(nucl) (**Table 1**).
2. The surface section of the nucleus.

This requires taking random pictures of nucleus or using portions of nucleus already in the pictures taken to establish the first counting (**Table 1**). Again at least 10–20 pictures should be counted till converging results are obtained (*see* **Subheading 4.6.**).

4.2.2. The Point-Hit Method (*13*; **Fig. 3A**).

This method is used to determine the surface section of an object. A grid (a series of vertical lines spaced by a distance "*d*," intersecting a series of horizontal lines spaced by the same distance "*d*") is needed. Each intersection defines a point that is surrounded by a space "d^2." By randomly overlapping the grid with the object of interest, it is possible to count the number of grid points, P(object), falling within and on the boundaries of this object. Each point representing a space d^2, the number of points P multiplied by d^2 gives an estimate of the surface section of the object. Magnification should be taken into account. So it comes

$$\text{Surface section S(object)} = \text{P (object)} \times d^2/\text{mag}^2 \text{ (in } \mu m^2)$$

In theory, at least 10 points should fall into the boundaries, so the density of the grid depends on the size of the structure and the magnification of the pictures.

When the object is the nucleus, P(nucl) and therefore S(nucl) can be estimated. The background LD is the number of gold particles falling on the nucleus boundary Ng(nucl) divided by the surface section of nucleus S(nucl) on which the gold particles have been counted.

A

Point hit method

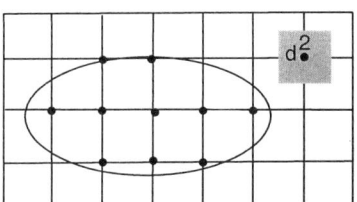

grid: 1 cm space (d)
10 points (P)
35 k (mag)

$$\text{Surface} = \frac{P \times d^2}{\text{mag}^2} = \frac{10 \times (1 \times 10^{-4})}{(35 \times 10^3)^2} \ \mu m^2$$

B

Intersection method

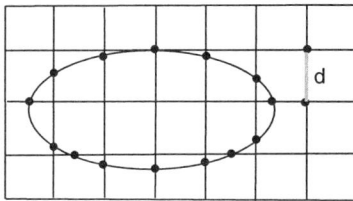

grid: 1 cm space (d)
14 intersections (I)
35 k (mag)

$$\text{Length} = \frac{I \times d}{\text{mag}} = \frac{14 \times (1 \times 10^{-4})}{(35 \times 10^3)} \ \mu m$$

Fig. 3. The point-hit and the intersection methods. *(A)* The surface of a specimen is estimated by counting grid points (10) falling within the boundaries of the specimen. The gray square represents the space d^2 specified by each point. The equation indicates how to calculate the surface. *(B)* The perimeter length of the same specimen is estimated by counting intersections (14) between a grid and the boundaries of the same object. The gray line represents the length d specified by each intersection.

Background LD = Ng(nucl)/S(nucl) (in gold particles/μm^2)

The background LD should be around one to two gold particle/$\mu m2$.

4.2.3. Examples

The background LD in each of the five patterns has been estimated and recorded in **Table 2A** and **B** (second column). The 11 points of a grid whose "d" is 1 cm

Table 2
(A) Estimation of the Surface Section of Organelle as Depicted in Fig. 2 using the Point-Hit Method

	nucl[a]	ER	mito	Golgi Stack	CGN	TGN	endo/ lyso	cyto
P	11	18	7	19	20	44	24	78
S(org) (μm^2)	0.44	0.72	0.28	0.76	0.8	1.76	0.96	3.12

grid = 1 cm; mag = 50 K.
[a]The boxed second columns shows data related to the nucleus used to estimate the background.

(B) Estimation of the Labeling Density for Each Organelle, LD(org) = Ng(org) × S(org)

Patterns	nucl	ER	mito	Golgi Stack	CGN	TGN	endo/ lyso	cyto
1	2.2	0	0	**10.5**	0	0.6	1.0	1.9
2	2.2	1	1	1.3	0	0	0	0
3	2.2	1.2	3.6	1.3	0	0.6	1.0	0.3
4	6.8	5.0	3.6	**26.3**	6.2	5.7	8.3	4.2
5	2.2	2.7	3.6	**13.1**	8.7	6.8	**10.4**	1.6

nucl, Nucleus; ER, Endoplasmic Reticulum; mito, Mitochondria; CGN, Cis Golgi Network; TGN, Trans Golgi Network; endo/lyso, the endosomal/lysosomal compartment; cyto, Cytoplasm.

at a magnification of 50K fall on the nucleus in **Fig. 4**. That gives a surface section of nucleus of 0.44 μm^2. In patterns 1, 2, 3, and 5, there is only one gold particle on the nucleus, so the background LD is 2.2 gold particles/μm^2. In pattern 4, there are three gold particles. The background LD is thus 6.8 gold particles/μm^2.

4.3. Specific Gold Labeling

4.3.1. Estimation of the Organelle Labeling Density, LD(org)

In order to know whether the gold labeling on an organelle is significantly higher than the background, the labeling density on the organelle has to be estimated. The same point-hit method is applied. As for the nucleus, the number of points from a grid falling into the boundary of organelles as drawn in **Fig. 4** is estimated. That is recorded in **Table 2A**, as well as the conversion of these numbers into surface section of each organelle S(org). Since the number

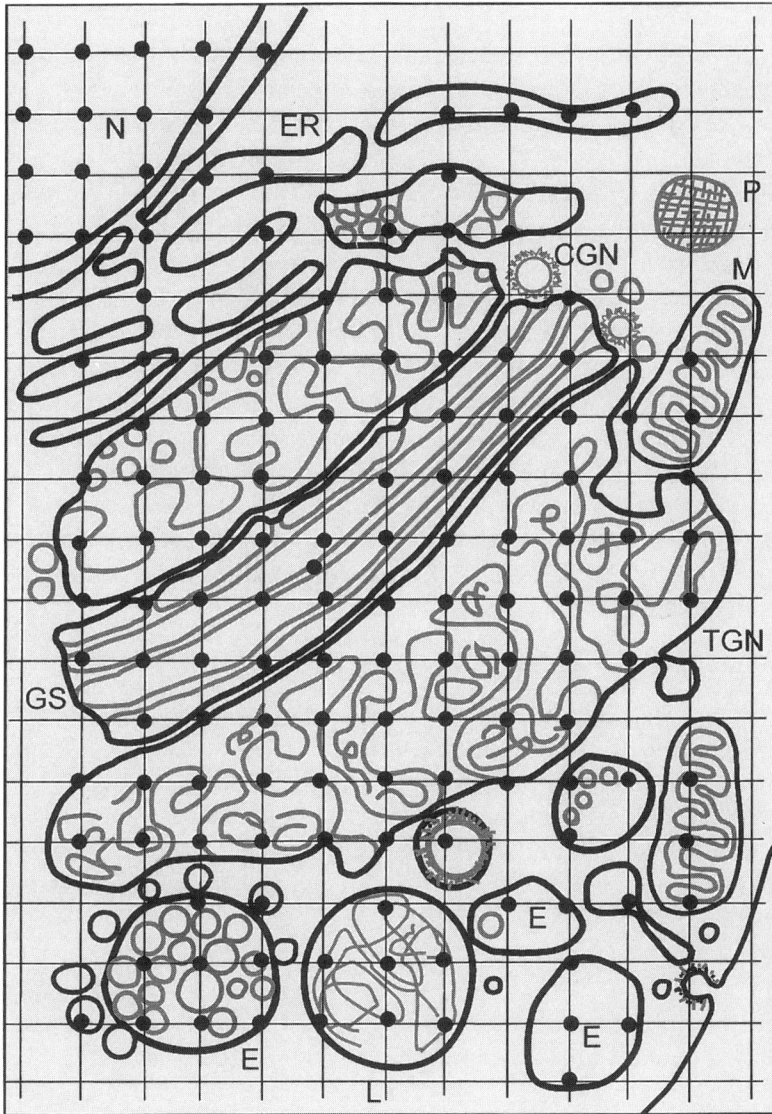

Fig. 4. Point-hit method on the intracellular compartments. The profile in **Fig. 2** has been overlain with a grid ($d = 1$ cm). The magnification is estimated to be 50K. The number of points falling within the boundaries of compartments is recorded in **Table 2**. (*See* Color Plate 5 following p. 180.)

of gold particles falling on each organelle Ng(org) has been established in **Table 1**, it is easy now to calculate the labeling density LD(org) = Ng(org)/S(org) as recorded in **Table 2B**.

4.3.2. Examples

There are some comments to be drawn out of these estimations. In pattern 1, the background is 2.2 gold particles/μm^2 whereas the LD on the Golgi stack is 10.5. The LD on the endosomal/lysosomal compartment, TGN or cytoplasm are similar to the background and therefore are not significant. The labeled protein is likely to be localized on the Golgi stack. In pattern 2, the LD on Golgi apparatus is low but because this organelle is always labeled, it could be significant. However, the situation needs to be improved. In pattern 3, The LD calculation reveals what mere observation has already revealed. No organelle exhibits an LD higher than the background. There is no specific labeling. In pattern 4, the background is high but so is the LD on Golgi stack (23.6 gold particles/μm^2). All the other compartments have an LD very close to background except the endosomal/lysosomal compartment which does exhibit an LD slightly higher than background. It could be potentially significant and should be investigated further. In pattern 5, the LD on the Golgi stack, the CGN, the TGN and the endosomal/lysosomal compartment is significantly higher than background. That is an unexpected situation in which the specific gold labeling extends much more than the limit of the Golgi stack boundary.

4.3.3. Calculation

Once the LD for each organelle as well as the background has been established, it is possible to subtract the background from the LD of each organelle and calculate the specific LD for each organelle, LDspec(org) (**Table 3**) LDspec(org) = LD(org) – background (in gold particles/μm^2) and to calculate the number of specific gold particles on each organelle, Ng, spec(org) (**Table 3**). Ng,spec(org) = LDspec(org) × S(org) (in number of specific gold particles per organelle).

4.4. Relative Distribution

Once the number of specific gold particles for each organelle has been established, the relative distribution (RD) of gold particles over cell structures (**Table 3**) can be calculated. The RD indicates where the bulk of the gold particles is and therefore, in most cases, where the bulk of the antigen is. That is not a reflection of the concentration because it does not take into account the size of the compartment.

The relative distribution expresses that out of the total number of specific gold particles, $x\%$ is located in one compartment, $y\%$ in a second, and $z\%$ in a third, etc. In pattern 1, the total number of specific gold particles is 6 and they are all (100%) located on the Golgi stack. In pattern 4, 17 gold particles are specific, 88% of them are associated with the Golgi stack and 12% with the

Table 3
(A) Estimation of the Specific Labeling Density for for Each Organelle,
LD spec(org) = LD(org)–Background (gold particle\μm2

Patterns	nucl	ER	mito	Golgi Stack	CGN	TGN	endo/lyso	cyto
1	0	0	0	**8.3**	0	0	0	0
2	0	0	0	0	0	0	0	0
3	0	0	1.2	0	0	0	0	0
4	0	0	0	**19.5**	0	0	1.5?	0
5	0	0.5	0.1	**10.9**	**6.5**	**4.6**	**8.2**	0

(B) Estimation of the Number of Specific Gold Particles for Each Organelle, Ng,spec(org) = LDspec(org) × S(org), and the Relative Distribution (RD) of Gold Particles over the Different Organelles Relative Distribution, (RD,%)

Patterns	nucl	ER	mito	Golgi Stack	CGN	TGN	endo/lyso	cyto
1	0	0	0	**6** **100%**	0	0	0	0
2	0	0	0	0	0	0	0	0
3	0	0	0	0	0	0	0	0
4	0	0	0	**15** **88%**	0	0	1–2 12%	0
5	0	0	0	8 27%	5 17%	8 27%	**8** 27%	0

nucl, Nucleus; ER, Endoplasmic Reticulum; mito, Mitochondria; CGN, Cis Golgi Network; TGN, Trans Golgi Network; endo/lyso, the endosomal/lysosomal compartment; cyto, Cytoplasm.

endosomal/lysosomal compartment. In pattern 5, 29 gold particles are specific, 27% are on the Golgi stack, 17% in the CGN, 27% in the TGN and 27% in the endosomal/lysosomal compartment.

The procedure to calculate the relative distribution could seem long and tedious but when the background is as low as one to two gold particles/μm², the situation is much easier and the LD is very close to LDspec. It is usually the case. However, when the background is higher and one is to compare two organelles whose surface sections are very different (one very large compared with the other one), then the background gold particles will affect very differently the number of gold particles associated with each compartment. Without estimation of the specific LD, the relative distribution will not be accurate. It will be biased toward the large compartment.

4.5. Linear Density

The determination of the specific labeling density (LD) gives an estimate of the concentration of gold particles per μm² of section. That is a meaningful parameter for the soluble (nonmembrane-bound) antigens which are located in the lumen of the compartment or in the cytoplasm.

For the membrane-bound antigens, the determination of the labeling density only (but very importantly) helps in the estimation of the relative distribution. Their linear density (the number of gold particles per length unit of membrane, in μm) should be estimated. For instance, the calculation of the relative distribution of gold particles in pattern 5 has shown that four organelles are labeled, the CGN, the Golgi stack, the TGN, and the endosomal/lysosomal compartment. It could be interesting to know where the linear density of gold particles is the highest.

Again, that requires the establishment of two parameters:

1. The number of specific gold particles, Nspec(org) associated with the membrane within a distance of 20 nm (which is the length of the antibody and protein A gold).
2. The length of membrane. The Intersection method (*13*; **Fig. 3B**) should be applied. Membranes of interest are drawn and randomly overlaid with a grid. Each intersection (I) of the grid with the membrane is counted.

The space between each line of the grid is "d," so it comes

$$\text{Length } L(\text{org}) = I \times d/\text{mag (in μm)}$$

The linear density is the specific number of gold particles associated with the membranes of interest divided by the length of membrane.

$$\text{Linear density} = \text{Nspec(org)}/L(\text{org}) \text{ (in gold particles/μm)}$$

Determination of the linear density can be very time-consuming. For intracellular organelles, high magnification pictures (90–120 K) and a fine grid (0.2–0.5 cm) should be used. In case of TGN and CGN, which are networks of membranes, the measurement of length will be underestimated because not all membranes are visible. Consequently the linear density can be overestimated. However, for the plasma membrane, it is a perfect technique.

4.6. Presentation of the Results (12)

Given one pattern, there will be variations from profile to profile within experiments and from experiment to experiment. The problem is to know when to stop counting and how to express the results. Pattern 4 of **Fig. 1** will be used to illustrate this point.

Table 4
(A) Ng, spec and Relative Distribution (RD) per Profile

Profiles		a	b	c	d	e	f	g	h
Ng,spec	Golgi stack	15	12	1	20	15	13	30	10
	endo/lyso	2	9	3	0	5	3	8	2
RD	% in Golgi stack	88	57	25	100	75	81	78	83
	% in endo/lyso	12	43	75	0	25	19	22	17

(B) Cumulative (Cum) Ng,spec and RD

Profiles		a	a+b	..+c	..+d	..+e	..+f	..+g	..+h
Cum Ng,spec	Golgi stack	15	27	28	48	63	76	106	116
	endo/lyso	2	11	14	14	19	22	30	32
Cum RD	Cum % in Golgi stack	88	71	66	77	77	77	78	78
	Cum % in endo/lyso	12	29	34	23	23	23	22	22

4.6.1. Cumulative Presentation

The cumulative presentation allows to determine the convergence of the results. As presented in **Table 4A**, the relative distribution (RD) of specific gold particles between the Golgi stack and the endosomal/lysosomal compartment has been established for each of the eight profiles. The same could be done for the linear density or the labeling density.

Next, the number of specific gold particles falling on either compartment (Golgi stack and the endosomal/lysosomal compartment) are added for profile a and profile b ($a + b$, **Table 4B**). From these newly generated numbers, a cumulative RD can be calculated. Then the number of gold particles of profile c are added ($a + b + c$) and a new cumulative RD is calculated, and so on. We observe that even if the relative distribution in individual results did not seem very convergent, the cumulative presentation indicates that they converges readily after four profiles; 78% of the gold particles are located in the Golgi stack. That is good indication that the counting is meaningful, but two other parameters should be established.

4.6.2. The Standard Deviation (SD)

The standard deviation (SD) indicates the dispersion of the results, i.e., the variation from profile to profile within one experiment.

From N number of profiles (eight in our example), there are N number of results (RD in the Golgi stack in our example). "xi" is the result of the profile i

(**Table 4A**). "*xm*" is the algebraic mean of the all the results and is the sum of the results divided by the number of results. Here the mean RD in the Golgi stack is 73.4%. To calculate the standard deviation, we apply

$$SD = \sqrt{1/(N-1)} \; \Sigma(xi-xm)^2.$$

So the results are expressed as xm±SD. In our example, the RD in Golgi stack is 73.4±22.7. If the odd result of profile c is left out, it comes 80±12 which is acceptable.

4.6.3. The Standard Error of the Mean

The standard error of the mean (SEM) describes the uncertainty (due to sampling error) in the mean value of the data, *xm*. That indicates how confident one can be with it, within one experiment or from one experiment to the next.

When the SEM is calculated within one experiment, it is the SD calculated in this experiment divided by the square root of the number of results equals 8. In our example it comes as $22.7/(8)^{-2}=8.02$. So the results are expressed as *xm*±SEM, 73.4±8.

The SEM can be calculated to measure the uncertainty of the mean from E number of different experiments. Each experiment *i* is expressed as *xmi*±SD*i*. We can calculated the mean of the *xm* and the SD*m* on this calculation.

$$SEM = SDm/\sqrt{E}.$$

The final results will be expressed as

$$RD = mean \; xm \pm SDm/\sqrt{E}.$$

The SEM should be between 5 and 10%.

For instance, together with the results of the experiment presented in **Table 4** which give RD = 73.4±22.7, three other similar experiments have given the following results 75±15; 71±10 and 79±20. The mean of the mean±SDm is 74.6±3.36. And the mean of the mean ±SEM is $74.6\pm3.36/\sqrt{4} = 74.6\pm1.68$. This experiment is perfect.

5. Limitations and Alternatives

There are two other requirements that immunolabeling should fulfill

1. Possibility of performing multiple labeling of different antigens in one specimen (*see* also **Subheading 3.1.**).
2. Equal accessibility of antigen molecules at their different locations.

However, cryoimmunolabeling presents problems related to these two requirements whereas resin sections could be an alternative to meet those requirements, but not without other additional problems.

5.1. Multiple Labeling

Performing multiple labeling using cryosections is possible and extremely useful in the identification of compartments (*see* also **Subheading 3.1.**). However, there are two kinds of problems that need to be overcome.

5.1.1. Technique

One problem concerns the sequential labeling in which the first antibody is visualized by a gold conjugate of one size, followed by the second antibody and the second gold conjugate of the second size. Different techniques have been developed to ensure that the second gold conjugate does not react with the first antibody and that the first gold conjugate does not react with the second antibody. The use of free protein A after the protein A complex is meant to quench free antibodies and solve the first problem. The aldehyde fixation after the first labeling is meant to solve the second problem *(14)*. But those extra steps could affect the labeling density of the second antigen. It is advisable to check multiple labeling in single labeling including the extra steps. Ideally, the LD should be the same. In case of a double labeling involving a mouse antibody (for instance, a monoclonal antibody) and a rabbit antibody, it is also possible to use antimouse IgG, and antirabbit IgG for the sequential labeling, thus avoiding the problem of cross reactivity *(15)*.

5.1.2. Steric Hindrance

If the labeling of the second antigen greatly overlaps that of the first one, it is likely that its labeling density will be lower than that observed in a single labeling. This is likely because of steric hindrance by the first antibody/gold conjugate. Therefore, some cautions should be taken because some parameters could be affected, such as the labeling density and possibly the relative distribution if the decrease in labeling density is not reduced uniformly in all compartments.

However, double labeling has been successfully used for quantitative purposes *(16,17)* given a certain amount of controls and standardisation of the labeling.

Resins sections are the alternative if the resin embedding itself has not damaged the antigenicity. Indeed, the resins sections could be labeled on both sides (one antigen per side preventing the steric hindrance) by floating them on the different solutions before mounting them on grids. It is a wonderful technique which has been applied successfully by the group of Bendayan *(18)* and Slot *(4)*.

5.2. Organelle Density and Antibody Penetration

In general, when a compartment has "n" times more gold particles than another compartment, it has "n" times more antigenic molecules. That is true

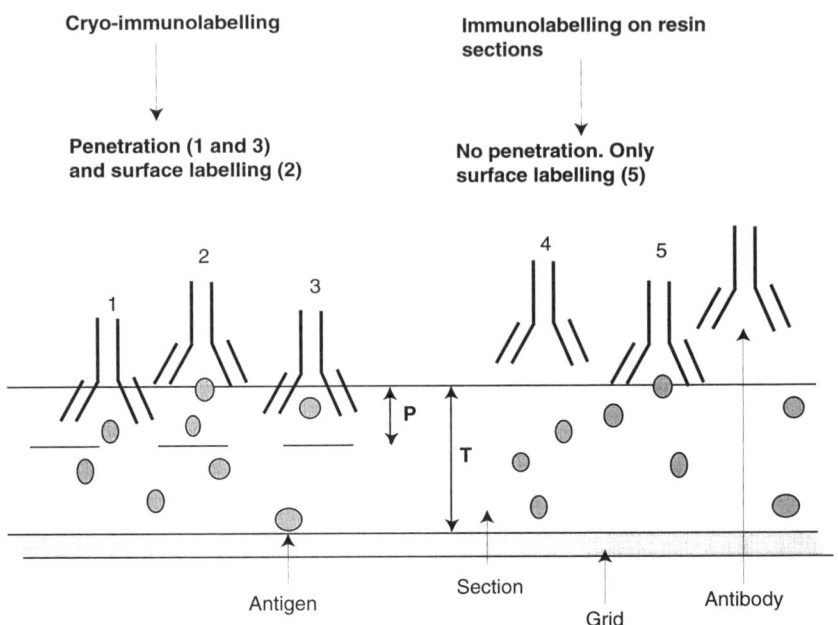

Fig. 5. Antibody penetration. The antibodies and gold conjugates can penetrate a certain fraction *P* (represented by the dashed line) of the thickness *T* of a cryosection. So some of the antigens situated within this penetration zone *P* could be labeled (1 and 3) as well as being labeled in surface *(2)*. In resin sections, however, there is no antibody penetration within the thickness of the section, so that only the surface antigens could be labeled *(4)*. As a result, the gold labeling in resin sections is likely to be reduced.

except when the density of the organelles is very different. The comparison of number of gold particles between compartments is only possible if the compartments have similar density. But some are very loose and not very concentrated as the ER. Some are very concentrated and dense such as the peroxisomes or the secretory granules. The comparison of gold labeling (calculation of the relative distribution of the gold particles) between these two types of compartments cannot be performed using cryosections. Antibodies penetrate within the thickness of cryosections (**Fig. 5**). Variations in antigen accessibility (penetration) owing to differences in density of cellular material affect the labeling efficiency (*see* **Subheading 1.2.1.**). to the extent that in dense secretory granules, there is almost only surface labeling (comparable to resin section, **Fig. 5**) whereas the antibody penetration is high in the endoplasmic reticulum and would be greatly reduced in resin sections. This causes problems for quantitation. Estimating the relative distribution of gold particles present on dense organelles by using cryosections would greatly underestimate them *(19)*

Resin sections offer solutions to overcome this problem. Because there is no penetration of antibodies at all in any compartment, comparison is possible *(4)*.

5.3. Absolute Quantitation

This last aspect is a matter for the future and it is the last challenge of cryoimmunoEM. How to relate the number of gold particles to the actual amount of protein in one particular compartment?

Some attempts made by the group of Slot *(4,20,21)* have been successful for soluble antigens. It consists in the embedding in a known volume of gelatine of a known amount of soluble antigen, which will be processed together with the tissue, in which the antigen is actually located. For instance, it has been performed in various ways for amylase in exocrine pancreas and concentrated in zymogene granules. Both blocks were mounted together, labeled, and viewed. The labeling density could then be related to the concentration of antigen. The group of Griffiths have also succeeded in absolute quantitation of membrane antigens *(22)*.

The second approach is to reduce the thickness of cryosections as much as possible so that the labeling owing to the antibody penetration is almost equal to the surface labeling. That has not been achieved yet and is an issue for the future.

Acknowledgment

I wish to thank Drs. Eija Jämsä, Barbara Svetstrup and Jim Shorter for critically reading this manuscript. I would also like to thank Dr. Jan Slot and Professor Hans Geuze for their teaching of this technique, and the example and inspiration they provided in using it.

References

1. Milstein, C. (1981) 12th Sir Hans Krebs Lecture. From antibody diversity to monoclonal antibodies. *J. Biochem.* **118,** 429–436.
2. Needman, J. (1936) Order and life.
3. Hobot, J. A. (1989) Lowicryls and low temperature embedding for colloidal gold methods, in *Colloidal Gold: Principles, Methods and Applications*, vol. 2 (Hayat M. A., ed.), Academic, San Diego, pp. 75–95.
4. Oprins, A., Geuze, H. J., and Slot, J. W. (1994) Cryosubstitution dehydration of aldehyde-fixed tissue: A favorable approach to quantitative immunocytochemistry. *J. Histochem. Cytochem.* **42,** 497–503.
5. Tokayasu, K. T. (1980) Immunocytochemistry on ultrathin frozen sections. *Histochem. J.* **12,** 381–390.
6. Geuze, H. J., Slot, J. W., and Tokayasu, K. T. (1979) Immunocytochemical localisation of amylase and chymotrypsinogen in the exocrine pancreatic cell with special attention to the Golgi complex. *J. Cell Biol.* **82,** 697–704.
7. Slot, J. W. and Geuze, H. J. (1981) Sizing of protein A-colloidal gold probes for immunoelectron microscopy. *J. Cell Biol.* **90,** 533–536.

8. Griffiths G., McDowall, A., Back, R. and Dubochet, J. (1984) On the preparation of cryosections for immunocytochemistry. *J. Ultrastruct. Res.* **89,** 65–73.

9. Liou, W., Geuze, H.J. and Slot, J.W. (1996) Improving structural integrity of cryosections forimmunogold labeling. *Histochem. Cell. Biol.* **106,** 41–58.

10. Shima, D. T., Haldar, K., Pepperkok, R., Watson, R., and Warren, G. (1997). Partitioning of the Golgi apparatus during mitosis in living HeLa cells. *J. Cell Biol.* **137,** 1211–1228.

11. Geuze, H. J., Slot, J. W., van der Ley, P. A., and Scheffer, R. C. T. (1981) Use of colloidal gold particles in double labelling immuno-electron microscopy of ultrathin frozen tissue sections. *J. Cell Biol.* **89,** 653–661.

12. Samuels, M. L. (1981) Statistics for the life sciences, in *Mathematics and Statistics*, Maxwell MacMillan Internat. Ed., Canada.

13. Weibel, E. R. (1979) A stereological methods. I., in *Practical Methods for Morphometry*, Academic, London, England.

14. Slot, J. W., Geuze, H. J., Gigengack, S., Lienhard, G. E., and James, D. E. (1991) Immuno-localisation of the insulin regulatable glucose transporter in brown adipose tissue of the rat. *J. Cell Biol.* **113,** 123–132.

15. Nilsson, T., Pypaert, M., Hoe, M., Slusarewicz, P., Berger, E. G., and Warren, G. (1993) Overlapping distribution of two glycosylatransferases in the Golgi apparatus of HeLa Cells. *J. Cell Biol.* **120,** 5–13.

16. Baas, E. J., van Santen, H. M., Kleijmeer, M. J., Geuze, H. J., Peters, P. J., and Ploegh, H. L. (1992) Peptide-induced stabilization and intracellular localization of empty HLA class I complexes. *J. Exp. Med.* **176,** 147–156.

17. Rabouille, C., Strous, G. J., Crapo, J. D., Geuze, H. J., and Slot, J. W. (1993) The differential degradation of two cytosolic proteins as a tool to monitor autophagy in hepatocytes by immunocytochemistry. *J. Cell Biol.* **120,** 897–908.

18. Bendayan, M. (1982) Double immunocytochemical labelling applying the proteinA-gold technique. *J. Histochem. Cytochem.* **30,** 81–89.

19. Posthuma, G., Slot, J. W., Geuze, H. J. (1987) The usefulness of the immunogold techniques in quantitation of the a soluble protein in ultrathin sections. *J. Histochem. Cytochem.* **35,** 405–410.

20. Posthuma, G., Slot, J. W., Geuze, H. J. (1986) A quantitative immunoelectron microscopical study of amylase and chymotrypsinogen in peri and tele insular cells of the rat exocrine pancreas. *J. Histochem. Cytochem.* **34,** 203–211.

21. Slot, J. W., Posthuma, G., Chang, L. Y., and Crapo, J. D. (1989) Quantitative aspects of immunogold labelling in embedded and in non embedded section. *Am. J. Anat.* **185,** 271–280.

22. Griffiths, G. and Hoppeler, H. (1986) Quantitation in immunocytochemistry: Correlation of immunogold labelling to absolute number of membrane antigen. *J. Histochem. Cytochem.* **34,** 1389–1397.

23. Rabouille, C., Hui, N., Hunte, F., Kieckbush, R., Berger, E. G., Warren, G., and Nilsson, T. (1995) Mapping the distribution of Golgi enzymes involved in the construction of complex oligosaccharides. *J. Cell Sci.* **108,** 1617–1627.

9

Microwave Processing Techniques for Electron Microscopy: *A Four-Hour Protocol*

Rick T. Giberson and Richard S. Demaree, Jr.

1. Introduction

The use of microwave energy to reduce processing times during fixation for electron microscopy (EM) is well established in the literature *(1–5)*. The microwave protocol described in this chapter is based on the work of Giberson et al. *(6,7)* and differs from the microwave literature in that each step (fixation, dehydration, embedding, and resin polymerization) in specimen processing is done in the microwave. The time savings gained from this approach are considerable and the results, based on our experience and that of others *(8,9)*, are equal or better than those achieved from routine processing protocols.

Anyone who has used a microwave oven, either for cooking or science, recognizes that it is an efficient tool for heating fluids or food. The reliance on the microwave oven's best attribute, rapid heating, does not appear to be the best strategy when it comes to scientific applications. In our experience *(5–7)*, rapid heating (5–20 s) and high final temperatures after fixation (40–55°C) are not necessary to achieve excellent fixation results (**Fig. 1A–D**). The protocol, with modifications from earlier work *(6,7)*, allows for precise temperature control during each step of microwave processing.

The successful use of this protocol relies to a great extent on the technology being used rather than the type of reagents. A diverse range of specimens (kidney, tumors, liver, brain, nematodes, plants, bacteria, embryos, yeast, etc.) have been successfully processed using this protocol. There is no evidence, as of this writing, that successful results (good fixation, dehydration, embedding, and polymerization) hinge on reagent choices. That is not to say that the use of certain reagents (fixatives, buffers, etc.) and/or extending different steps (fixa-

From: *Methods in Molecular Biology*, vol. 117: *Electron Microscopy Methods and Protocols*
Edited by: N. Hajibagheri © Humana Press Inc., Totowa, NJ

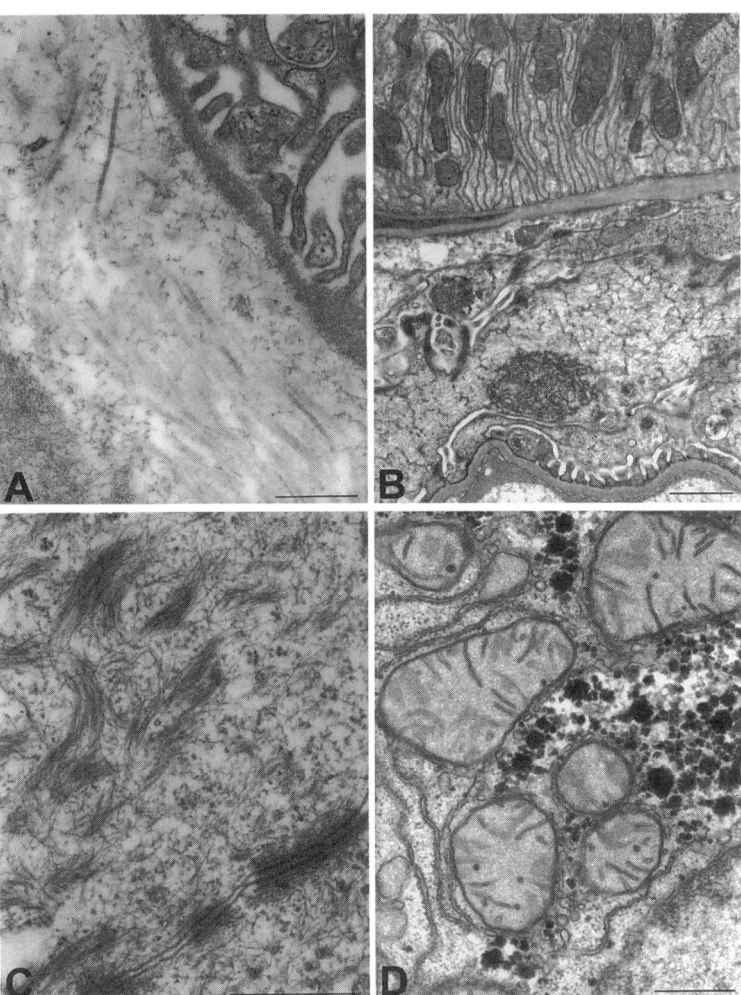

Fig. 1. **(A)–(D)** were processed using the four hour protocol. Each tissue was in aldehyde fixative for varying lengths of time prior to starting microwave-assisted processing.
(A) Human kidney biopsy (tissue in fixative ≤20 min prior to microwave processing) showing fibrous deposits in Bowman's capsule. The starting point of processing was **Subheading 3.2., step 6.** Bar, 0.5 μm. **(B)** Mouse kidney (tissue in fixative ≤20 min prior to
microwave processing) processed starting at **Subheading 3.2., step 6.** Note the absence of
any urinary space between Bowman's capsule and the underlying glomerulus. This finding
appears to be valid and is currently under study (unpublished research). Bar, 1μm. **(C)**
Patient biopsy having a clinical diagnosis of a Thymoma. The tissue was in a modified
Karnovsky's fixative for five mo prior to micorwave processing. The starting point of processing was **Subheading 3.2., step 4.** Bar, 0.5 μm. **(D)** Mouse liver which was profusion
fixed prior to starting at **Subheading 3.2., step 4.** Glycogen, rough endoplasmic reticulum
and mitochondria are well preserved in this sample. Bar, 0.5 μm.

tion, dehydration, and embedding) will not improve results when direct comparisons are made.

1.1. Microwaves and Immunocytochemistry

In addition to improving processing times, microwave fixation retains antigen preservation similar to that observed in conventionally fixed material (*see* refs. *11–13*). For light microscopy immunochemistry, microwave irradiation has been used on paraffin sections to demonstrate improved labeling efficiency, or antigen retrieval (*see* refs. *14–16*). Although the results obtained varied with the antibodies tested, there did seem to be an enhanced labeling effect after microwave treatment. It seems that at least some of these effects are not the result of heating alone suggesting that a microwave effect is present. One result of this seems to be to improve accessibility of antigens to antibodies.

Although it would seem logical to assume that improvement of antigen accessibility by microwave irradiation is possible for sections prepared for EM, there has been no quantitative study yet reported. A report indicating that heavy metal staining of epoxy resin sections for TEM is enhanced in the microwave (*17*) suggest that thin sections are not harmed by microwaves, and unpublished results (Webster, personal communication) suggest that microwaves do improve labeling efficiency of antibodies in some systems. However, as the studies examing the effects of microwaves on thin sections for TEM examination are still preliminary, there is little that can yet be concluded.

A final comment, the use of the microwave for tissue processing, prior to immunolabeling, has worked very well for localization of diffusible antigens in botanical tissue (unpublished results)."

2. Materials

Note: The Model 3450 Laboratory Microwave Processor and microwave accessories came from Ted Pella, Inc., Redding, CA (TPI) (**Fig. 2**).

2.1. The Microwave Processor and Creation of a Cold Spot within the Microwave Cavity

1. Use a microwave processor designed for laboratory use. The Model 3450 (TPI #3450) Laboratory Microwave Processor used for this protocol has the following features:
   a. automatic magnetron prewarming (important during fixation and dehydration);
   b. a temperature restrictive probe (necessary for dehydration and embedding);
   c. software and RS232 Port for temperature and time monitoring via a PC;
   d. water recirculator/chiller for the water load in the microwave cavity;
   e. safety exhaust vent for the microwave processor cavity;
   f. comes complete with all accessories;
   g. programmable for time and power level

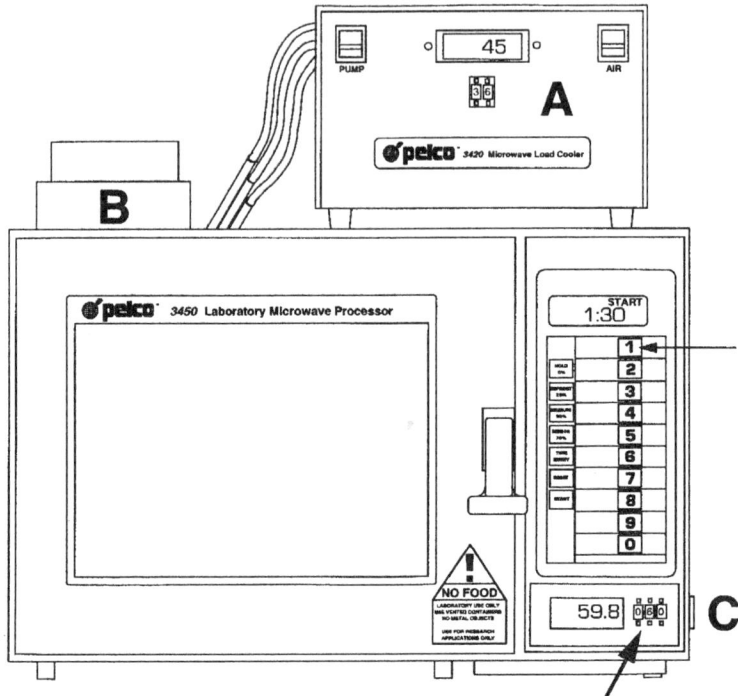

Fig. 2. The Model 3450 Microwave processor showing the water recirculator (**A**), exhaust vent (**B**) and RS232 port (**C**) for attachment of the microwave to a PC. Software is provided for the monitoring and recording of temperature and time data during a microwave processing run. The small arrow indicates one of the 10 programmable key pads. The large arrow indicates where the temperature restriction is set (dehydration and embedding steps).

   2. Neon bulb array (TPI #36141) for cold spot localization.
   3. Grid map (TPI #36110) for marking water load and sample locations.
   4. Two 400 mL plastic beakers (Tri-Pour® or similar).

2.2. Microwave Fixation

   1. Aldehyde fixative: 2.5% glutaraldehyde in 0.1 *M* sodium cacodylate buffer at pH 7.3. A 25% or less stock solution of glutaraldehyde is used to make up the fixative (*see* **Note 1**).
   2. Osmium fixative: 2% aqueous osmium tetroxide.
   3. 1.7 mL polypropylene microcentrifuge tubes (National Scientific Supply Co., Inc., #CN1700-GT P) or similar which are placed into a teflon holder (TPI #36134) for microwave irradiation. The tubes are used for the fixation steps and the buffer rinse between fixatives.

4. Buffer: 0.1 *M* sodium cacodylate, pH 7.3.
5. Crushed ice to cool the fixatives.

2.3. Microwave Dehydration

1. 50 mL volumes of 50, 70, 90, and 100% acetone (this is enough to do from 1–7 samples in the 60 × 15 mm polypropylene Petri dishes—*see* **Subheading 3.6., step 2**).
2. For the start of microwave dehydration the specimens are transferred from the microcentrifuge tubes to flow-through baskets (TPI #36142-1) which are held in the bottom half of a 60 × 15 mm polypropylene Petri dish (TPI #36135) containing 15–18 mL of 50% acetone.
3. Use ultrapure water to make the acetone dilutions (*see* **Note 2**).

2.4. Microwave Embedding

1. A mixture of Epon and Spurr's resin (*see* **Note 3**), adapted from Bozzola and Russell *(10)*, is composed of the following: 10.0 g of ERL 4206; 6.0 g of DER 736; 26.0 g of NSA; 24 drops of DMAE (from a glass pasteur pipet); 25.0 g of any Epon 812 substitute (i.e., Eponate 12, Polybed 812, EMbed 812), 13.0 g DDSA, 12.0 g NMA, 32 drops DMP-30 (from a glass pasteur pipet). This makes enough resin for approximately 50 BEEM® 00 capsules after all embedding steps have been completed.
2. A 1:1 mixture of 100% acetone and Epon/Spurr's (~20 mL).

2.5. Microwave Polymerization

1. A rectangular plastic tray (TPI #36133), 2" deep × 4.5" wide × 9" long, which will hold approximately 1000 mL of water (*see* **Note 11**).
2. Size 00 polyethylene embedding capsules (BEEM® or similar available from electron microscopy [EM] supply companies).
3. A Teflon® capsule holder (TPI #36131) for securing BEEM® or similar type 00 embedding capsules.
4. ParafilmM® (available from VWR and EM supply companies) to seal capsules prior to polymerization.

3. Methods
3.1. Creation of a Cold Spot within the Microwave Cavity

1. Tape the grid map to the bottom of the microwave cavity ~2.5 cm from the front.
2. Two 400 mL plastic beakers are filled with ~375 mL of water. The tubing inside the microwave from the water recirculator is placed in the beaker on the left and the recirculator turned on. The water volume is adjusted to 375 mL with the recirculator running.
3. The other water load is placed on the right side of the microwave cavity with the neon bulb array (**Fig. 3**) positioned between the two loads where the cold spot/sample area is indicated in **Fig. 4**.

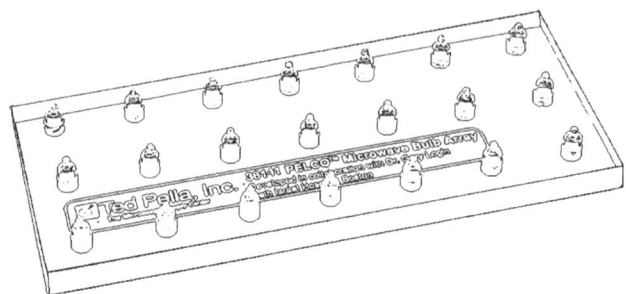

Fig. 3. The neon bulb array contains 21 neon bulbs and has the following dimensions: 7.6 × 17.8 cm.

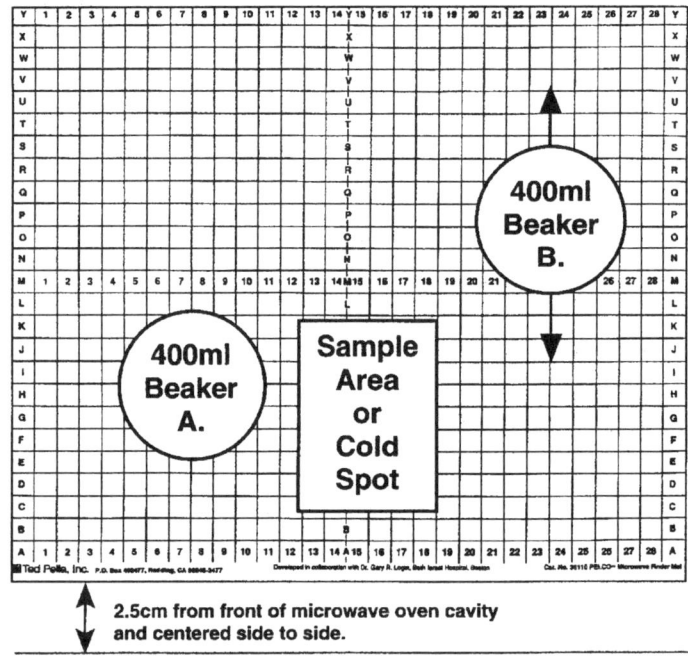

2.5cm from front of microwave oven cavity
and centered side to side.

Fig. 4. The grid map indicates the water load locations and the cold spot/sample area which is configured using the neon bulbs shown in **Fig. 3**. Beaker A is serviced by the water recirculator and not moved during cold spot determination. Beaker B is moved foward or backward (as shown by the arrows) in the oven cavity to configure the cold spot/sample area.

4. The water loads are positioned (*see* 5 below) so that an area (~4 × 8 cm) of at least six adjacent bulbs does not come on during 10 s of microwave irradiation at 100% power. This area is called the cold spot and becomes the sample placement area (**Fig. 4**).

5. Move the beaker on the right forward or backward to create the cold spot (**Fig. 4**). Leave the beaker on the left with the tubes from the recirculator in a fixed position during this process.
6. Mark the water load locations and the sample placement area on the grid map with a permanent marker (i.e., Sharpie®).
7. Check the cold spot parameters by placing the Teflon holder with three microcentrifuge tubes containing 600 μL of water each in the cold spot and microwaving the tubes for 40 s at 100% power. The change in temperature (ΔT) should be ≥10°C and ≤15°C after microwave irradiation.
8. If the ΔT is <10°C decrease the volume of the water loads.
9. If the ΔT is >15°C increase the volume of the water loads (larger beakers) or add a third smaller water load on the right side.

3.2. Aldehyde Fixation

Note: All reagents should be made and ready for use prior to starting fixation.

1. Specimens for processing should be ≤2.0 mm^3.
2. Place the specimens (from 1 to 7 tissue blocks) in a 1.7-mL microcentrifuge tube containing 600 μL of fixative.
3. Cool the fixative and sample to between 10 and 15°C prior to microwave irradiation (this step is not required for #4 below). This step will insure that the final temperature after microwave irradiation will not exceed 30°C. Record the temperature immediately prior to placing the specimens and Teflon® holder into the cold spot in the microwave (**Fig. 5**).
4. If the tissue has been in fixative for ≥1 h microwave the specimens for a time interval of 10-20-10. The time interval 10-20-10 (10 s at 100% power – 20 s at 0% power – 10 s at 100% power) is programmed into one of the numbered key pads on the front of the microwave processor (**Fig. 2**). Use the temperature probe to record the temperature immediately before and after microwave irradiation (*see* **Note 5**). No cooling is required for the 10-20-10 interval.
5. If the tissue has been in fixative for ≤1 h microwave the sample for 40 s at 100% power in the cold spot. Record the temperature after microwave irradiation. Let the sample sit in fixative outside the microwave oven for 5 min after microwave irradiation before going to the buffer rinse (*see* **Notes 6** and **7**).
6. If the tissue has been in fixative for ≤20 min microwave the sample for 40 s at 100% power in the cold spot. Record the temperature after microwave irradiation. Let the sample sit in fixative outside the microwave oven for 5 min after microwave irradiation and then repeat steps 3 and 5.

3.3. Buffer Rinse

1. Remove the fixative from the microcentrifuge tube and add buffer and then immediately replace the buffer with fresh buffer (*see* **Note 8**).
2. Let the specimens sit in the second change of buffer 5 min outside the microwave at room temperature.

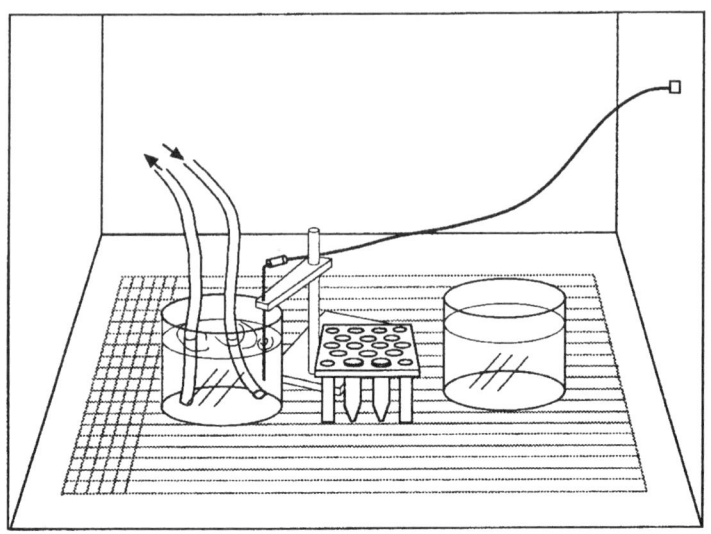

Fig. 5. The placement of the temperature probe (in beaker A, *see* **Fig. 4**) and Teflon®
holder (cold spot/sample area) with microcentrifuge tubes is shown for the fixation
steps.

3.4. Osmium Fixation

1. Working in a fume hood remove the buffer and add 600 μL of fixative.
2. Cool the fixative and sample to between 10 and 15°C prior to microwave irradiation. This step will ensure that the final temperature after microwave irradiation
   will not exceed 30°C. Record the temperature immediately prior to placing the
   specimens and Teflon® holder into the cold spot in the microwave.
3. Microwave at 100% power for 40 s and immediately record the temperature after
   microwave irradiation.
4. Let the sample sit in fixative in the fume hood for 5 min after microwave irradiation and then repeat steps 2 and 3.

3.5. Water Rinse

1. In a fume hood, remove the osmium and replace with water.
2. Replace the water with a smaller volume (~100 μL) to facilitate transfer of the
   specimen from the microcentrifuge tubes to the flow-through baskets (**Fig. 6**).
3. The flow-through baskets for this exchange are contained in the bottom half of
   the 60 × 15 mm Petri dish which contains 15–18 mL of 50% acetone.

3.6. Acetone Dehydration

1. The acetone dehydration series is composed of the following:1 × 50%; 1 × 70%;
   1 × 90%; 2 × 100%.

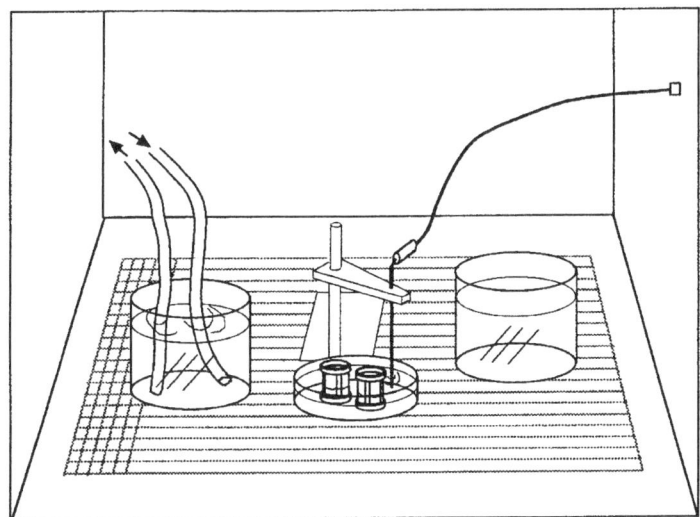

Fig. 6. The placement of the temperature probe (in the plastic Petri dish, but not in a basket) and Petri dish with baskets (cold spot/sample area) is shown for the dehydration and embedding steps.

2. The Petri dish with baskets (up to seven baskets can be placed in a dish) and acetone is placed in the cold spot in the microwave.
3. The temperature probe is placed in the dish (not in a basket) (**Fig. 6**) and the temperature restriction on the front of the microwave is set at 37°C (**Fig. 2**).
4. The samples are irradiated at 100% power for 40 s.
5. Have a second Petri dish ready with the next acetone change and transfer the baskets using a curved Kelly-type hemostat.
6. Repeat steps 3 and 4 until all dehydration steps have been completed.

3.7. Embedding

1. The embedding series is composed of the following microwave steps:1:1 100% acetone—resin × 15 min; 100% resin 2 × 15 min.
2. Transfer the baskets as described for dehydration to a Petri dish containing a 1:1 mixture of 100% acetone and resin.
3. The Petri dish with baskets is positioned in the cold spot.
4. The temperature probe is placed in the dish (not in a basket) and the temperature restriction on the front of the microwave is set at 45°C (**Fig. 2**) for all embedding steps (*see* **Note 10**).
5. Microwave at 100% power for 15 min.
6. Transfer the baskets to a Petri dish containing 100% resin and microwave at 100% power for 15 min. Make sure the temperature probe is in the resin before starting the microwave (*see* **Notes 9** and **12**).
7. Repeat step 6.

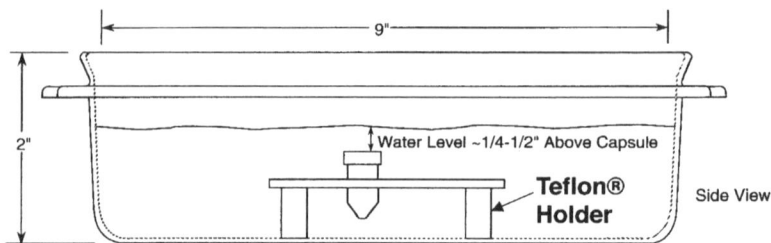

Fig. 7. The Teflon® holder with 00 embedding capsules is shown in the rectangular plastic tray covered with water for the polymerization step.

3.8. Polymerization

1. Transfer the specimens from the flow-through baskets to 00 size polyethylene capsules which are partially filled (~1/3) with resin. The capsules should be in the Teflon® holder shown in **Fig. 7**.
2. Slightly overfill the capsules with resin and place the ParafilmM® in the cap as shown in **Fig. 8** and cap each capsule. Make sure the tab between the capsule and the cap has been broken. If not the cap will pop open during polymerization.
3. Place the Teflon® holder with capsules into the plastic tray and remove the water load on the right.
4. Place the tray in the microwave and fill to within 1/4" of the top with tap water (**Fig. 7**).
5. Microwave at 100% power for 75 min. The water load serviced by the water recirculator remains in the microwave during polymerization. The temperature probe is not needed for polymerization. It can be removed from the microwave or used to monitor the temperature of the water surrounding the capsules. If the probe is used to monitor water temperature the temperature restriction must be set above 100°C.

4. Notes

1. The protocol described in this chapter has been subjected to many different laboratory settings and widely diverse tissues. The choice of the type of fixative (glutaraldehyde, modified Karnovsky's, paraformaldehyde alone, buffered, and unbuffered osmium tetroxide) and buffer (sodium cacodylate, phosphate buffers, HEPES), in our experience, should be based on what is used routinely in the laboratory for processing the tissue in question. This is the best starting point when trying microwave-assisted rapid processing for the first time. The addition of $MgCl_2$ and/or $CaCl_2$ to the fixative buffer will increase the ΔT outside the 10–15°C parameter described in **Subheading 3.1.**, step 7. If there are preferences, they are in the use of acetone during the dehydration steps and the resin combination of Epon/Spurr's for the embedding and polymerization steps.
2. Acetone eliminates the need for propylene oxide as a transition solvent, works well with all epoxy resins used to date (EMbed 812, Eponate 12, Polybed 812,

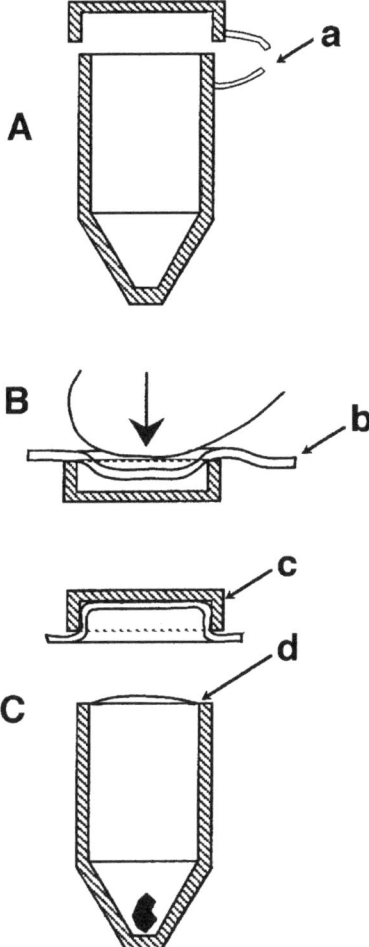

Fig. 8. **(A)** The tab on the 00 embedding capsule is broken free from the capsule (a). **(B)** ParafilmM® (b) is placed in the cap. **(C)** The capsule is overfilled slightly with resin (d) and the cap (c) is placed on the capsule.

LX112, Araldite) and is quite satisfactory when used with LR White and UNICRYL™. There is no reason why ethanol cannot be used in place of acetone for dehydration. If this is done, then acetone or propylene oxide can be used for the transition solvent. The 100% propylene oxide steps are done identically to the 100% acetone steps in the microwave.

3. The resin combination, Epon/Spurr's, has proven to be an excellent choice for microwave processing. That is not to say that any of the epoxies listed in **Note 2** and Spurr's cannot be used alone for microwave processing. However, in our experience, and that of others who have been using the Epon–Spurr's mixture for

microwave embedding and polymerization, the following comments appear to be valid:

   a. after polymerization the resin is easier to trim with a razor blade and section (1μm or ultrathin) than other epoxy resins;

   b. the resin appears to be more stable in the electron beam than other epoxy resins;

   c. the infiltration and crosslinking properties in difficult samples (botanical and yeast, for example) are superior to either resin used alone.

4. Once the cold spot is established that location will remain constant over time as long as the water load volumes and locations remain fixed.

5. During the fixation steps the temperature probe is placed in the water load on the left side of the microwave cavity and the temperature restriction on the front of the microwave is set at 45°C or above. This ensures that the magnetron will run continuously during the 40 s time interval and that the ΔT will be in the 10–15°C temperature range, the indication that the sample has received enough microwave energy for reproducible fixation. If for any reason the ΔT is less than 8°C, repeat the step and check temperature settings, water load locations, and size. Line voltage will also affect ΔT. Low line voltage (100–105 V) will typically produce ΔT's <8°C and line voltages above 120 V can produce ΔT's above 15°C. If the ΔT is above 15°C and the final temperature below 40°C after microwave irradiation no harm has been done.

6. In the osmium fixation step and occasionally in the aldehyde step, difficult to fix tissues like plant, certain bacteria and peripheral nerves will require an additional microwave fixation step. This is especially true of peripheral nerves.

7. If en bloc staining with uranyl acetate is desired treat it as a 40 s fixation step. Cool the sample with the uranyl acetate and microwave for 40 s at 100% power. Make sure to note the ΔT. It should be the same as a fixation step for aqueous uranyl acetate.

8. Increasing the ionic strength of a buffer or the addition of $MgCl_2$ or $CaCl_2$ to fixatives or buffers can push the ΔT outside the established parameters. As long as the final temperature is not above 40°C it will not be a problem.

9. The water load on the right can be changed when it gets hot to the touch (>50°C). This is done usually once or twice during the entire protocol.

10. Make sure the temperature probe is always in the resin or 1:1 mixture during each microwave step of embedding. If it is not, the resin will polymerize rapidly (<2–3 min) and the microwave processing run will be ruined.

11. The volume and dimensions of the plastic tray are important. The volume controls the heating rate of the water which is critical during the first 10 min of the polymerization step. If the water column is too deep then all microwave energy will be absorbed by the water and none by the resin filled capsules.

12. When LR White is polymerized in the microwave the total time drops to 45 min instead of 75 min. When UNICRYL™ is polymerized in the microwave the total time is 60 min and a temperature restriction of 95°C is used throughout the time period.

13. The water level in the tray should be checked periodically (about every 20 min) and water added (cold tap water is fine) when the level is approaching the top of the capsules. Even if the water level drops significantly that does not hurt the sectioning quality of the blocks, although the block appearance will probably suffer. The loss of a cap after the first 15 min of the polymerization step does not indicate disaster.

References

1. Giberson, R. T. and Demaree, R. S., Jr. (1995) Microwave fixation: Understanding the variables to achieve rapid reproducible results. Microsc. Res. Tech. **32,** 246–254.
2. Benhamou, R., Noel, S., Grenier, J., and Asselin, A. (1991) Microwave energy fixation of plant tissue; an alternative approach that provides excellent preservation of ultrastructure and antigenicity. *J. Electron Microsc. Tech.* **17,** 81–94.
3. Login, G. R. and Dvorak, A. M. (1985) Microwave energy fixation for electron microscopy. *Am. J. Pathol.* **120,** 230–243.
4. Login, G. R., Stavinoha, W. B., and Dvorak, A. M. (1986) Ultrafast microwave energy fixation for electron microscopy. *J. Histochem. Cytochem.* **34,** 381–387.
5. Login, G. R., Dwyer, B. K., and Dvorak, A. M. (1990) Rapid primary microwave-osmium fixation. I. Preservation of structure for electron microscopy in seconds. *J. Histochem. Cytochem.* **38,** 755–762.
6. Giberson, R. T., Demaree, R. S., Jr., and Nordhausen, R. W. (1997) Four-hour processing of clinical/diagnostic specimens for electron microscopy. *J. Vet. Diagn. Invest.* **9,** 61–67.
7. Giberson, R. T., Smith, R. L., and Demaree, R. S. (1995) Three hour microwave tissue processing for transmission electron microscopy: From unfixed tissues to sections. *Scanning* **17(suppl. 5),** 26–27.
8. Rassner, U. A., Crumrine, D. A., Nau, P., and Elias, P. M. (1997) Microwave incubation improves lipolytic enzyme preservation for ultrastructural cytochemistry. *Histochem. J.* **29,** 387–392.
9. Madden, V. J. and Henson, M. M. (1997) Rapid decalcification of temporal bones with preservation of ultrastructure. *Hearing Research,* **111,** 76–84.
10. Bozzola, J. J. and Russell, L. D. (1992) *Electron Microscopy: Principles and Techniques for Biologists,* Jones and Bartlett, Boston, MA, figure legends, p. 26.
11. Login, G. R., Galli, S. J., Morgan, E., Arizona, N., Schwartz, L. B., and Dvorak, A. M. (1987a) Rapid microwave fixation of rat mast cells. I. Localization of granule chymase with an ultrastructural postembedding immunogold technique. *J. Lab. Invest.* **57(5),** 592–599.
12. Login, G. R., Schnitt, S. J., and Dvorak, A. M. (1987b) Methods in laboratory investigation: Rapid microwave fixation of human tissues for light microscopy immunoperoxidase identification of diagnostically useful antigens. *J. Lab. Invest.* **57(5),** 585–591.
13. Login, G. R., Galli, S. J., and Dvorak, A. M. (1992) Immunocytochemical localization of histamine in secretory granules of rat peritoneal mast cells with conven-

tional or rapid mcirowave fixation and an ultrastructural post-embedding immunogold technique. *J. Histochem. Cytochem.* **40(9),** 1247–1256.

14. Leong, A. S.-Y., Milios, J., and Duncis, C. G. (1988) Antigen preservation in microwave-irradiated tissues: A comparison with formaldehyde fixation. *J. Pathol.* **156,** 275–282.

15. Cuevas, E. C., Bateman, A. C., Wilkins, B. S., Johnson, P. A., Williams, J. H., Lee, A. H. S., Jones, D. B., and Wright, D. H. (1994) Microwave antigen retrieval in immunocytochemistry: A study of 80 antibodies. *J. Clin. Pathol.* **47,** 448–452.

16. Shi, S., Key, M. E., and Kalra, K. L. (1991) Antigen retrieval in formalin-fixed, parafin-embedded tissues: An enhancement method for immunohistochemical staining based on microwave oven heating of tissue sections. *J. Histochem. Cytochem.* **39(6),** 741–748.

17. Estrada, J. C., Brinn, N. T., and Bossen, E. H. (1985) A rapid method of staining ultrathin sections for surgical pathology TEM with the use of the microwave oven. *Brief Scientific Reports* **83(5),** 639–641.

10

Electron Microscopic Enzyme Cytochemistry

Nobukazu Araki and Tanenori Hatae

1. Introduction

The primary aim of enzyme cytochemistry is elucidation of intracellular localization of enzymes. Enzyme cytochemical reactions are based on the enzymatic conversion of a substrate, followed by the deposition of electron dense products at the enzyme site. So, enzyme cytochemistry reveals not only the location of enzyme but also the intensity of its catalytic activity. The data obtained by enzyme cytochemistry can be easily compared to biochemical enzyme assay in test tubes. Immunocytochemistry detects the enzyme molecule itself, but gives no information about enzyme activity. Instead, immunocytochemistry is more correlative to immnological techniques such as Western blotting. Thus, enzyme cytochemistry shows distinct biological significance from immunocytochemistry.

Enzyme cytochemistry is a useful tool to identify an organelle by detecting its maker enzyme, and it can be applied to three-dimensional (3-D) electron microscopy (EM) of intracellular organelles (**Fig. 1**) *(1,2)*. Although the solo use of enzyme cytochemistry seems to be less attractive nowadays, the importance of enzyme cytochemistry would increase when combined with other techniques. Enzyme cytochemistry is being used as the final markers in immunocytochemistry and hybridocytochemistry *(1)*. Combination of enzyme cytochemistry with immunocytochemistry is also a useful tool to analyze the relationship between one enzyme reaction and another immunoreaction (**Fig. 2**) *(1,3)*. Thus, enzyme cytochemistry may form a disciplinary bridge between morphology, biochemistry and molecular biology.

Among various enzyme cytochemical reactions, phosphatase enzyme cytochemistry is well established and one of the most common techniques. The original method for the cytochemical demonstration of phosphatases was the lead-based

From: *Methods in Molecular Biology, vol. 117: Electron Microscopy Methods and Protocols*
Edited by: N. Hajibagheri © Humana Press Inc., Totowa, NJ

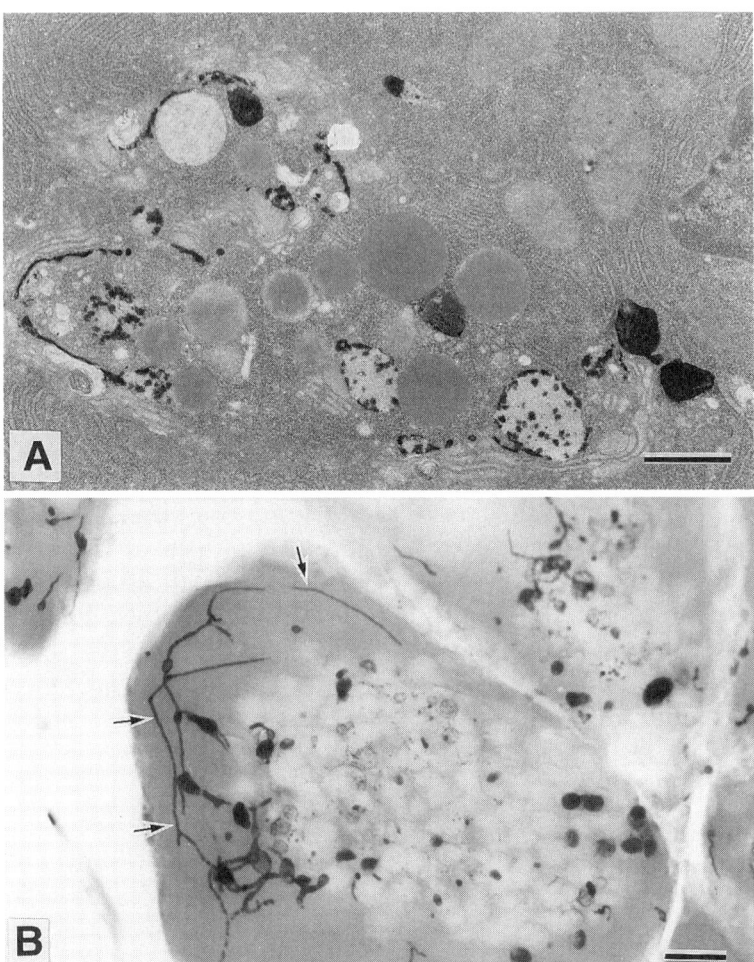

Fig. 1. Enzyme cytochemical demonstration of acid phosphatase in rat pancreatic exocrine cells by the lead-based method. (**A**) Reaction products are seen in lysosomes, trans-Golgi cisternae and immature secretory granules. (**B**) Two-hundred kilovoltage transmission electron micrograph of 2 μm-thick section of rat pancreatic exocrine cells. Thick section enables us to observe three-dimensional structure of nematolysosomes (tubular lysosomes) showing acid phosphatase activity (arrows). Bars = 1 μm

method of Gomori (1952) *(4)*. Phosphatases hydrolyze monoesters of phosphoric acid and release phosphate ions. Lead ions contained in the incubation medium precipitate as lead-phosphate where enzyme reaction occurs. This method is still widely used, since the lead phosphate can be readily observed under a light microscope after converting into lead sulphide with ammonium sulphide solution.

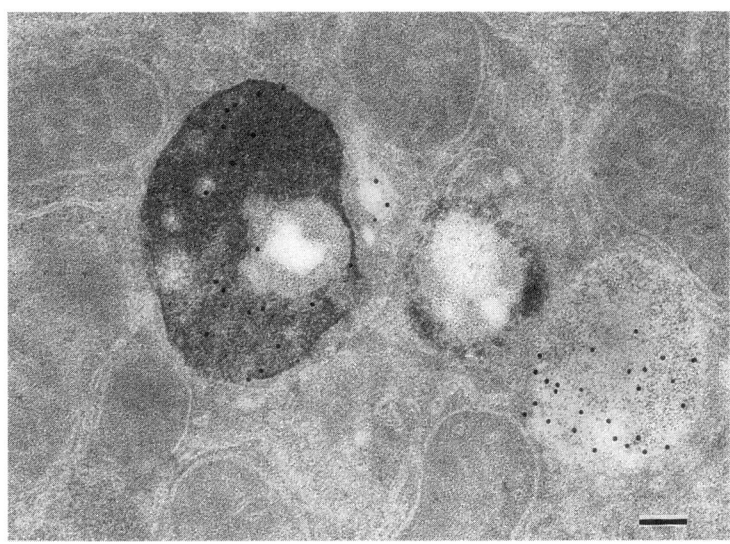

Fig. 2. Electron micrograph of rat liver doubly labeled with acid phosphatase enzyme cytochemistry and postembedding immunocytochemistry for saposin C, a sphingolipid activator protein. Acid phosphatase was detected by the cerium-based method. Immunogold particles within acid phosphatase-positive compartments strongly prove that saposin C is localized in lysosomes. Bar = 0.1 µm

Recently, cerium has been increasingly employed as a capture agent for phosphate, instead of lead *(5)*. The cerium-based method is advantageous for combination with postembedding immunocytochemistry, since cerium phosphate precipitation is more stable than lead phosphate *(3)*.

As basic examples, we introduce here the lead-based method and cerium-based method for cytochemical demonstration of acid phosphatase activity. In addition, a procedure for combination of enzyme cytochemistry and postembedding immunocytochemistry is described.

Although the number of enzymes whose reaction products can be cytochemically visualized is limited, various methods have been established to detect about 100 enzymes such as phosphatases, esterases, oxidases, and reductases. One can detect many other enzyme activities by similar procedures simply by changing reaction media. Lists of composition by of other reaction media can be found elsewhere *(6,7)*.

2. Materials
2.1. Enzyme Cytochemistry of Acid Phosphatase

1. 5X Buffer stock: 0.15 *M* pipes buffer, pH 7.2 (*see* **Note 1**). Dissolve 18 g pipes in 300 mL distilled water during neutralizing with NaOH, then adjust pH to 7.2.

Bring the final volume to 400 mL with distilled water. This stock solution can be stored up to a few months at 4°C.

2. 8% paraformaldehyde stock solution:Add 8 g paraformaldehyde powder to 60 mL distilled water. Heat solution to 60–70°C with stirring in a fume hood. Add a few drops of 1 N NaOH until milky solution become clear. Bring the final volume to 100 mL with distilled water. Cool and filter with a Whatman No. 1 filter paper. This solution can be divided into aliquots, and stored at –20°C up to several months. Thaw the necessary amount of aliquots when preparing a fixative. The just-thawed solution is cloudy, but it clears again after heating up to 40°C.

3. Fixative: 2% paraformaldehyde and 0.25% glutaraldehyde in 30 mM pipes buffer, pH 7.2 containing 8% sucrose. Make fresh from stock solutions. (*see* **Note 2**).

4. Rinse buffer DMSO plus: 30 mM pipes buffer, pH 7.2 containing 10% dimethyl sulfoxide (DMSO) and 8% sucrose. Store at 4°C.

5. Rinse buffer: 30 mM pipes buffer, pH 7.2 containing 8% sucrose. Store at 4°C.

6. Incubation medium: Incubation medium either of a) or b), and control medium c) should be freshly made just prior to incubation.

   a. Lead-based medium: Prepare 50 mL of 0.05 M acetate buffer, pH 4.7, by combining 6.25 mL of 0.2 M acetic acid (1.15 mL/100 mL) , 6.25 mL of 0.1 M sodium acetate (1.64 g/100 mL) and 37.5 mL of distilled water. Dissolve 4.0 g sucrose and 50 mg of lead nitrate completely in 0.05 M acetate buffer, pH 4.7. Add 5 mL of 3% sodium β-glycerophosphate slowly while stirring, so that the medium should be clear (*see* **Note 3**). Final pH should be adjusted to 5.0 using 0.1 N HCl or NaOH.

   b. Cerium-based medium: Prepare 25 mL of 0.2 M acetate buffer, pH 5.0, by combining three portions of 0.2 M acetic acid and seven portions of 0.2 M sodium acetate. Add 20 mL distilled water. Dissolve 15 mg sodium β-glycerophosphate and 2.5 g sucrose. Add 5 mL of 20 mM CeCl$_3$ (75 mg/mL) slowly during stirring. 25 μL of 0.15% Triton X-100 may be optionally added to increase permeabilization.

   c. Cytochemical control media: As a cytochemical control, sodium β-glycerophosphate, an enzyme substrate, can be omitted from the medium. In another control, add a specific inhibitor such as 10 mM sodium fluoride to the full medium of a) or b) (*see* **Note 4**).

7. 1% osmium tetroxide in 30 mM pipes buffer.

8. Ethanols (50, 70, 80, 90, 95, and 100%).

9. Propylene oxide.

10. Embedding medium: Spurr's resin.

11. Embedding mold: gelatin capsules or silicon molds.

12. Copper grids (200–300 mesh).

2.2. Combination of Enzyme Cytochemistry with Immunocytochemistry

1. 0.25% NH$_4$Cl in 30 mM pipes buffer, pH 7.2 containing 8% sucrose.

2. LR white resin kit (London Resin, U.K.).

3. Dimethylformamide.
4. UV polymerizer.
5. Nickel grids (200 mesh).
6. To combine with postembedding immunocytochemistry, additional materials such as blocking solutions, primary antibodies, and colloidal gold conjugates would be necessary (*see* Chapters 5, 6, and 7).

3. Methods

3.1. Enzyme Cytochemistry for Detecting Acid Phosphatase Activity

1. Fixation: Fix cells or tissue with 2% paraformaldehyde and 0.1–0.25% glutaraldehyde in 30 mM pipes buffer, pH 7.2 containing 8% sucrose for 30–60 min at room temperature (*see* **Note 2**).
2. Rinse the specimen in 30 mM pipes buffer, pH 7.2 containing 8% sucrose and 10% DMSO for 30 min by changing the buffer several times.
3. Slice: If the specimen is tissue, freeze the tissue and cut into 30–50 μm-thick sections with a cryostat or a freezing microtome. Then, return the sections into the buffer.
4. Incubate the sections in a incubation medium as described in **Subheading 2.** for 30 min at room temperature or 37°C while agitating.
5. Following incubation, rinse the sections in the rinse buffer without DMPO by changing the rinse buffer several times, and postfixed with 1% osmium tetroxide in 30 mM pipes buffer for 30 min at 4°C.
6. Embedding: Dehydrate the specimen in a graded series of ethanol solutions and propylene oxide for 10 min each and substitute with the Spurr's resin. Then, place the specimen in an appropriate mold filled with the resin and polymerized in an oven at 70°C overnight.
7. Sectioning: Prepare ultrathin sections using an ultramicrotome and mount on grids. The section may be stained with 1% uranyl acetate. Observation of 1–2 μm-thick sections by a high or intermediate voltage electron microscope operated at 200–1000 kV would be useful to analyze 3-D structure of reaction-positive compartments (**Fig. 1B**).

3.2. Combination of Enzyme Cytochemistry with Postembedding Immunogold Cytochemistry

1. Fix the tissue or cells with 4% paraformaldehyde and 0.1% glutaraldehyde in 30 mM pipes buffer, pH 7.2, containing 8% sucrose for 1 h at room temperature.
2. Following fixation, rinse the specimen three to five times over a period of 30 min in the buffer.
3. Rinse the specimen with 0.25% NH$_4$Cl in the buffer for 10 min to quench free aldehyde groups.
4. Perform steps 2–4 as described **Subheading 3.1.** The cerium-based incubation medium should be employed (*see* **Note 7**).

5. Following enzyme cytochemical incubation, rinse in the buffer thoroughly. Skip the postfixation with osmium (*see* **Note 8**).
6. Dehydrate in a graded series of dimethylformamide at 4 or –20°C, and then substitute LR white for dimethylformamide.
7. Embed in LR white containing accelerator 0.5 µL/mL resin and polymerize by UV irradiation at 4 or –20°C for 24–48 h (*see* also Chapters 6 and 7).
8. Prepare ultrathin sections and mount on nickel grids.
9. Perform the postembedding immunogold cytochemistry as described elsewhere (*see* also Chapters 6 and 7).

4. Notes

1. Although sodium cacodylate has been widely used as a buffer reagent for fixative and rinse solution for enzyme cytochemistry, pipes is more preferable than cacodylate to preserve both enzyme activity and ultrastructure.
2. The condition of fixation is very important to obtain good results. In earlier studies, higher concentrations of glutaraldehyde than 2% were frequently used as fixative. Even if enzyme reaction could be cytochemically detected, such fixation tends to make false negative reaction in some compartments having weak enzyme activity, owing to partial enzyme inactivation by glutaraldehyde. Mixture of 2–4% paraformaldehyde and lower concentrations of glutaraldehyde than 0.25% is generally advisable, although the most appropriate condition should be determined depending on the kind of enzymes and cell types. If possible, perfusion fixation should be applied to tissues. The adequate fixation allows the cytochemical demonstration of acid phosphatase activity in nematolysosomes (tubular lysosomes) where it had been hardly detected (**Fig. 1**) *(2)*.
3. It is important to prepare a clear incubation medium. Use of a cloudy incubation medium may result in nonspecific reactions or a diffusion of reaction products from an appropriate reaction site. The purity of the reagents and distilled water is also critical.
4. The reaction specificity should be assessed by cytochemical controls using substrate-free medium and/or known specific inhibitors for the enzyme. If any reaction products would be observed in cytochemical controls, the reaction should be considered as results of other nonspecific enzymes or nonenzymatic artifacts.
5. In earlier studies, use of nonfrozen sections by a Vibratome or a tissue chopper was recommended for electron microscopic cytochemistry, since it has been believed that frozen sections lead to poor ultrastructure. However, an addition of 10% DMSO into the buffer has enough cryoprotection effect to preserve ultrastructure at the electron microscopic level (**Fig. 1A**). Moreover, DMSO and the freeze-thawing process improve membrane penetration of the enzyme substrate and the capture agent into cells and organelles.
6. Incubation time is dependent on the type of enzyme and intensity of enzyme activity in the cell. Longer incubation than 40 min should be avoided, since artifacts such as nuclear staining may occur.

7. The composition of fixative may vary depending on the antigen. If the antigen is very sensitive to glutaraldehyde, try to omit the glutaraldehyde from the fixative.
8. Lead phosphate precipitation, reaction products by the lead-based method, will be mostly lost from the specimen during postembedding immunoreaction *(3)*.
9. The antigenicity of most antigen would be not retained by the osmium postfixation.

References

1. Araki, N., Yokota, S., Takashima, Y., and Ogawa, K. (1995) The distribution of cathepsin D in two types of lysosomal or endosomal profiles of rat hepatocytes as revealed by combined immunocytochemistry and acid phosphatase enzyme cytochemistry. *Exp. Cell Res.* **217,** 469–476.
2. Araki, N. Matsubara, H., Takashima, Y., and Ogawa, K. (1989) Three-dimensional distribution of acid phosphatase-positive thread-like structures in rat pancreatic exocrine cells. *Acta Histochem. Cytochem.* **22,** 585–590.
3. Araki, N., Yokota, S., and Takashima, Y. (1993) The cerium-based method is advantageous in combination with post-embedding immunocytochemistry. *Acta Histochem. Cytochem.* **26,** 569–576.
4. Gomori, G. (1952) *Microscopic Histochemistry, Principle and Practice*, University Chicago Press, Chicago, IL, pp. 189.
5. Robinson, J. M. and Karnovsky, M. J. (1983) Ultrastructural localization of several phosphatases with cerium. *J. Histochem. Cytochem.* **31,** 1197–1208.
6. Ogawa, K. and Barka, T. (eds.) (1993) *Electron Microscopic Cytochemistry and Immunocytochemistry in Biomedicine*, CRC Press, Boca Raton, FL.
7. Stoward, P. J. and Pearse, A. G. E. (ed.) (1991) *Histochemistry*, vol. 3, 4th Ed., Churchill Livingstone, Edinburgh, Scotland.

11

In Situ Molecular Hybridization Techniques for Ultrathin Sections

Jean-Guy Fournier and Françoise Escaig-Haye

1. Introduction

Several methods in molecular biology have now found a wide application in the morphological science domain allowing *in situ* detection of nucleic acids. It first became possible to visualize molecules in their natural environment 30 years ago, by adaptating nucleic acid hybridization techniques to histocytological preparations *(1)*. To date, many other molecular procedures whether or not derived from hybridization techniques can be used for *in situ* applications: polymerase chain reaction *(2)*, nick-translation *(3)*, tranferase-end labeling *(4)*, and reverse transcription *(5)*. Thus, the *in situ* hybridization technique has opened a new field of research that we can refer to as molecular histology *(6)*. The technique of combining the molecular approach at the histological level is a powerful tool and is now widely used to examine the anatomical analysis of gene function. However, though it is possible to identify which type of cell contains nucleic acid sequences of interest and in which specific region of the cell they are preferentially found *(7–12)*, *in situ* hydridization does not permit observation of the exact position of nucleic acid molecules in relation to the fine structure of the cell and its organelles. Such analysis is only possible if one is able to observe molecule detection with a high resolution. In order to do this, in the last 10 years, several efforts have been made to adapt *in situ* hybridization to electron microscopy (EM). This task was greatly helped by the development of nonisotopic labeled probes such as those which have incorporated biotinylated nucleotides *(13,14)*. These allow for rapid and high resolution visualization of detected molecules, especially when the biotin is identified directly with ligands or antibodies conjugated to gold particles. Whatever the ultrastructural strategies used: preembedding *(15–17)*, postembedding *(18–21)*,

From: *Methods in Molecular Biology*, vol. 117: *Electron Microscopy Methods and Protocols*
Edited by: N. Hajibagheri © Humana Press Inc., Totowa, NJ

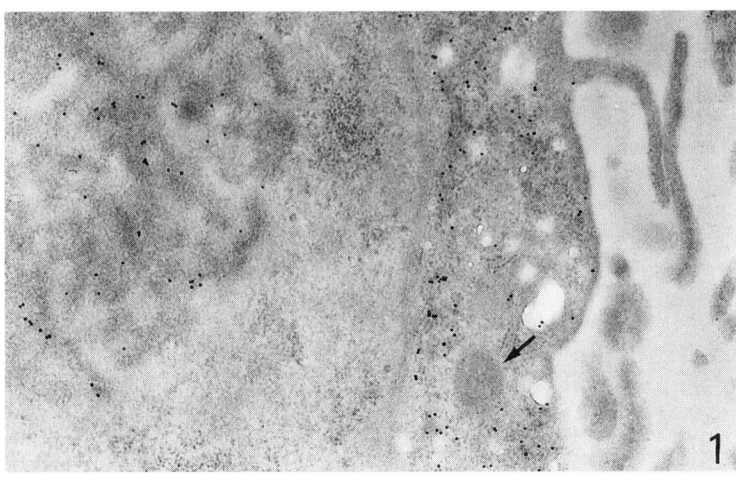

Fig. 1. Lowicryl K4M-embedded Vero cell ultrathin-section hybridized with biotinylated-nrDNA probe specific for nuclear ribosomal genes and revealed by direct method using streptavidin complexed with 10 nm gold particles. nrRNA ultrastructural immunogold labeling is observed over the nucleolus and cytoplasmic ribosomal particles. Mitochondria spared by the hybridization reaction do not display gold particles (arrow) (×40,000).

cryoultramicrotomy*(22)*, or whole-mounted cell deposited on grids *(23–25)*, they should all follow the fundamental rules of the molecular hybridization reaction. This consists of establishing—in appropriate conditions—a base specific association according to the Watson–Crick criteria, between the nucleotide sequences of the genetic probe and the complementary sequences present within the cell. The aim here is to obtain a high sensitivity of that reaction combined with good cell ultrastructural preservation. Cryoultramicrotomy fulfills the first criterion, but with poor cell integrity, whereas the preembedding protocol provides a high quality of morphology, but is associated with a low sensitivity of hybridization reaction. The third postembedding approach offers a compromise in which detection sensitivity and morphology preservation are acceptable *(26)*. In this procedure, the use of acrylate-methacrylate hydrosoluble resins is a prerequisite for the successful hybridization of the nucleic acid sequences exposed only at the surface of the ultrathin sections. Among the resins available, Lowicryl K4M is the most popular, but other Lowicryl types such as HM20 *(27)* and K11M *(28)* or acrylic resins such as LR Gold *(29)*, LR White *(30)*, and Unicryl *(31)* are also efficient in permitting *in situ* molecular hybridization at the electron microscope level. The choice of the resin should be oriented to the one providing the best morphology of the material used.

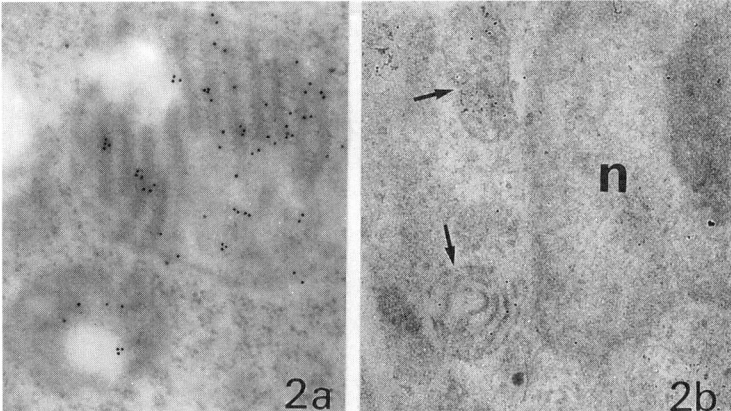

Fig. 2. Lowicryl K4M-embedded CEM cell ultrathin section hybridized with biotinylated-mtrDNA probe specific for mitochondrial ribosomal genes revealed by indirect method using GAR-IgG complexed with 10 nm-gold particles. (**A**) mtrRNA ultrastructural immunolabeling is observed over mitochondria in close association with intraorganelle membrane cristae. (×50,000). (**B**) mtrRNA ultrastructural immunolabeling after posthybridization treatment with RNase H nuclease. Reduced labeling over mitochondria (arrows) certifies the specificity of the hybridization reaction (*see* **Note 13**). Nucleus (n) (×7,000).

In the present chapter, the authors describe an ultrastructural postembedding hybridization method for the detection of cellular and viral RNA sequences, using Lowicryl K4M resin and biotinylated DNA probes revealed by a colloidal gold immunocytochemistry marker technology. They also describe how to conduct simultaneous double detection on the same ultrathin section (RNA/protein or RNA/RNA). The targeted RNA sequences (ribosomal RNA of nuclear or mitochondrial origin and HIV RNA), exemplify the necessity of visualizing the molecules in relation to the submicroscopic cell structures. Indeed, to reach a precise level of information about structure-function in this case, the electron microscope is the only means of analyzing the subcompartments of highly organized structures such as mitochondria (**Fig. 2 and refs. *18,32–34*)**, nucleolus (**Fig. 1 and refs *19,20,35)***, and virus structure (**Figs. 3 and 4 and refs. 36–38**).

2. Materials

2.1. Fixation

1. Stock solution A:0.4 *M* NaH$_2$PO$_4$.
2. Stock solution B:0.4 *M* Na$_2$HPO$_4$12 H$_2$O.
3. Stock solution C :16 mL A + 84 mL B pH 7.4 (double strength buffer).

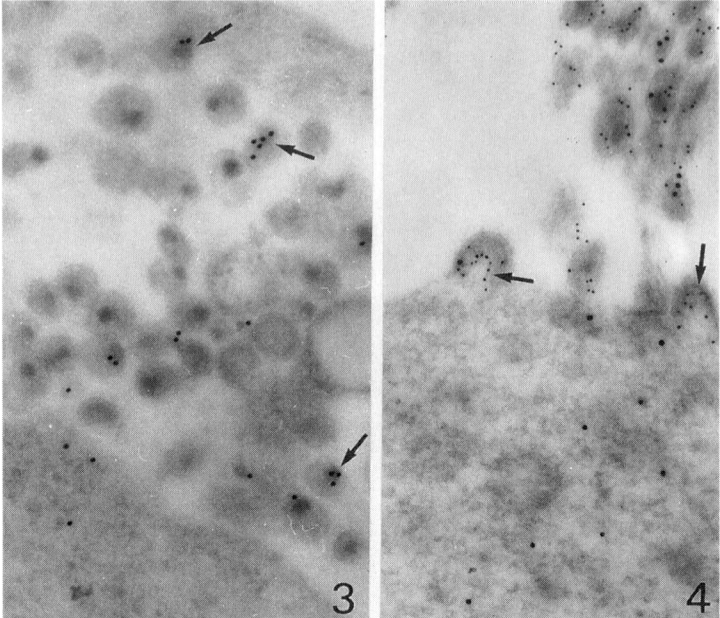

Fig. 3. Lowicryl K4M-embedded infected CEM cell ultrathin section hybridized with biotinylated-HIVDNA probe specific for HIV-1 genome, revealed by indirect method using GAR-IgG complexed with 10 nm gold particles. Viral RNA ultrastructural immunolabeling is observed over cytoplasm and the core of mature virions (arrows) (×70,000).

Fig. 4. Simultaneous detection of viral RNA, visualized as in **Fig. 3** and viral protein (p24) visualized by indirect immunocytochemical method using GAM-IgG complexed with 5 nm gold particles. RNA 10 nm gold signal is identical to **Fig. 3**, whereas viral protein 5 nm gold signal is only localized over mature virions and viral buddings (arrows) (×80,000).

5. Paraformaldehyde (powder).
6. Glutaraldehyde (25%).

2.2. Embedding

Lowicryl resins are available in kits (BALTEC) containing three components.

1. Crosslinker A.
2. Monomer B.
3. Initiator C.
4. 100% Ethanol.

2.3. Solutions and Reagents for Preparation of Labeled Probes

1. Recombinant plasmid probe (*see* **Note 1**).

2. Nick-translation buffer X10:50 mM Tris-HCl, 10 mM MgCl$_2$, 100 mM DTT, 50 mg/mL BSA, pH 7.2.
3. 100 mM (dATP, dCTP, dGTP) (Pharmacia) stock .
4. DNase I (Boehringer) 5 mg/mL diluted in distilled water with 50% glycerol.
5. DNase polymerase *E. Coli* (Boehringer).
6. Bio-11-dUTP (Enzo Biochemicals).
7. 2 M sodium bisulfite (PBS-Organics).
8. 1 M methylhydroxylamine (PBS-Organics).
9. 500 mM EDTA, pH 8.0.
10. 100% Ethanol.
11. Sonicated salmon sperm DNA (Sigma) 10mg/mL in distilled water.
12. *E. Coli* RNA (Sigma) 20 mg/mL in distilled water.
13. 3 M sodium acetate.

2.4. Hybridization

1. Deionized formamide (*see* **Note 2**).
2. 20X sodium chloride citrate buffer: 3 M NaCl, 0.3 M Na citrate, pH 7.0.
3. Dextran sulfate (*see* **Note 3**).
4. Sonicated salmon sperm DNA (SIGMA) 10mg/mL in distilled water.
5. RNase A (Sigma).
6. RNase H (Boehringer).
7. RNase H buffer: 100 mM KCl, 20 mM Tris-HCl, pH 7.5, 1.5 mM MgCl$_2$, 150 mg/mL BSA, 1 mM DTT, 0.7 mM EDTA, 13 mM HEPES.

2.5. Immunocytochemical Visualization

1. 0.01 M Phosphate-buffered saline (PBS).
2. Bovine serum albumin (BSA) (Sigma fraction V).
3. Tween-20.
4. Rabbit antibiotin (ENZO diagnostic).
5. Mouse antisulfone antibody (PBS-Organics).
6. Goat antirabbit antibody (GAR) conjugated to 10 nm gold particles (Amersham).
7. Goat antimouse antibody (GAM) conjugated to 5 nm gold particles (Amersham).
8. Streptavidin conjugated to 10 nm gold particles (Amersham).

2.6. Staining for Electron Microscopy

Uranyl acetate in aqueous saturated solution.

3. Methods

3.1. Fixation of Tissues and Cells

3.1.1. Fixatives

1. Double phosphate buffer 1/1 in distilled water before being used for fixation.
2. Dissolve 2 g of paraformaldehyde in 25 mL distilled water by heating to 65°C.

3. Add 1 *N* sodium hydroxide until the solution clears.
4. After cooling the fixative is prepared by mixing 5 mL 8% paraformadehyde, 5 mL 0.2 *M* phosphate buffer and 0.1 mL 10% glutaraldehyde (*see* **Note 4**).

3.1.2. Cells

1. Cells are harvested after trypsinization then washed in culture medium using centrifugation at 1200 rpm (800 *g*) for 20 min.
2. The medium is removed and the fixative added.
3. Keep for 2 h at 4°C (*see* **Note 5**).

3.1.3. Tissues

1. Small fragments of tissue (1 mm³) are immersed in fixative overnight, then rinsed in 0.1 *M* phosphate buffer.
2. Tissue fragments can be stored several months in buffer complemented with 0.3 *M* sucrose and 0.01% azide.

3.2. Lowicryl Resin Preparation

1. Weigh out in a brown glass container the crosslinker A (2.7 g) and the monomer B (17.30 g).
2. Gently mix the compounds with a glass rod.
3. Add the initiator C (0.10 g).
4. Mix until complete dissolution in the resin (*see* **Note 6**).

Safety: Vapors of Lowicryl resins are highly toxic, and care must be taken when handling mixtures. Use a fume hood and wear gloves for subsequent steps involving use of the resins, because contact with the skin may cause eczema.

3.3. Dehydration and Infiltration

Cells and tissue samples are prepared for Lowicryl resin embedding according to the progressive lowering temperatures (PLT) technique *(39)*.

3.3.1. Dehydration

1. 30% Ethanol at 0°C 30 min.
2. 50% Ethanol at –20°C 60 min.
3. 70% Ethanol at –20°C 60 min.
4. 90% Ethanol at –20°C 60 min.
5. 100% Ethanol at –20°C 60 min.

3.3.2. Infiltration

1. 100% Ethanol/K4M (1/1) at –20°C 60 min.
2. 100% Ethanol/K4M (1/2) at –20°C 60 min.
3. 100% Lowicryl at –20°C 60 min.
4. 100% Lowicryl at –20°C overnight.

5. The samples are next put in gelatine or BEEM capsules (*see* **Note 7**).

3.4. Polymerization

1. Capsules are placed under UV lamps (15W emitting 360 nm Wavelength) for at least 48 h at −35°C (*see* **Note 8**).
2. Polymerization is continued for 2–3 d at room temperature

3.5. Ultrathin Sections

1. Ultrathin section are obtained from a Reichert Ultramicrotome using a diamond knife (*see* **Note 9**).
2. Section are collected on uncoated grids of 600 mesh.
3. The grids can be stored in dry atmosphere for several weeks before use.

3.6. Preparation of the Probe

Add to a microcentrifuge tube in the following in order.

1. 27 µL sterile distilled water.
2. 10 µL Bio-11-dUTP.
3. 5 µL nick-translation buffer X10.
4. 5 µL dNTP (CTP, GTP, ATP) diluted 1:100 in distilled water.
5. 1 µL probe (0.5–1 µg) .
6. 5 µL DNAse I.
7. Mix and touch spin.
8. Incubate for 2–3 min at 37°C
9. Stop the reaction on ice and add:
10. 2 µL DNA polymerase.
11. Gently agitate the mixed reagents (do not use a vortex).
12. Incubate for 75 min at 15°C.
13. Stop the reaction by the addition of 3 µL of EDTA (*see* **Note 10**).

3.7. Precipitation of Labeled-Probe

1. Add to a microcentrifuge tube.
   a. 1.5 µL sonicated salmon sperm DNA 10 mg/mL diluted in water.
   b. 2.5 µL E Coli RNA (Sigma) 20 mg/mL diluted in water.
   c. 3.7 µL sodium acetate.
2. Mix well and touch spin then add 140 µL cold ethanol (100%).
3. Incubate at −20°C overnight and then spin for 30 min.
4. Discard the supernatant.
5. Rinse the pellet four times with 70% ethanol without resuspension.
6. Allow to dry for 2–3 h at room temperature.

3.8. Hybridization Buffer

1. Add to a microcentrifuge tube containing the dried probe pellet.
   a. 25 µL deionized formamide.
   b. 10 µL 20X SSC, pH 7.0.

    c. 10 µL 50% dextran sulfate.

    d. 4.5 µL sonicated salmon sperm DNA 10 mg/mL.

2. Make sure the solution is homogeneous by mixing with a Pasteur pipet, then centrifuge 15 s at 12,000*g*.

3. Take the volume necessary for experiments (2–3 µL/grid) in a microcentrifuge tube the cap of which is perforated with a needle.

4. Denature the probe by plunging the tube in boiling water for 3 min and chill on ice.

5. The remaining probe hybridization solution is stored at –20°C for maximum of 4–5 d.

3.9. Thin-Section Hybridization Procedures

3.9.1. Pretreatment of Sections

1. Incubate the control sections with the RNase A diluted 1/10 in 2X SSC for 1 h at 37°C in a moist chamber then rinse in 2X SSC twice for 2 min.

2. Rinse in distilled water for 2 min and then air dry the grids.

3.9.2. Incubation of Sections with Probes

1. Deposit 2–3 µL of denatured hybridization Solution in one well of the multiwell porcelain plaque (Pelanne Instrument) (*see* **Note 11**).

2. Cover the hybridization solution drop with the grid section-side down onto the liquid.

3. The plaque is then covered with the cap and sealed with parafilm.

5. Incubate overnight at 37°C.

3.9.3. Washings

Float the grids, at room temperature, successively on:

1. 100 µL 100% non deionized formamide diluted 1/1 in 8X SSC twice for 5 min.

2. 100 µL 4X SSC twice for 5 min.

3. 100 µL 0.2X SSC twice for 5 min.

4. 100 µL distilled water once for 2 min.

3.9.4. Posttreatment of Sections

1. Incubate the sections in solution of RNase H at 16 U diluted in 1 mL of the buffer for 1 h at 37°C.

2. Rinse in buffer twice for 2 min.

3.10. Immunocytochemical Visualization of Hybridized Probe

All the steps are done at room temperature and, during all steps of the procedure the sections should be kept wet.

3.10.1. Direct Method

1. Incubate grids on blocking solution PBS/BSA 1% for 15 min.

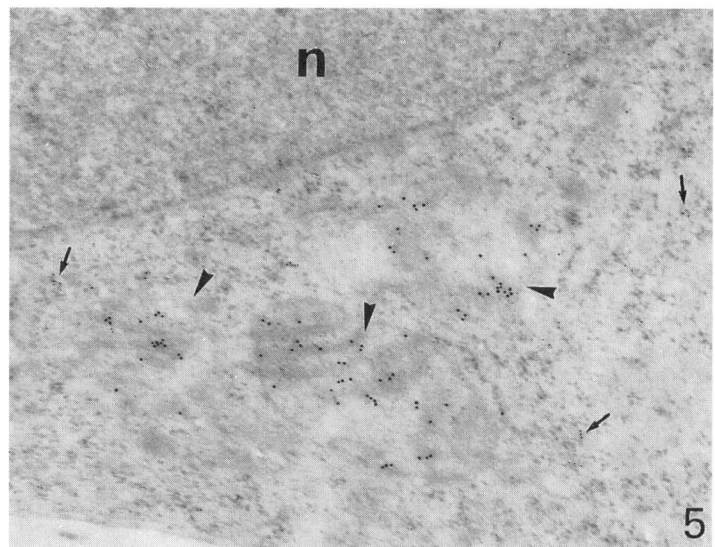

Fig. 5. Lowicryl K4M-embedded CEM cell ultrathin section hybridized simultaneously with biotinylated-mtrDNA probe and sufonated-nrDNA probe revealed by indirect method using GAR-IgG complexed with 10 nm gold particles, and GAM-IgG complexed with 5nm gold particles, respectively. The 10 nm gold labeling is concentrated over mitochondria (arrow heads), whereas the 5 nm gold labeling is spread over the cytoplasm (arrows). Nucleus (n) (×50,000).

2. Incubate grids on gold conjugated streptavidin diluted 1/50 in PBS/BSA 1% for 45 min.
3. Rinse the grids in PBS/Tween 0.05% and distilled water for 10 min each.

3.10.2. Indirect Method

1. Incubate sections on blocking solution (1% BAS in PBS) for 60 min.
2. Incubate sections on a drop of antibiotin diluted to 1% in blocking solution for 60 min.
3. Rinse grids in 0.05% Tween in PBS twice for 10 min.
4. Incubate girds on GAR (gold antirabbit) diluted 1/50 in blocking solution for 60 min.
5. Rinse in PBS/Tween 0.05% and distilled water for 10 min each.

Three examples of indirect immunocytochemical methods are shown in **Figs. 3, 4,** and **5**.

3.11. Double Labeling

3.11.1. RNA/RNA (Fig. 5)

The hybridization mixture contains the two probes the 10-20µg/mL and visualization is done by simultaneous immunocytochemistry as described in **Subheadings 3.9.2., 3.9.3.,** and **3.10.2.**

3.11.1.1. PREPARATION OF SULFONATED PROBE

1. Add to a microcentrifuge tube 2 µL nuclear rDNA probe (0.5–1 µg) and 6 µL distilled water then place for 1 min in boiling water.
2. Cool it in ice and then add 4 µL sodium metasulfite (solution A) and 1 µL methylhydroxylamine (solution B) and keep overnight at room temperature.
3. Precipitate in ethanol and then wash in 70% ethanol four times and lyophylize the sample.
4. Add the hybridization buffer containing the biotinylated mitochondrial rDNAprobe prepared according to the **Subheading 3.8**.

3.11.1.2. HYBRIDIZATION AS DESCRIBED IN **Subheading 3.9.2**.

3.11.1.3. WASHINGS AS DESCRIBED IN **Subheading 3.9.3**.

3.11.1.4. VISUALIZATION BY IMMUNOCYTOCHEMISTRY AS DESCRIBED IN **Subheading 3.10.2**.

In step 2, the sections are incubated simultaneously with the two primary antibodies (antibiotin and antisulfone) diluted 1/100 in 1% BAS in PBS.

3.11.2. RNA/Protein

1. Hybridization with the HIV probe as described in **Subheading 3.9.2.** steps.
2. Wash as described in **Subheading 3.9.3.** steps.
3. Visualization by immunocytochemistry as described in **Subheading 3.10.2**. The biotinylated probe, and the p24 viral antigen are detected using in step 2, the two primary antibodies (antibiotin and anti-p24 diluted 1/100 and 1/20 respectively) and in step 4, the two secondary antibodies (GAR gold and GAM gold) diluted 1/50.

3.12. Staining

1. Allow the sections to stain for 15 min in saturated aqueous solution of uranyl acetate.
2. Wash in distilled water for twice 5 min and allow the sections to dry for 30 min before observation in the electron microscope (*see* **Note 12**).

4. Notes

1. Three probes are used in the present case. The first is the nuclear ribosomal probe corresponding to the 6.7 kb *E. coli* fragment of mouse genomic DNA inserted into pBR-322 *(19)*. The second is the mitochondrial ribosomal probe corresponding to the 1.5 kb XbaI 1-2 DNA fragment of human mitochondrial genome inserted into pUC-19 *(32)*. The third is the 9 kb DNA fragment of HIV-1 genome inserted into pUC-18 *(36)*. Among the different types of probe available, recombinant probes appear the most appropriate for postembedding hybridization in terms of sensitivity. It has been demonstrated that the vector is a powerful amplification factor of signal creating a network that increase the number of molecules reported at the hybridization site *(40)*.

2. Formamide is deionized using MB 1 Amberlit monobed resin (Biorad). A hundred milliliter of formamide is treated with 3 g of resin by gentle stirring for 30 min in darkness. After filtration through Whatman 3M paper, formamide is aliquoted in a microcentrifuge tube and stored at –20°C. We recommend using freshly deionized formamide, for the first experiments. The quality of this reagent is essential for the *in situ* hybridization reaction

3. Here, we propose a rapid and easy preparation of Dextran sulfate solution at the exact concentration of 50%.
   a. Weigh out 50 mg in microcentrifuge tube.
   b. Add 65 μL of distilled water.
   c. Mix with a plugged Pasteur pipet until complete dissolution.
   d. Centrifuge for 30 s.
   e. The solution is very viscous and should be taken with a tip cut 0.5 cm from its extremity.

4. The glutaraldehyde solution is prepared at the time of use from 25% stock glutaraldehyde diluted in distilled water. Glutaraldehyde fixes molecules strongly inducing carbon bridges which can prevent the accessibility of the probe. The concentration of this aldehyde should be tested. Generally 0.1% offers the best compromise. Some detections work at 2% and others are completely inhibited with 0.1%.

5. To accelerate fixation, after 1 h the pellet is detached from the bottom of the tube with a Pasteur pipet and fresh fixative added.

6. Since the resin components have a low viscosity, the mixture is stirred gently to avoid air bubbles and subsequent oxygen dissolution in the resin that can interfere during the polymerization. Make sure that the mixture is completely homogeneous before use.

7. Capsules should be completely filled with the resin and capped in order to avoid trapping air bubbles.

8. There are several systems for performing polymerization. An apparatus is provided by Leica Company which offers security and easy handling. However, it is possible to construct a polymerization chamber at low cost with high efficiency. Whatever the system chosen, the important thing is that it should permit indirect ultraviolet (UV)-irradiation on all sides of the capsules filled with resin and containing the samples. The capsules are held in a 0.5-cm thick plexiglass plaque drilled with several holes of a diameter slightly smaller than that of the capsules. The plate is held to the cross-piece support by a gripper, covered with aluminium foil and is placed 30 cm below the UV lamps. The system is set up in a freezer with a chamber, the walls of which are covered with aluminium foil, thus permitting the reflection of UV-light by the capsules. The temperatures and times we use for embedding in Lowicryl are slightly different from those generally recommended. This indicates that the resins are sufficiently versatile to allow embedding at different temperatures and under different conditions, which should be adapted by each investigator.

9. In the case where the blocks are difficult to cut because of humidity content, they can be kept in a oven at 60°C for 2–3 d.

10. Avoid the use of phenol chloroform extraction, generally recommended to improve the quality of probe, as the lipophylic properties of the reporter molecules incorporated in the probe promote its solubility in the organic phase.

11. Alternatively, the moist chamber can be constituted by a Petri dish in which a filter paper saturated with 4XSSC is placed in the bottom and covered with a parafim sheet. Grids are deposited on the surface of a drop of reagent. For prolonged incubation (overnight) a second larger Petri dish containing the first one and paper saturated with 4XSSC is recommended.

12. To increase the contrast of the image, it is recommended viewing at an accelerating voltage of 60 KV.

13. Several checks should be carried out to assess the specificity of the hybridization reaction. They routinely include 1) hybridization with a unrelated probe (for example vector plasmid), and hybridization buffer alone and 2) RNase A pretreatment to abolish all RNA sequences within cells. To this, we add RNase H posttreatment because as pointed out previously *(9)* this control is more rigorous because only the cellular RNA sequences that had specifically established hydrogen bonds to the complementary sequences in the DNA probe are digested.

References

1. Gall, J. C. and Pardue, M. L. (1969) Formation and detection of RNA-DNA hybrid molecules to cytological preparations. *Proc. Natl. Acad. Sci. USA* **63,** 378–383.

2. Nuovo, G. J. (1992) PCR in situ hybridization, Raven, New York.

3. Wijsman, J. H., Jonker, R. R., Keijzer, R., Van de velde, C. J., Cornelisse, C. J., and Van Dierendonck, J. H. (1993) A new method to detect apoptosis in paraffin sections: in situ end labelling of fragmented DNA. *J. Histochem. Cytochem.* **41,** 7.

4. Gorczyca, W., Bruno, S., Darzynkiewicz, R., Gonj, J., and Darzynkiewicz, Z. (1992) DNA strand breaks occurring during apopotosis their early in situ detection by terminal deoxynucleotidyl transferase and nick-translation assays and prevention by serine protease inhibitors. *Int. J. Oncol.* **1,** 639–648.

5. Longley, J., Merchant, M. A., and Kacinski, B. M. (1989) In situ transcription and detection of CD1a mRNA in epidermal cells: an alternative to standard in situ hybridization techniques. *J. Invest. Dermatol.* **93,** 432.

6. Fournier, J. G. (1994) *Molecular Histology.* Lavoisier TEC and DOC, Paris, France.

7. Fournier, J. G., Kessous, A., Tiollais, P., Simard, R., Brechot, C., and Bouteille, M. (1982) Hepatitis B virus genome expression detected by in situ hybridization in a human hepatoma cell line. *Biol. Cell* **44,** 197–200.

8. Fournier, J. G., Rozenblatt, S., and Bouteille, M. (1983) Localization of measles virus acid nucleic sequences in infected cells by in situ hybridization. *Biol. Cell.* **49,** 287–290.

9. Lawrence, J. B. and Singer, R. H. (1986) Intracellular localization of messenger RNAs for cytoskeletal proteins. *Cell* **45,** 407–415.

10. Meyer, J. L., Fournier, J. G., and Bouteille, M. (1986) Expression of integrated hepatitis B virus DNA in PLC/PRF/5, Hep 3B, and L6EC3 cell lines detected by in situ hybridization. *Med. Biol.* **64,** 367–371.

11. Lawrence, J. B., Singer, R. H., and Marselle, L. M. (1989) Highly localized tracks of specific transcripts within interphase nuclei visualized by in situ hybridization. *Cell* **57,** 493-502.

12. Fournier, J. G., Prevot, S., Audoin, J., Letourneau, A., and Diebold, J. (1990) Fine analysis of HIV-1 RNA detection by in situ hybridization, in *Modern Pathology of AIDS and Other Retroviral Infections* (Haase, H., Gluckmann, J. C., and Racz, P., eds.), Karger, Basel, pp. 62–68.

13. Langer, P. R., Waldrop, A. A., and Ward, D. C. (1981) Enzymatic synthesis of biotin-labelled polynucleotides novel nucleic acid affinity probes, *Proc. Natl. Acad. Sci. USA* **78,** 6633–6637.

14. Singer, R. H. and Ward, D. C. (1982) Actin gene expression visualized in chicken muscle tissue culture by using in situ hybridization with biotynilated nucleotide analogue. *Proc. Natl. Acad. Sci. USA* **79,** 7331–7335.

15. Trembleau, A., Calas, A., and Fevre-Montange, M. (1990) Ultrastructural localization of oxytocin mRNA in the rat hypothalamus by in situ hybridization using a synthetic oligonucleotide. *Mol. Brain Res.* **8,** 37–45.

16. Guitteny, A. F. and Bloch, B. (1989) Ultrastructural detection of the vasopressine messenger RNA in the normal and Brattleboro rat. *Histochemistry* **92,** 277–281.

17. Tong, Y., Zhao, H., Simard, J., Labri, F., and Pelletier, G. (1989) Electron microscopic autoradiographc localization of prolactin mRNA in a rat pituitary. *J. Histochem. Cytochem.* **37,** 567–571.

18. Binder, M., Tourmente, S., Roth, J., Rebaud, M., and Gehring, W. J. (1986) In situ hybridization at the electron microscope level: localization of transcripts on ultrathin sections of Lowicryl K4M embedded tissue using biotinylated probes and protein A-gold complexes. *J. Cell Biol.* **102,** 1646–1653.

19. Escaig-Haye, F., Grigoriev, V., and Fournier, J. G. (1989) Detection ultrastructurale d'ARN ribosomal par hybridation in situ à l'aide d'une sonde biotinylée sur coupes ultrafines de cellules animales en culture. *C. R. Acad. Sci., Paris, D* **429,** 429–434.

20. Thiry, M. and Thiry-Blaise, L. (1989) In situ hybridization at the electron microscope. Improved method for precise localization of ribosomal DNA and RNA. *Eur. J. Cell Biol.* **50,** 235–243.

21. Puvion-Dutilleul, F. and Puvion, E. (1989) Ultrastructural localization of viral DNA in thin sections of herpes simplexe virus type 1 infected cells by in situ hybridization. Eur. *J. Cell Biol.* **49,** 99–109.

22. Morel, G., Dihl, F., and Gossard, F. (1989) Ultrastructural distribution of GH mRNA and GH intron I sequences in rat pituitary gland: effects of GH releasing factor and somatostatin. *Mol. Cell Endocrinol.* **65,** 81–90.

23. Hutchinson, N., Langer-Safer, P., Ward, D., and Hamkalo, B. (1982) In situ hybridization at the electron microscope level:hybrid detection by autoradiography and colloidal gold. *J. Cell Biol.* **95,** 609–618.

24. Radic, M. Z., Lundgren, G., and Hamkalo, B. A. (1987) Curvature of mouse satellite DNA and condensation of heterochromatine. *Cell* **50,** 1101–1108.

25. Singer, R. H., Langevin, G. L., and Lawrence, J. B. (1989) Ultrastructural visualization of cytoskeletal messenger RNAs and their associated protein using double label in situ hybridization. *J. Cell Biol.* **108,** 2343–2353.

26. Leguellec, D., Trembleau, A., Pechoux, C., and Morel, G. (1992) Ultrastructural non-radioactive in situ hybridization of GH mRNA in rat pituitary gland: pre-embedding vs ultra-thin sections vs post-embedding. *J. Histochem. Cytochem.* **40,** 979–986.

27. Fournier, J. G. and Escaig-Haye, F. (1993) Electron microscopy of rRNA in situ hybridization on Lowicryl sections, in *Hybridization Techniques for Electron Microscopy* (Morel, G., ed.), CRC, Boca Raton, FL, pp. 243–268.

28. Leguellec, D., Frappart, L., and Desprez, P. Y. (1991) Ultrastructural localization of mRNA encoding for the EGF receptor in human breast cell cancer line BT 20 by in situ hybridization. *J. Histochem. Cytochem.* **39,** 1–6.

29. Mc Fadden, G. I., Cornish, E. C., Boning, I., and Clarke, A. E. (1988) A simple fixation and embedding method for use in situ hybridization histochemistry of plant tissues. *Histochem. J.* **20,** 575–586.

30. Wenderroth, M. A. and Eisenberg, B. R. (1991) Ultrastructural distribution of myosin heavy chain mRNA in cardiac tissue: a comparison of frozen and LR White embedment. *J. Histochem. Cytochem.* **39,** 1025–1033.

31. Cenacchi, G., Musiani, M., Gentilomi, G., Righi, S., Zerbini, M., Chandler, J. G., Scala, C., La Placa, M., and Martinelli, G. N. (1993) In situ hybridization at the ultrastructural level: localization of cytomegalovirus DNA using digoxigenin labelled probes. *J. Submicrosc. Cytol. Pathol.* **25,** 341–345.

32. Escaig-Haye, F., Grigoriev, V., Peranzi, G., Lestienne, P., and Fournier, J. G. (1991) Analysis of human mitochondrial transcripts using electron microscopic in situ hybridization. *J. Cell Sci.* **100,** 851–862.

33. Tourmente, S., Savre-Train, I., Berthier, F., and Renaud, M. (1990) Expression of six mitochondrial genes during Drosophila oogenesis: analysis by in situ hybridization. *Cell Diff. Dev.* **31,** 137–149.

34. Lecher, P., Petit, N., Beziat, F., and Alziari, S. (1996) Localization by ultrastructural in situ hybridization of mitochondrial transcript in epithelial cells of a Drosophila sub-obscuradeletion mutant. *Eur. J. Cell Biol.* **71,** 423–427.

35. Puvion-Dutilleul, F. and Puvion, E. (1996) Non-isotopic electron microscope in situ hybridization for studying the functional sub-compartmentalization of the cell nucleus. *Histochem. Cell Biol.* **106,** 59–78.

36. Escaig-Haye, F., Grigoriev, V., Sharova, I. A., Rudneva, V., Buckrinskaya, A., and Fournier, J. G. (1992) Ultrastructural localization of HIV-1 RNA and core proteins: Detection using double immunogold labelling after in situ hybridization and immunocytochemistry. *J. Submicrosc. Cytol. Pathol.* **24,** 437–443.

37. Mandry, P., Murray, B. A., Rieke, L., Becker, H., and Hofler, H. (1993) Post-embedding ultrastructural in situ hybridization on ultrathin cryosections and LR white resin sections. *Ultrastruct. Pathol.* **17,** 185–194.

38. Ukimura, A., Deguchi, H., Kitaura, Y., Fujioka, S., Hirasawa, M., Kawamura, K., and Hirai, K. (1997) Intracellular viral localization in murine coxsackievirus-B3

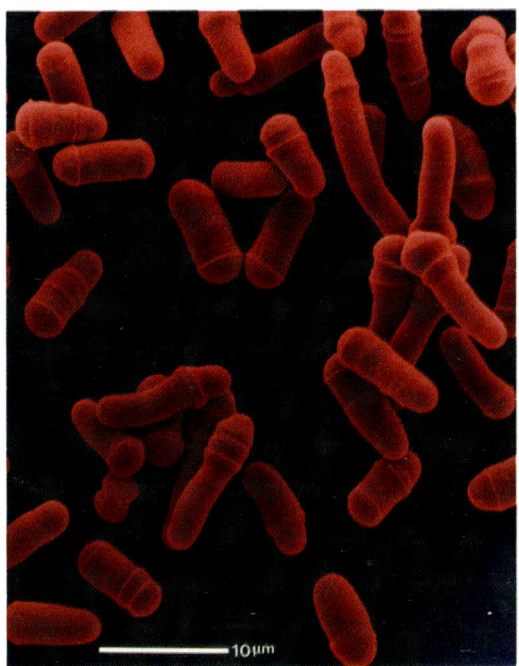

Plate 1, Fig. 1; (*see* discussion in Chapter 12, p. 184) Scanning electron micrograph showing a population of fission yeast *S. pombe.*

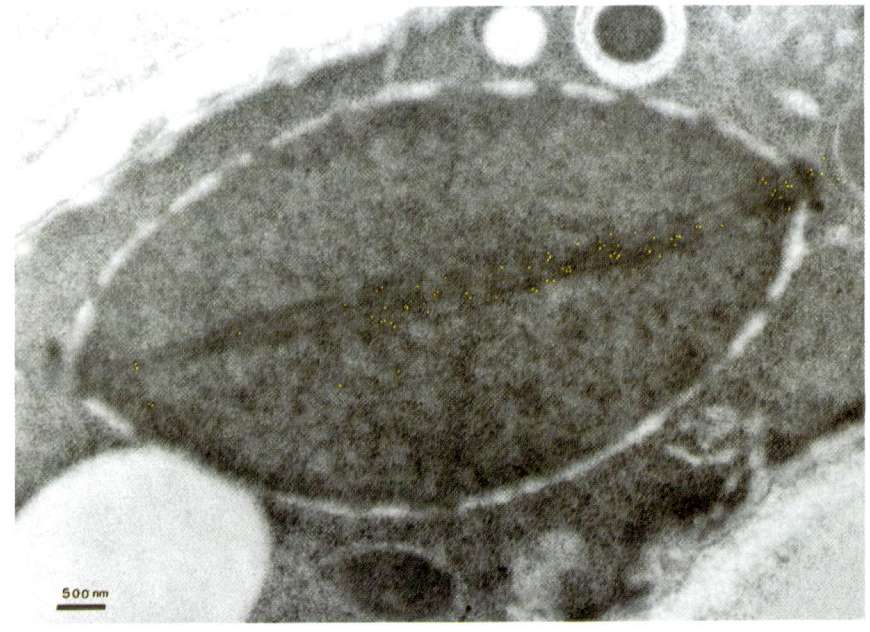

Plate 2, Fig. 9; (*see* discussion in Chapter 12, p. 199) Immunolocalization of Tubulin on a mitotic spindle of fission yeast.

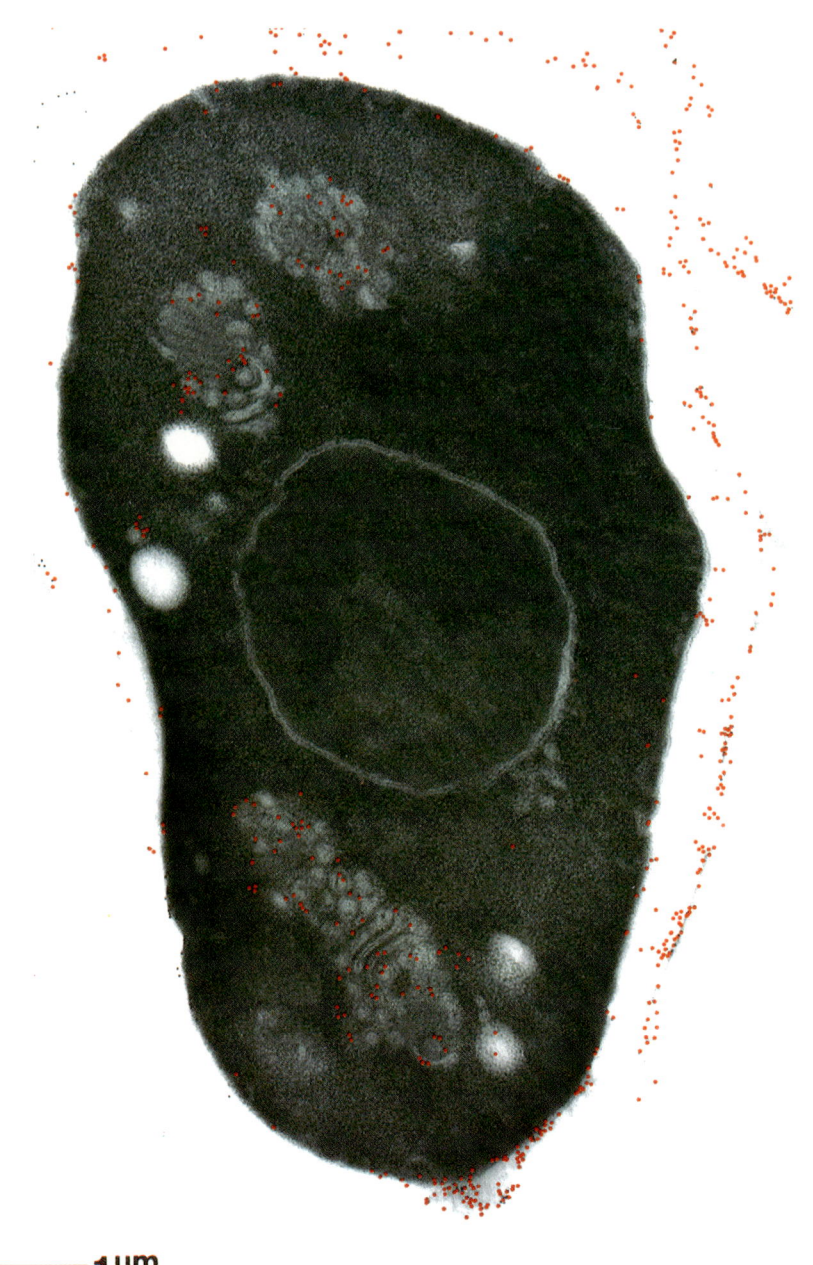

Plate 3, Fig. 2; *(see* discussion in Chapter 12, p. 192) Ultrathin section of Lowicryl embedded *S. Pombe* incubated with polyclonal rabbit antiserum raised against purified galactosyltransferase, showing labelling of the Golgi stack and cell wall.

Plate 4, Fig. 2; (*see* discussion in Chapter 8.) The boundaries of 7 different compartments (the same as depicted in Fig. 8-1) have been drawn. The nucleus is pink, the endoplasmic reticulum (ER) is yellow, the mitochondria is orange, the Cis Golgi Network (CGN) in red, the Golgi stack is in blue, the trans Golgi network (TGN) in green and the endosomal/lysosomal compartment in violet. Note that the boundaries of the Golgi stack, the CGN and the TGN do not exactly follow the membranes. The boundaries of the Golgi stack were drawn by following the outmost membranes of the cis and trans cisternae. The boundaries of the CGN and TGN were drawn by follwing the interface between the outmost membranes and the more amorphous cytoplasm **(23)**.

Plate 5, Fig. 4; (*see* full caption on p. 135 and discussion in Chapter 8.) The profile in Fig. 2 has been overlaid with a grid (d = 1 cm). The magnification is estimated to be 50K. The number of points falling within the boundaries of compartments is recorded in Table 2.

myocarditis. Ultrastructural study by electron microscopic in situ hybridization. *Am. J. Pathol.* **150,** 2061–2074.

39. Carlemalm, E., Villiger, W., Hobot, J. A., Acetarin, J. D., and Kellenberger, E. (1985) Low temperature embedding with Lowicryl resins: two new formulations and some applications. *J. Microsc.* **140,** 55–72.

40. Singer, R. H., Lawrence, J. B., and Rashtchian, R. N. (1987) Towards a rapid and sensitive in situ hybridization methodology using istopic and non-isotopic probe, in *In Situ Hybridization: Application to the Central Nervous System* (Valentino, K., Eberwin, J., and Barchas, J., eds.), Oxford University Press, New York, pp. 71–96.

12

Preparation of the Fission Yeast
Schizosaccharomyces pombe for Ultrastructural
and Immunocytochemical Study

**M.A. Nasser Hajibagheri, Kenneth Sawin, Steve Gschmeissner,
Ken Blight, and Carol Upton**

1. Introduction

The fission yeast *Schizosaccharomyces pombe* serves as a model system for investigating a wide variety of problems in eukaryotic cellular and molecular biology. Although probably most widely studied in relation to the control of the eukaryotic cell cycle *(1)* and the events of cell division *(2,3)*, *S. pombe* has also received significant attention in studies of protein trafficking and secretion *(4)*, cell wall biosynthesis *(5)*, cellular and molecular aspects of meiosis, genetic recombination, and spore formation *(6)*, signal transduction pathways and control of gene expression *(7–10)* and cell growth *(11,12)*, cell morphology *(13,14)* and the cytoskeleton *(15,16)*. In many ways we can think of *S. pombe* as essentially a typical eukaryote trapped inside a cell wall, and the resulting constraints on cell morphology confer on it a stereotyped cellular architecture, well suited to ultrastructural analysis. The details of this architecture at both the light- and electron-microscopic level can be found in reference *(17)* ; here we will only highlight certain features.

Fission yeast are roughly cylindrical cells with hemispherical ends, with a diameter of about 3.5 µm (*see* **Fig. 1**); they grow in length only, by extension of cell ends, from approximately 7 µm at "birth" to about 12–14 µm at cell septation, which occurs precisely in the middle, which is also the position of the interphase nucleus *(18)*. Extensile growth of cells, and concomitant deposition of new cell wall material, is asymmetric; early in the cell cycle, growth is exclusively at the "old" end (i.e., that which was an end in the mother cell

From: *Methods in Molecular Biology*, vol. 117: *Electron Microscopy Methods and Protocols*
Edited by: N. Hajibagheri © Humana Press Inc., Totowa, NJ

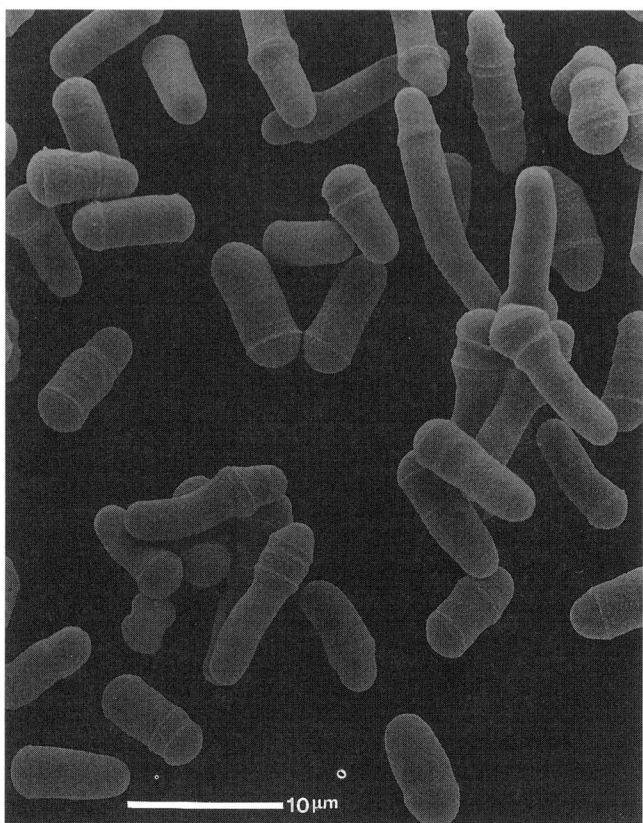

Fig. 1. Scanning electron micrograph showing a population of fission yeast *S. pombe*. (*See* Color Plate 1, following p. 180.)

during the previous cell cycle), while later in the cell cycle growth occurs at both the old end and the "new" end (i.e., that which was most recently formed by septation) *(12)*. The distribution of filamentous-actin matches the patterns of cell growth, including the formation of an actin ring at the cell middle prior to septation *(15)*. Tubulin, the other major cytoskeletal polymer, forms cytoplasmic microtubules that extend the length of the cell during interphase (although microtubule dynamics and polarity are as yet unknown), while during mitosis they form an intranuclear spindle, and some astral microtubules extend from the nuclear surface to the ends of cells *(16,19)*. Upon completion of mitosis, cytoplasmic microtubules exhibit a transient distribution, emanating from the septum (the so-called "post-anaphase array") before returning to a typical interphase array. Intracellular membrane systems also show stereotyped patterns. Most tubules of the endoplasmic reticulum are present near the cell surface *(17)*. Golgi stacks/clusters are also found, although not in the highly concentrated organization typical of mammalian cells *(35)*.

When deprived of nutrients fission yeast differentiate and enter a meiotic cycle *(20)*. Initially, via the action of specific signalling peptide "pheromones" *(21)*, cells of opposite mating type arrest progression through the mitotic cycle *(22)*, and begin to grow towards each other *(23,24)*, eventually fusing to form a zygote. Cell fusion is followed by nuclear fusion, and subsequently the result- ant diploid nucleus replicates its DNA and undergoes a series of characteristic movements during which time the (nuclear) spindle pole body, normally asso- ciated with chromosomal centromeres during vegetative growth, becomes associated with telomeres *(25)*. Genetic recombination also takes place at this time, prior to reductional and equational meiotic divisions, followed by spore formation around each of the four daughter nuclei.

Fission yeast is used successfully in studies of the cell cycle, but the degree of chromatin condensation during normal fission yeast cell division is not really sufficient to allow identification of individual chromosomes or chro- mosomal structures at the ultrastructural level. In addition, the large cell length means that exhaustive serial section work may be difficult, in contrast to flattened mammalian cells growing on a substrate. Also, the presence of a cell wall may complicate specific special procedures where access to the cytoplasm prior to sectioning may be desired. On the other hand, it is impor- tant to note that one of the principal strengths of *S. pombe* as a model cell is that it is well established as a genetic system *(26)*, with several hundred clas- sically defined loci, as well as ordered cosmid and P1 maps covering the genome *(27)*, the sequencing of which is expected to be complete by 2001. Simple methods exist for creating gene deletions *(28)* and manipulating gene expression *in vivo (29)*. Thus, any morphologic or cytologic analysis, whether at the light- or electron-microscopic level, is enhanced by the possibility of doing parallel experiments in different genetic backgrounds. With specific regard to using genetic systems for cell biological research, it is worth noting that although in some aspects of its molecular genetics *S. pombe* is not as well-developed as its distant relative budding yeast *(S. cerevisiae)*, the two yeasts are no more phylogenetically related than either is to mammals *(30)*. Therefore, identifying a conserved process, function, or gene in both yeasts often provides strong evidence that a similar process occurs in human cells. Moreover, in spite of its simple, nonmotile lifestyle, in many aspects of its cell biology *S. pombe* may more strongly resemble multicellular eukaryotes than does *S. cerevisiae*. These differences, ranging from the ability of *S. pombe* to splice mammalian introns *(31)* and the complex nature of its origins of DNA replication and kinetochores *(32,33)*, to the pronounced changes in the orga- nization of the cytoskeleton during cell division characteristic of higher eukaryotes *(15)*, suggest that much of what we learn from fission yeast will be applicable to mammalian cells as well.

Efforts to understand the molecular biology of the cell cycle and of development in this organism has resulted in the production of many antibodies against specific gene products and their localization by immunofluorescence light microscopy. Their localization by EM immunocytochemistry has lagged behind, in part because preservation of this organism for even conventional EM is difficult. In order to analyze the detailed structures of subcellular compartments that have not been detected by either immunofluorescence or light microscopy, it is necessary to take advantage of the increased resolving power of the electron microscope by using immunogold labelling of ultrathin sections. We have used fission yeast cells in a number of projects during the last ten years to study the morphological preservation and immunolocalization of protein by three different methods; High pressure freezing/freeze-substitution, ultrathin cryosectioning, and progressive low temperature (PLT) embedding to establish a reliable method for morphological preservation and immunogold labelling of fission yeast protein. Our aim in this chapter is to highlight advantages and disadvantages of the above techniques and to help a new researcher in this field to choose the most desirable method for the project at hand.

2. Materials

2.1. Media

In our laboratory the standard rich liquid medium (YE) is based on yeast extract: 0.5% (w/v) Oxoid yeast extract, 3% glucose.

A completely defined medium (EMM2, Edinburgh minimal medium) is often used to avoid any possible variability between batches of yeast extract:

3 g/L potassium hydrogen phthallate	(14.7mM)
2.2 g/L Na_2HPO_4	(15.5 mM)
5 g/l NH_4Cl	(93.5 mM)
2% (w/v) glucose	(111 mM)
20 ml/L salts	(stock × 50)
1 ml/L vitamins	(stock × 1000)
0.1 ml/L minerals	(stock × 10,000)

Salts × 50

52.5 g/L $MgCl_2 \cdot 6H_2O$	(0.26 M)
0.735 mg/L $CaCl_2 . 2H_2O$	(4.99 mM)
50 g/L KCl	(0.67 M)
2 g/L Na_2SO_4	(14.1 mM)

Vitamins × 1000

1 g/L pantothenic acid	(4.20 mM)
10 g/L nicotinic acid	(81.2 mM)
10 g/L inositol	(55.5 mM)
10 mg/L biotin	(40.8 μM)

Minerals × 10,000

5 g/L boric acid	(80.9 mM)
4 g/L $MnSO_4$	(23.7 mM)
4 g/L $ZnSO_4.7H_2O$	(13.9 mM)
2 g/L $FeCl_2.6H_2O$	(7.40 mM)
0.4 g/L molybdic acid	(2.47 mM)
1 g/L KI	(6.02 mM)
0.4 g/L $CuSO_4.5H_2O$	(1.60 mM)
10 g/L citric acid	(47.6 mM)

After autoclaving, a few drops of preservative is added.
(chlorobenzene:dichloroethane:chlorobutane, 1:1:2)

For growth of auxotrophic mutants, both YE and EMM2 can be supplemented with adenine, histidine, leucine, uracil, and/or lysine hydrochloride as needed, all at 225 mg/mL. Solid media is made by adding 2% Difco Bacto Agar. Media are sterilized by autoclaving at 10 psi for 20 min; at this pressure very little caramelization of glucose takes place. To further avoid caramelization, EMM2 can be autoclaved without NH_4Cl, and supplemented with sterile-filtered NH_4Cl before use. The media are stored in 500 mL bottles, and agar is remelted in a microwave oven before using. Further details concerning unusual media, cell-generation times at different temperatures, etc. can be found in *(34)*. A wealth of additional information about fission yeast can be obtained from the Web site of the laboratory of S. Forsburg (Salk Institute, San Diego) and links therefrom (http://flosun.salkedu/users/forsburg/lab.html).

2.2. Chemicals and Solutions

1. 50 mM glycine in Sörensen's phosphate buffer, comprising 0.2 M sodium dihydrogen orthophosphate dihydrate and 0.2 M disodium hydrogen orthophosphate.
2. Fetal calf serum: 10% in PBS.
3. 2% paraformaldehyde in 0.1M Sörensen's phosphate buffer, pH 7.3.
4. 4% paraformaldehyde containing 0.1% monomericglutaraldehyde buffered in 0.1M Sörensen's phosphate buffer, pH 7.3.
5. 1% OsO_4 in 0.1M Sörensen's phosphate buffer.
6. Phosphate Buffered Saline (PBS).
7. 2.3 M sucrose in PBS.
8. Primary antibodies.

9. Protein A-gold. Protein A-gold from the Department of Cell Biology, University of Utrecht, 3584 CX Utrecht, The Netherlands.
10. Lowicryl resin (Agar Scientific Ltd., Stansted, UK).
11. 0.1% BSA (Sigma Type V) in PBS.
12. 1% glutaraldehyde in 0.1M Sörensen's phosphate buffer, pH 7.3.
13. Reynolds lead citrate.
14. 3% uranyl acetate in water.
15. 2% methyl cellulose in water (Use 25-centipoise viscosity methyl cellulose Sigma M6385).

2.3. Equipment

1. High pressure freezing machine (HPM 010, Bal-Tec AG, Liechtenstein).
2. Freeze-substitution unit (FUU 010, BAL-Tec).
3. Low temperature UV-polymerization device (TTP 010, BAL-Tec).
4. Low Temperature Embedding (Unit LTE 020, Bal-Tec).
5. Ultra-microtome equipped with cryo-attachment (Leica AG, Germany).
6. Transmission electron microscope.
7. Vacuum evaporator.
8. Cryoknife diamond.
9. Formvar/carbon coated specimen grids.
10. Aluminium specimen mounting pins.
11. Fine forceps.
12. Liquid nitrogen storage tank (for specimen storage).
13. Parafilm.

3. Methods
3.1. Growing Cells

1. From a single colony or "patch" growing on a solid agar-medium plate, make a "pre-culture" by adding cells to 10 mL liquid media with a sterile loop or toothpick.
2. Grow in flat-bottomed flask overnight in shaking air or water incubator at 32°C (for wild-type cells).
3. Dilute preculture in a larger volume of medium as appropriate, based on a generation time of 2 h 10 min in YE and 2 hr 30 min in EMM2 *(34)*. Grow overnight at 32°C to mid-log phase (i.e., to an OD$_{595}$ of 0.2–0.5, equivalent to 0.4–1.0 × 10^7 cells/mL). For growth at temperatures other than 32°C, generation times are: (YE) 3 h at 25°C, 2 h 30 min at 29°C, and 2 h 35.5°C; (EMM2) 4 h at 25°C, 3 h at 29°C, and 2 h 20 min at 35.5°C (*see* **Note 1**).

3.2. Progressive Lowering Temperature (PLT) and Low Temperature Embedding with Lowicryl HM20

1. Fixation: Two types of fixatives are used to preserve yeast cells for immunogold labelling studies.

    Fixation A: One h in 0.1% monomeric glutaraldehyde/4% paraformaldehyde (*see* **Note 2**) in 0.1M Sörensen's phosphate buffer, pH 7.3.

Fixation B: One h in 4% paraformaldehyde in 0.1*M* Sörensen's phosphate buffer, pH 7.3.

2. Rinse 3 × 5 min with 50 m*M* glycine in Sörensen's phosphate buffer, to quench unreacted aldehyde groups.

3. Dehydration: Dehydrate in graded methanol as follows:

50%	30 min	+ 4°C
70%	30 min	–20°C
80%	45 min	–20°C
90%	60 min	–45°C
95%	60 min	–45°C

4. Infiltration with Lowicryl resin at –45°C.

Lowicryl resin :	95% methanol	1:2	1 h
Lowicryl resin :	95% methanol	1:1	overnight
Lowicryl resin :	95% methanol	2:1	4 h

Pure fresh Lowicryl resin x 2 changes overnight

5. Polymerization: Polymerize in UV light (360 nm) for 24 h at –45°C, then harden in daylight for 2–3 days if necessary.

3.3. Cryosectioning Techniques

3.3.1. Morphology

1. Fix yeast cells in 4% paraformaldehyde/0.2% monomeric glutaraldehyde in 0.1*M* Sörensen's phosphate buffer, pH 7.3. for 1 h at room temperature on rotator.
2. Wash in 0.1*M* Sörensen's phosphate buffer for 5 min.
3. 2% paraformaldehyde in 0.1*M* Sörensen's phosphate buffer, pH 7.3., overnight.
4. Wash in 0.1*M* Sörensen's phosphate buffer for 3 × 30 min on rotator.
5. Post-fix in 1% OsO$_4$ in 0.1*M* Sörensen's phosphate buffer for 4 h at room temperature on rotator.
6. Wash in 0.1*M* Sörensen's phosphate buffer for 3 × 30 min on rotator.
7. Process as normal for cryosectioning of cells as described in Chap. 4.

3.3.2. Immunocytochemistry.

1. Fix yeast cells in 4% paraformaldehyde/0.2% monomeric glutaraldehyde in 0.1*M* Sörensen's phosphate buffer, pH 7.3. for 1 h at room temperature on rotator.
2. Wash in 0.1*M* Sörensen's phosphate buffer for 5 min.
3. 2% paraformaldehyde in 0.1*M* Sörensen's phosphate buffer, pH 7.3., overnight.
4. Wash in 0.1*M* Sörensen's phosphate buffer for 2 × 10 min on rotator.
5. Process as normal for cryosectioning of cells as described in Chap. 4.

3.4. Preparation of Yeast Cells by High Pressure Freezing

Detailed information on cell preparation by high pressure freezing/freeze substitution for both morphological and immunogold labelling studies *(36,37)* is presented in Chap. 5. Please refer to this chapter for a description and application of this technique to yeast.

3.5. Immunogold Labelling Procedure

Ultrathin sections are mounted on formvar/carbon-coated grids and are labelled (*see* **Note 4**) as follows.

All steps are carried out on strips of Parafilm in a moist chamber to prevent drying of the sections by evaporation of the reagent between the incubations. Incubations are carried out by floating grids on a 20 µl drop of reagent so as to expose only one side of the grid. Exposing both sides of the grid to the antibodies increases background.

1. Cryosections are usually collected on 2% gelatine in PBS (sodium phosphate buffer, pH 7.4, 0.1*M*) and it is necessary before labelling to remove gelatine by incubating cryosection at 37°C. for 30 min
2. 0.02*M* glycine in PBS.   3×1 min
3. 1% BSA (Sigma Type V) in PBS (PBS/BSA 0.1%).   1 min
4. Specific primary antibody diluted in PBS/BSA 1%   30 min
   (Drops of 5 µL containing 5–20 µg/mL specific antibody   room temper-
   sufficient to incubate one grid).   ature (RT), or
                                                                                                          overnight
                                                                                                          at + 4°C
5. Wash on PBS/BSA 0.1%.   4×1min
6. Incubate in secondary antibody in PBS/BSA 1% (*see* **Note 5**).   30 min
7. Protein A-gold, diluted as specified in PBS/BSA 1%
   (gold dilution are made fresh and used immediately).   30 min
8. Wash in PBS/BSA 0.1% (1×5 min and then wash in PBS).   (4×5 min)
9. Stabilize the reaction with 1% glutaraldehyde in PBS.   5 min
10. Wash on PBS.   2×5 min
11. In case of double labelling repeat steps 3–9 with different antibody, Protein A-gold combination after 5×2 min rinse with PBS/glycine.
12. Distilled water.   5×2 min
13. For resin embedded samples, the sections are contrasted with uranyl acetate (5 min) and lead citrate (4 min) as described in Chap. 7.
14. For cryosections, stain in saturate aqueous uranyl acetate for 5 min, distilled water (1 Dip) and ice cold 1.8% methyl-cellulose (25Ctp)/0.4% Uranylacetate (MC-UA).   5 min
15. Loop out the grids and reduce the MC-UA to an even thin film, and let it dry.

3.6. Applications

3.6.1. Immunogold Labelling of Yeast Cells Prepared by Progressive Lowering Temperature and Lowicryl Embeding Resin

Immunoelectron microscopy is the method of choice for the high resolution study of cellular antigens. A variety of methods exist for these investigation, choice largely being determined by the type of material, requirement of the study (qualitative or quantitative), and the equipment available.

Figs. 2 and **3** show electron micrographs of longitudinal and transverse thin sections of *S. pombe*. The cells contain several structures morphologically recognizable as Golgi. The distension of cisternae appears to be independent of processing , in that it is observed with both Lowicryl and frozen thin sections. The morphological preservation, and in particular that of Golgi membranes, was well defined in thin sections of yeast embedded in Lowicryl. Superior morphological preservation was obtained in samples embedded in Lowicryl HM20 as compared with K4M (data not shown). Polyclonal rabbit antiserum (**Note. 6**) raised against the purified galactosyltransferase labelled the Golgi stack and the cell surface but no intracellular compartments (**Figs. 2** and **3**). We assume that the antiserum recognizes both the galactosyltransferase and its oligosaccharides, the latter being responsible for the cell-surface staining observed (**Fig. 3**). The specificity of labelling is confirmed by its absence in a strain of *S.pombe* in which the enzyme had been deleted *(35,38)*.

3.6.2. Expression of Mammalian Protein Kinase C (39)

Mammalian protein kinase C (PKC) isotypes elicit a number of effects on expression in *Schizosaccharomyces pombe*. A small decrease in growth rate results from PKC-η expression, and treatment of these cells with phorbol esters leads to marked growth inhibition and vesicle formation *(39,40)*. Expression of PKC-η was accompanied by a massive accumulation of vesicles as shown in **Fig. 4.** These were fairly homogeneous in size, ranging from 70–90 nm in diameter and distributed throughout the cell cytoplasm. The origin of the vesicles was confirmed by labelling Lowicryl sections with a varieties of antibodies to known components of clathrin-coated vesicles (**Fig. 5**). Polyclonal antibody pep1a (against the N-terminal region of Lca) labelled accumulated vesicles and parts of the plasma membrane, which could represent the sites of budding vesicles. There was also occasional labelling of the vacuoles that might represent delivery sites for the endocytic vesicles. There was no labelling on Golgi, ER, or the nuclear envelope. The level of nonspecific labelling was low as shown by the absence of labelling over the nucleus and mitochondria. Concurrent addition of the pep1a epitope (0.63 mM) with antibody reduced labelling by 94% (Fig. 5B) as calculated by reduction in the density of gold particle per μm^2 of vesicles. Thus, binding was specific and the expression of specific mammalian PKC isotypes up-regulate endocytosis in *S. pombe*, providing a likely explanation for PKC-mediated receptor internalization in higher eukaryotes.

Fig. 6. illustrates ultrastructural localization of HA.CED-4 in *S. pombe* cells using an anti-HA monoclonal antibody. Cells were prepared as described in **Subheading 3.2.** and show that the expression of CED-9 caused the relocation of CED-4 to endoplasmic reticulum.

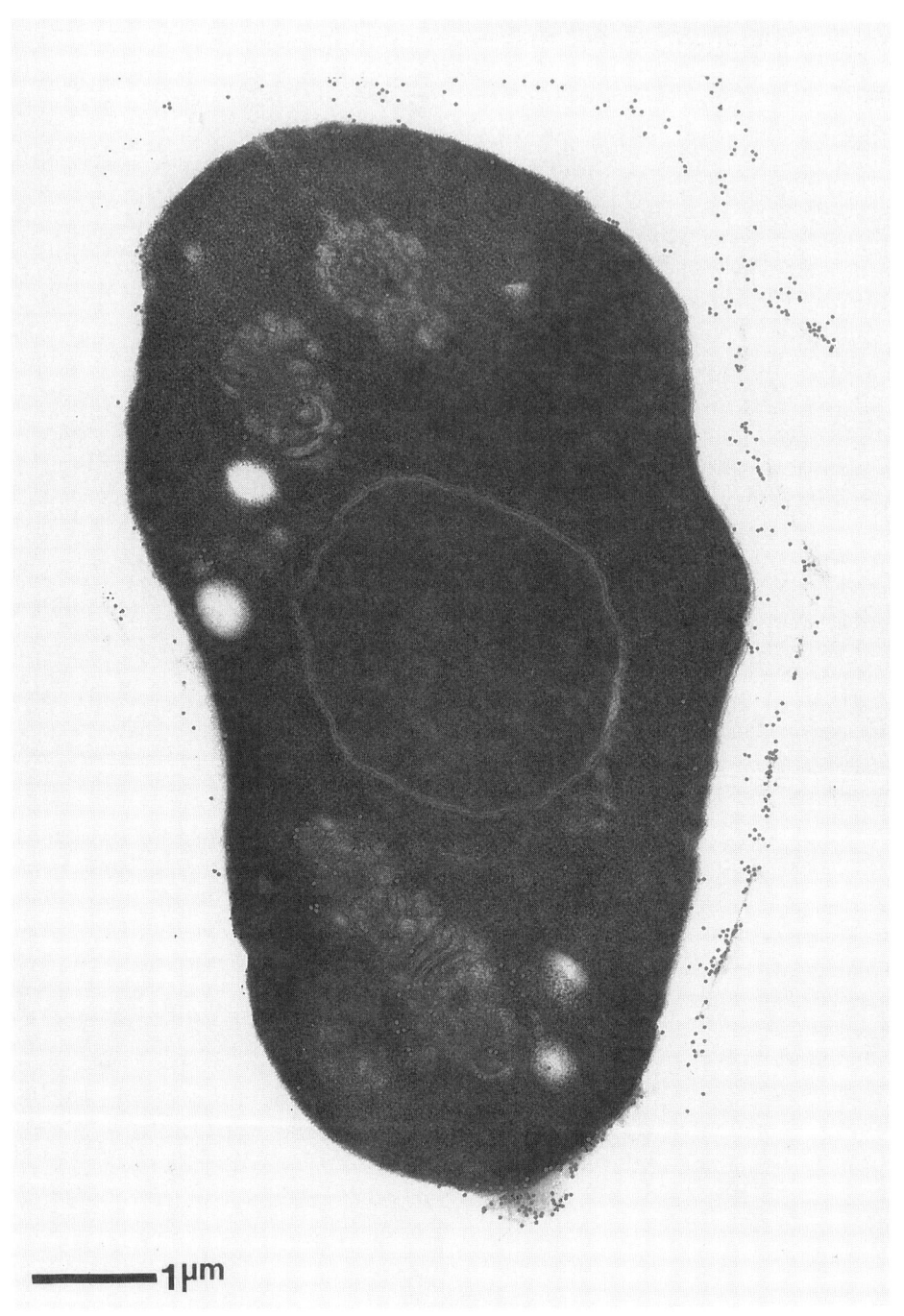

Fig. 2. Ultrathin section of Lowicryl embedded *S. pombe* incubated with polyclonal rabbit antiserum raised against purified galactosyltransferase, showing labelling of the Golgi stack and cell wall. (*See* Color Plate 3, following p. 180.)

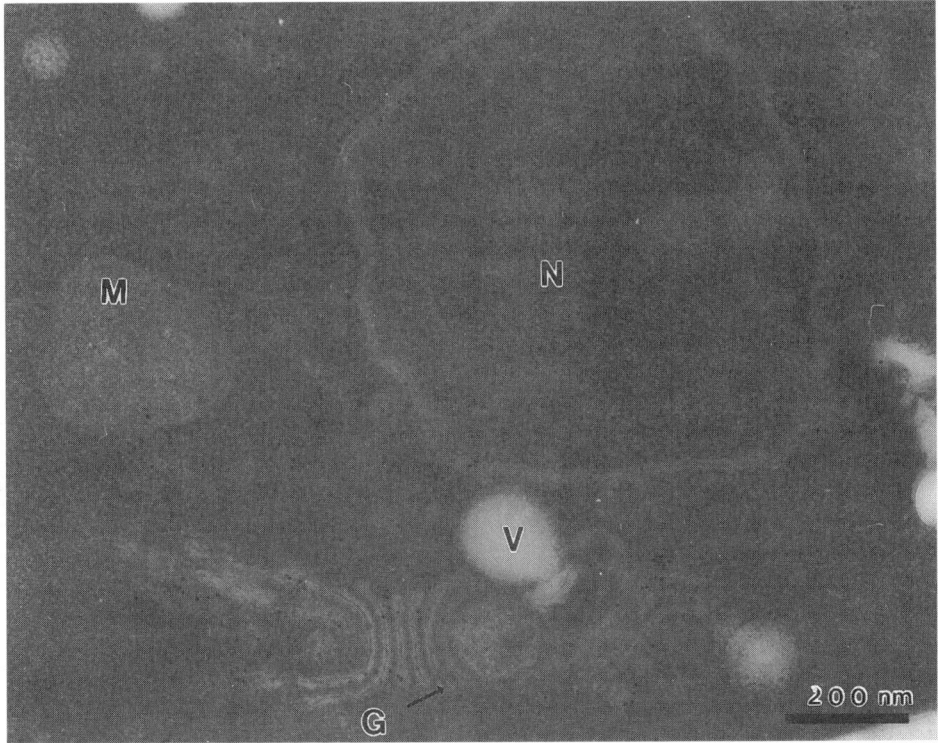

Fig. 3. Wild type *S. pombe* cells fixed with monomeric glutaraldehyde embedded in HM20 Lowicryl resin. Labelling demonstrates that the carbohydrate linkage created by a 1, 2-galactosyltransferase is constructed in the Golgi apparatus. *M*, mitochodrion; N-nucleus; V-vacuole; G., Golgi-apparatus.

3.6.3. High Pressure Frozen/Freeze Substitution Technique

Cryofixation is a physical technique of preserving cellular structure that is an important alternative to conventional chemical fixation. It is assumed that cryofixation compared to conventional chemical fixation provides images that are more likely to reflect the native structure of living cells. An additional benefit of cryofixation is that it may improve immunocytochemical localization.

The main advantages of cryofixation methods is the stabilization of cellular components by the much faster rate of fixation. The freezing method should produce amorphous ice from the specimen water (*see* Chap. 5 for more details). To achieve this, the specimens must be frozen as quickly as possible at a freezing rate not less than 10,000°C/sec. With the recent introduction of commercially available high-pressure freezing machine (Balzers AG [Liechtenstein], HPM010, and Leica, Reichert HPF, Wein, Austria), the routine cryofixation of

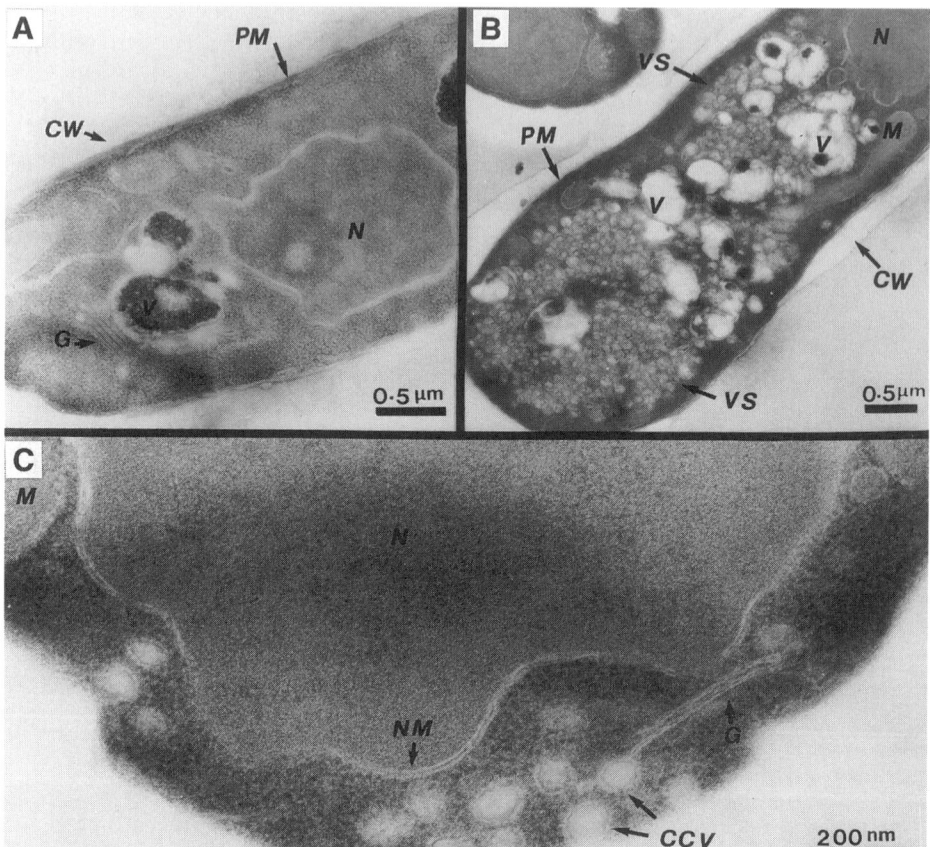

Fig. 4. PKC-η induces vesicle accumulation. (**A**) Mid-log phase PKC-η transformed cells, cultured in minimal selective medium with 1 μ*M* thiamine or without (**B,C**), were examined by electron microscopy as described in **Subheading 3.2.** PKC-η induces a large accumulation of vesicles (**B**) and active invagination from the membrane (**C**). CCV-putative budding clatherin-coated vesicles; CW-cell wall; G-Golgi apparatus; M-mitochondrion; N-nucleus; NM-nuclear membrane; PM-plasma membrane; V-vacuole; VS-vesicle.

large biological materials has finally become possible. Once frozen, specimens can be processed for (a) direct viewing of frozen specimens, (b) freeze-substitution, (c) freeze-fracture, and (d) cryomicrotomy. After substitution, the specimen can be embedded in epoxy resins for routine examination or in low temperature resins for immunogold labelling.

The major applications of this technique are to improve morphological preservation of cells such as yeast which are difficult to fix using standard chemicals, and to allow immunogold labelling in cases where the epitopes are

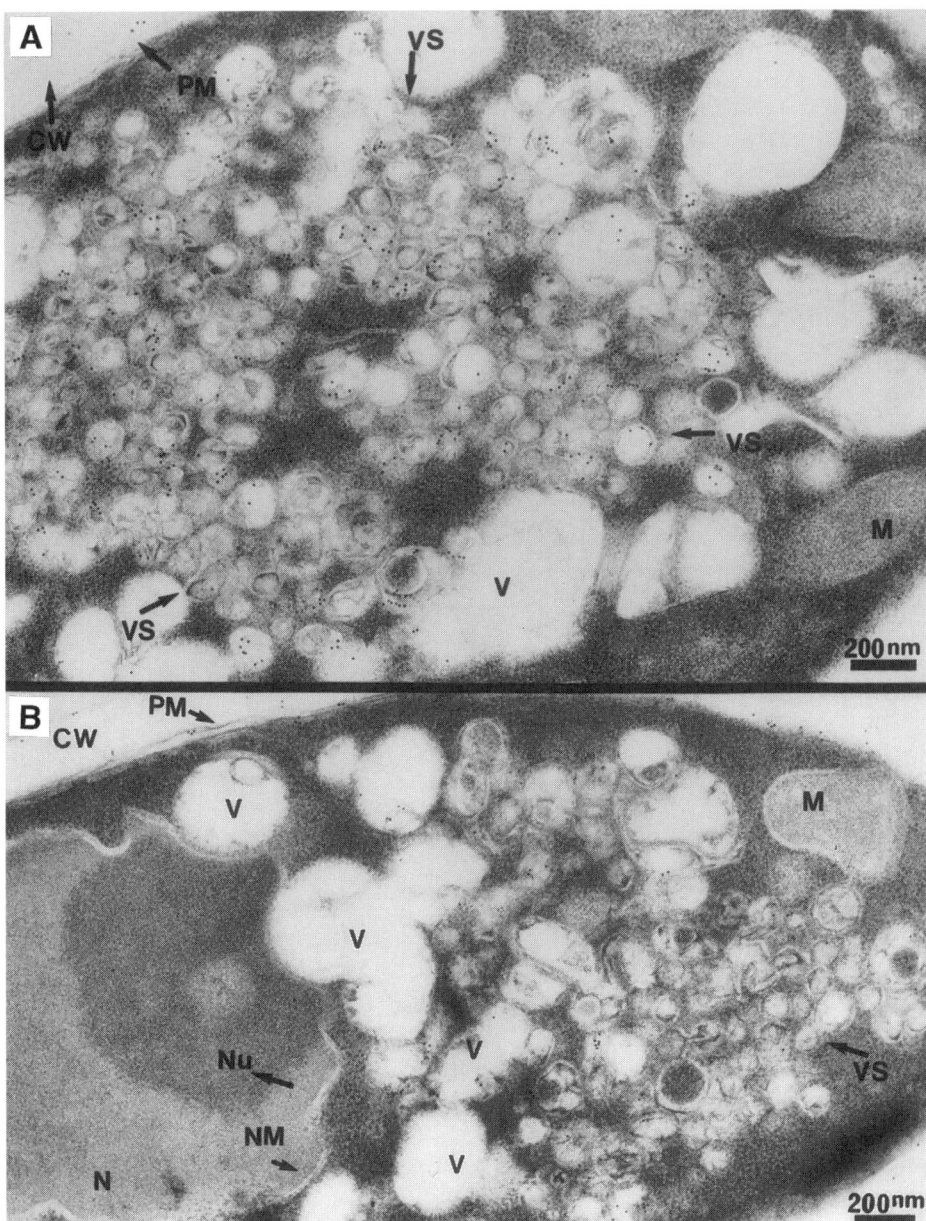

Fig. 5. Clathrin Lca localized on vesicles and the plasma membrane. (**A**) The gold particle are seen primarily on 70–90 nm vesicles and the plasma membrane (**B**) pep1a epitope-decreased labelling. Peptide 1a (0.63mM) was added to sections concurrently with antibody. CW-cell wall; M-mitochodrion; N-nucleus; Nu-nucleolus; PM-plasmas membrane; V-vacuole; Vs-vesicle.

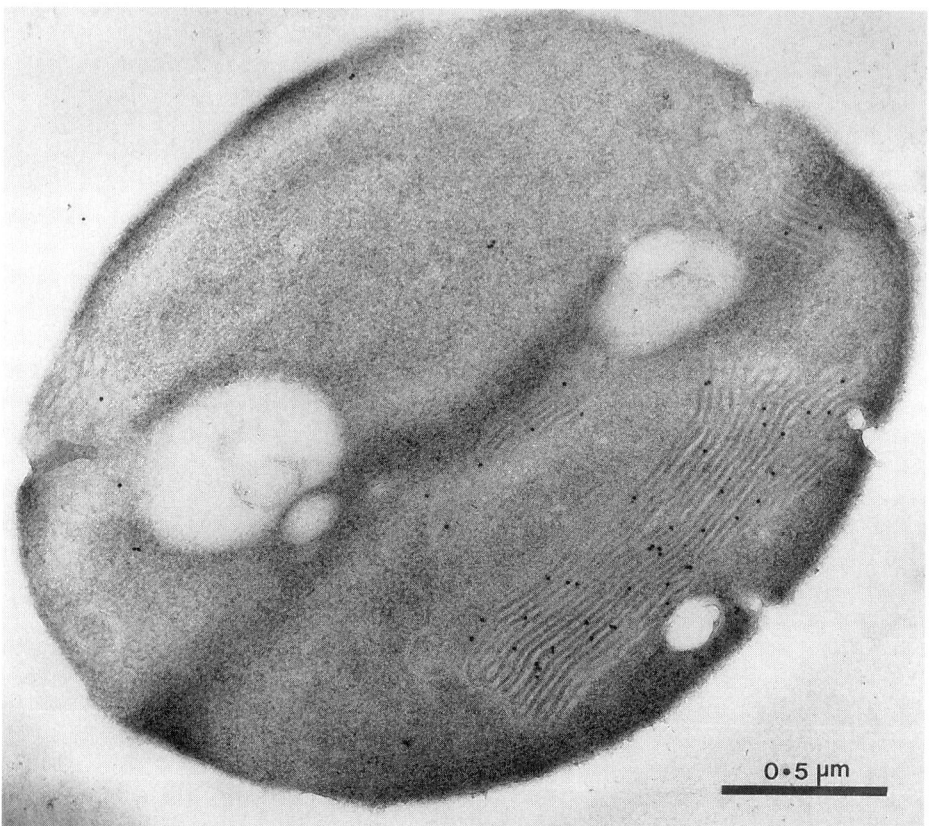

Fig. 6. Ultrastructure localisation of HA.CED-4 in *S. pombe* cells visualized by immunogold labelling using anti-HA monoclonal antibody. Cells were prepared as described in **Subheading 3.2.** Co-expression of CED-9 caused the relocation of CED-4 to endoplasmic reticulum.

destroyed by chemical fixation. Excellent results have been obtained in both morphological and immunogold labeling studies, and one can be confident therefore that this technique will have widespread uses for EM of both yeast cells and mammalian cells/tissues. One of the goals of this technique has been to develop methods of specimen preparation that will yield large numbers of well-fixed yeast cells that can be used for routine EM characterization or EM immunocytochemistry.

Use of the Balzers High-pressure freezer has allowed us to achieve this goal (Figure 7). Yeast cells were high-pressure frozen, then freeze-substituted in acetone for 3 days at - 90°C., warmed to –35°C and infiltrated with Lowicryl HM20 before UV polymerization. HPF in conjunction with freeze-substitution resulted

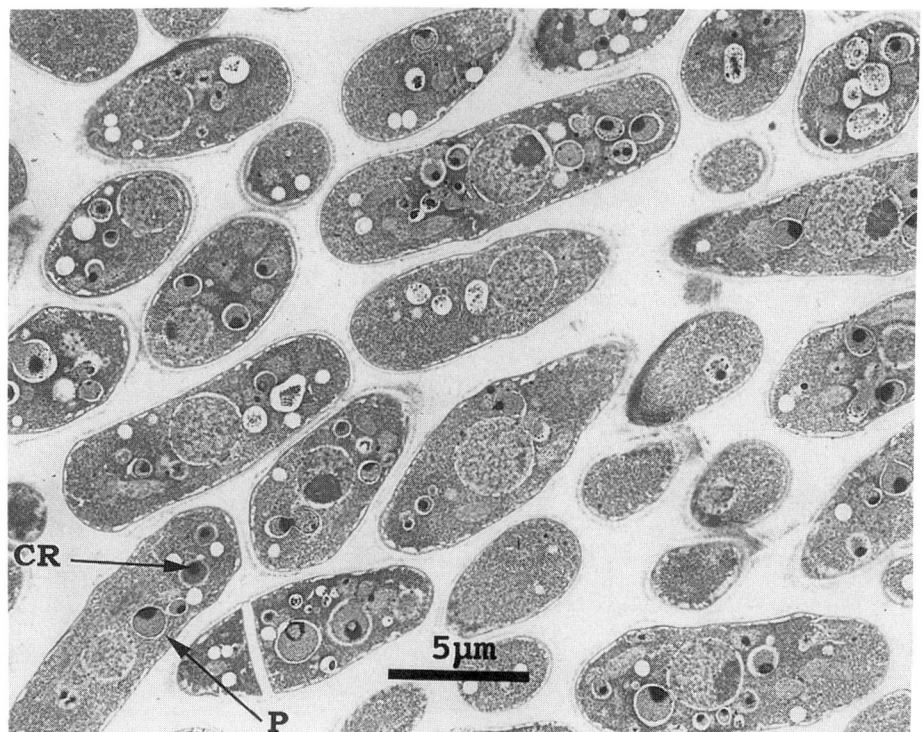

Fig. 7. High-pressure frozen *S. pombe*, freeze-substituted in acetone and embedded in resin, illustrating a very high yield of adequately frozen cells with good morphological preservation. The peroxisomes (P) in the yeast cell are well defined and contain a dense structure, the crystalloid (CR).

in excellent preservation of yeast ultrastructure (**Fig. 7**). In both *S. pombe* and *S. cerevisiae* (data not shown) improved preservation of membranes is visible. The peroxisomes in the yeast cell are well defined and contain a dense structure the so called crystalloid (*see* **Note 7**).

3.6.4. Pat1-114 Induced Meiosis in Fission Yeast.

The use of high-pressure freezing followed by freeze-substitution results in excellent morphological preservation of spore (SP) formation with pat-114 induced meiosis in fission yeast (**Fig. 8**). The pat1 gene in *S. pombe* encodes a protein kinase that prevents entry into meiosis. In diploid cells, signal transduction pathways sensing low nitrogen levels and the presence of both mating pheromones induce the production of the mei3 gene. The mei3 protein inhibits the pat1 kinase, leading to meiosis and spore formation. A pat1-114/pat1-114

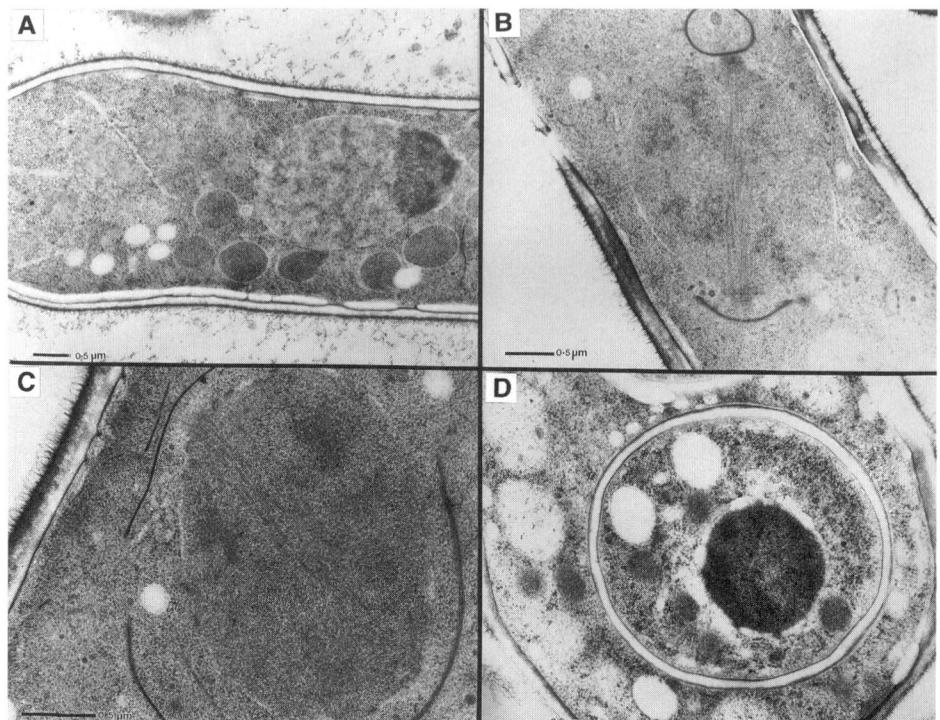

Fig. 8. High-pressure freezing followed by freeze-substitution results in excellent morphological preservation of spore formation with pat-114 induced meiosis in fission yeast. A, 0 h; B, 2 h; C, 4 h; D, 7h.

mutant diploid can be driven into meiosis by temperature inactivation of the kinase activity. This temperature-induced meiosis proceeds in a highly synchronous manner producing ascospores after 7 h.

In this experiment meiosis was induced with a temperature shift from 25°C to 34°C followed by a return to 25°C at 4 h. Samples were collected at hourly intervals during meiosis. Premeiotic DNA synthesis and meiosis I occur during the first 4 h. By the 4 h time point most of the cells are in meiosis II, spindle pole bodies (*see* **Notes 8** and **9**) have extended into the cytoplasm, and the prespore membranes have begun to form. By 5 h the prespore membranes are beginning to enclose the nuclei. After sealing off the nucleus, cytoplasm and organelles that will continue the new ascospore, the spore cell wall thickens during the last 2 h of the time course (unpublished data).

Other examples of immunogold labelling after high-pressure freezing /freeze-substitution are illustrated in **Figs. 9** and **10**. Postembedding immunogold labelling of mitotic spindles of *S. pombe* with 10 nm gold on Lowicryl section (**Fig. 9**)

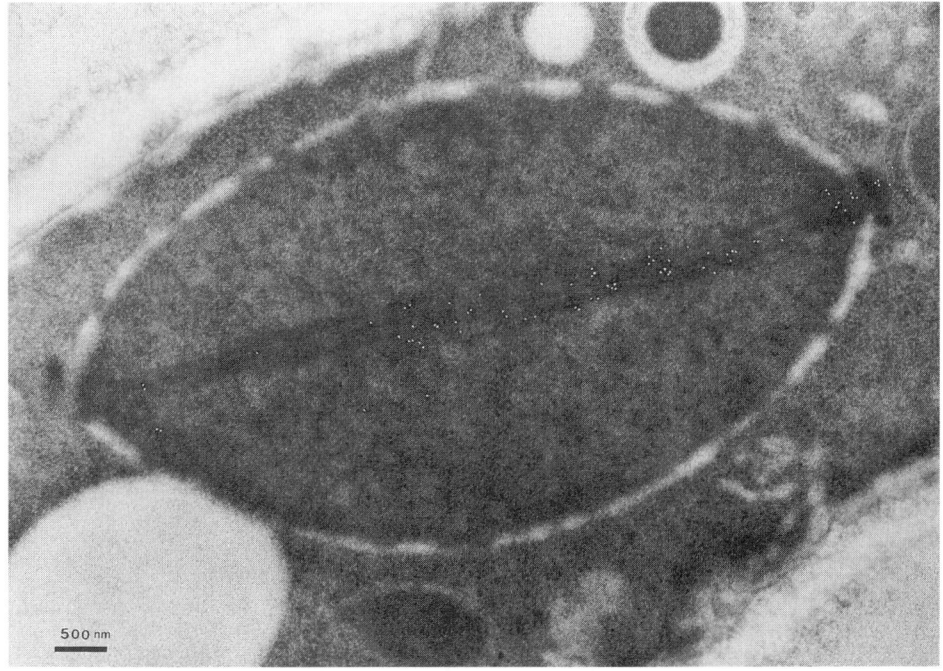

Fig. 9. Immunolocalisation of Tubulin on a mitotic spindle of fission yeast. (*See* Color Plate 2, following p. 180.)

using an anti tubulin antibody (*see* **Note 8**). **Fig. 10** shows labelling of the yeast cell nucleus after expression of cip 1 cyclin dependent kinase inhibitor in *S. pombe*: G2 arrest of the cells. Cells prepared by high-pressure freezing and freeze-substitution as described in Chap. 5.

3.6.5. Cryosectioning Technique

Ultrathin cryosectioning in combination with electron-dense gold markers offers the method of choice for the visualization and localization of antibody-antigen complexes in yeast cells. Different molecules can be visualized simultaneously whilst avoiding the accessibility problems of other methods. Ultrathin cryosectioning, well known and much used by scientists for mammalian cells or tissues *(41,42)*, has up to the present been little used for yeast . We have found that it is an excellent method for preservation of fine structure in yeast. The protocol described in **Subheading 3.3.** to prepare yeast cells for morphological examination of different strains or events such as apoptosis is now in routine use. **Figs. 11** and **12** show low and high magnifications of cryosections of *S. pombe* prepared in this way to illustrate the morphological

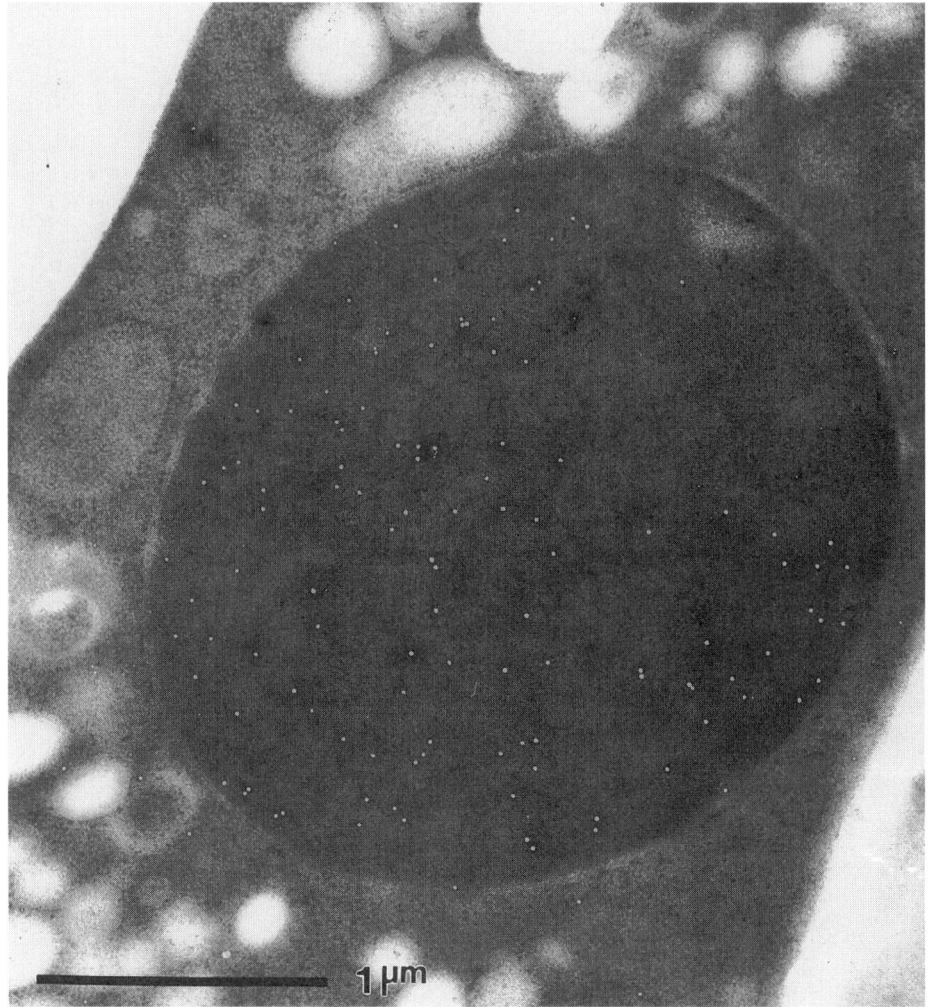

Fig. 10. Expression of cip-1-cyclin-dependent kinase inhibitor in *S. pombe*: G2-arrested cells. Cip1 is localized on the nucleus.

preservation of cell organelles in fission yeast. **Fig. 12B** shows an apoptotic cell in fission yeast. Induction of wild type ced4 expression in S.pombe resulted in rapid focal chromatin condensation and cell death *(43)*.

An example of immunogold labelling of a cryosection (**Note 10**) is presented in **Fig. 13** and shows labelling of pre-Golgi membranes with anti-GFP antibody in a yeast strain containing trunc NAGT1 + gfp fusion protein (unpublished data) . The cells were prepared as described in **Subheading 3.3.2.**

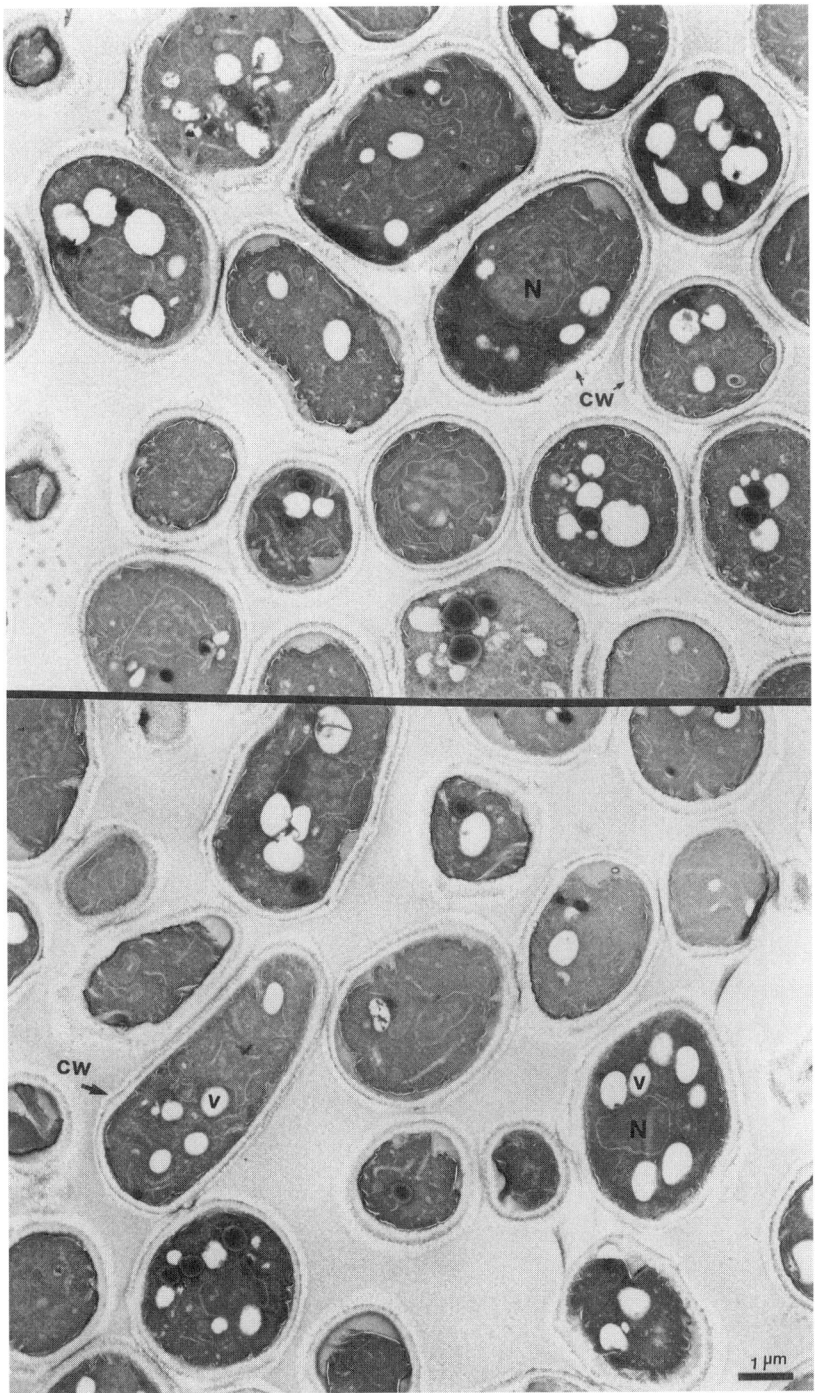

Fig. 11. Low-power magnification of cryosections of *S. pombe* prepared as described in **Subheading 3.3.**, illustrating the morphological preservation of cell organelles.

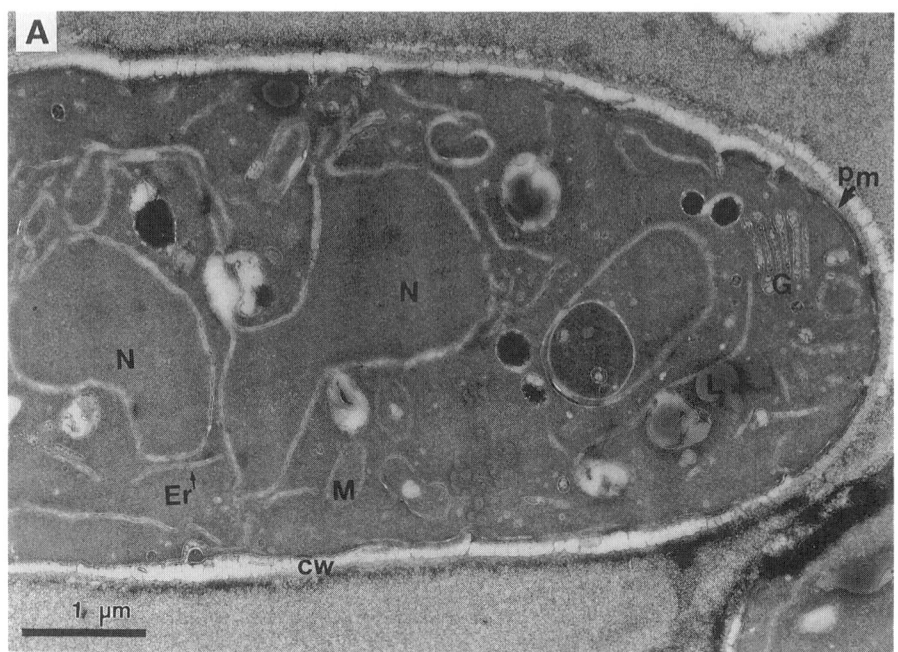

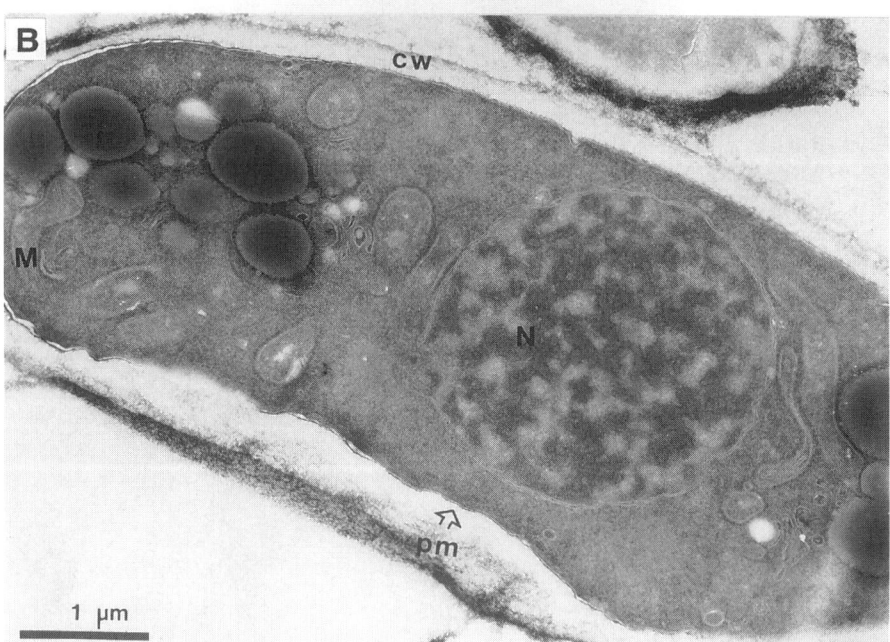

Fig. 12. High magnification of crysections of *S. pombe* prepared as described in **Subheading 3.3.**, illustrating the morphological preservation of cell organelles in normal (A) and an apoptotic cell (B). (N-nucleus; Er-endoplasmic reticulum; CW-cell wall; pm-plasma membrane; M-mitochondrion.)

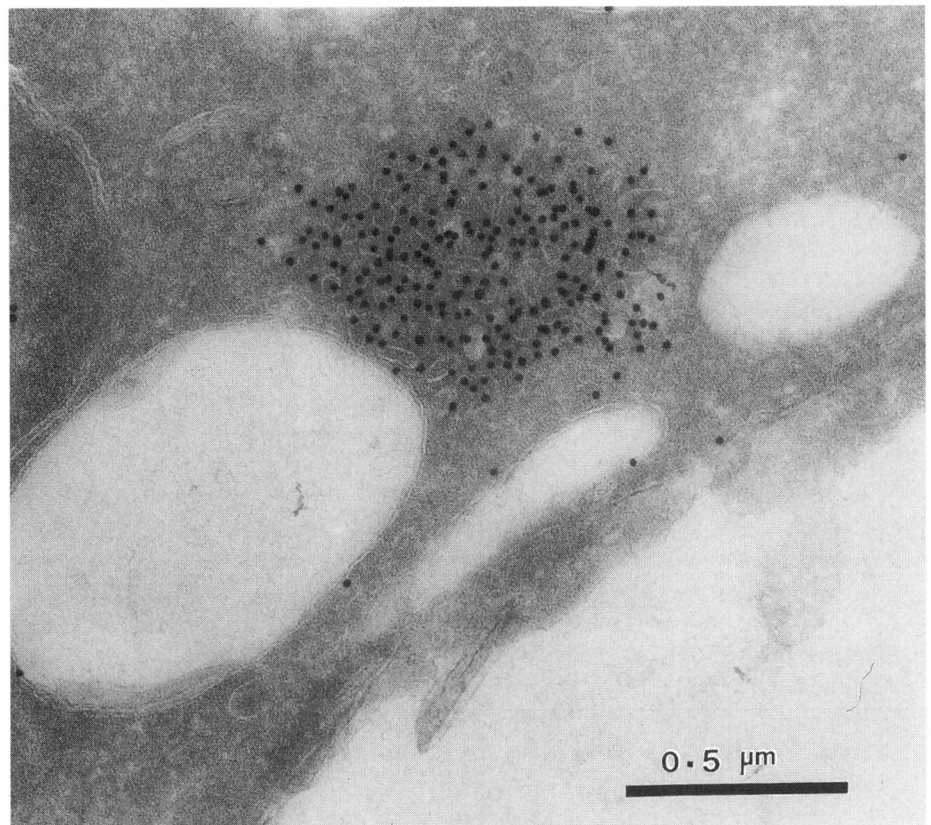

Fig. 13. Localisation of pre-Golgi membranes with anti-GFP antibody in yeast strain containing trunc NAGT1 + gfp fusion protein.

4. Notes

1. When using temperature-sensitive mutants in a study, it is important to note that temperature shifts alone in fission yeast can lead to transient perturbations of even wild-type cell physiology (e.g., filamentous actin is transiently disorganized by temperature shift from 25°C–36°C, requiring about 1–1.5 h to "adapt" to the new temperature). In the absence of a complete understanding of the reasons for such transient changes, such experiments require careful controls.

2. Maintenance of ultrastructure and retention of antigenicity are contradictory requirements of fixation and must be balanced carefully. Glutaraldehyde causes permanent cross-linking in proteins, whilst paraformaldehyde, especially in low concentrations, causes weaker reversible fixation. Full preliminary studies at light level are essential to determine if limitations in choice of fixatives exist. Localization at light level does not however guarantee success at EM level.

3. Each antibody must be titrated to determine the optimum dilution.
4. An essential feature of immunogold labelling is the demonstration of specificity. To check that nonspecific binding of either the primary or secondary antibody is not occurring it is necessary to carry out the following controls:
   A. Compare the immunogold labelling with that of a similar specimen lacking the antigen in question if possible.
   B. Treat the specimen with preimmune serum in place of the primary antibody.
   C. Either treat the section with the antibody preabsorbed with excess a soluble homogeneous form of the antigen (peptide) prior to labelling or label with primary antibody in presence of excess peptide.
   D. Treat the specimen with unrelated antibody to check for nonspecific labelling.
5. Only if specific antibodies without binding capacity for protein A are used, like sheep and goat IgG and some monoclonal, or when an enhancement of the gold signal is desired: Binding antibody (rabbit anti-goat or mouse; swine anti rabbit) diluted in PBS/BSA 1%.
6. A major problem with polyclonal antibodies is non-specific labelling of the cell wall. This can be overcome by preabsorbing the antibody with whole yeast cells as follows:
   A. Harvest the cells by centrifugation and wash once with PBS.
   B. Add sodium azide to 10 mM and let stand for 5 min.
   C. Wash the cells twice with PBS, dilute the antiserum 1:5 with PBS to 100 μL and add to 50 μL of washed yeast cells.
   D. Leave overnight at 4 °C, then centrifuge at 500 g for 15 min at 4°C.
7. The use of high-pressure freezing followed by freeze-substitution results in excellent preservation of peroxisomes. It is the only method which allows visualization of the crystalloid.
8. High-pressure freezing followed by freeze-substitution is the best preparation method for examining normal and aberration yeast mitotic spindles during the cell cycle.
9. Paraformaldehyde-and glutaraldehyde-based fixatives have proven useful for many studies of low-molecular-weight antigens and peptides; they are often reported to be less valuable with larger protein antigens (e.g., filaments). The latter are found by most authors to be better preserved by variants of noncrosslinking fixatives such as ethanol, acetone or methanol. Therefore, the best method for processing large protein antigen for EM examination may be by high-pressure freezing followed by freeze-substitution either in acetone or methanol.
10. In crysectioning, fixed cells or tissues are simply cryoprotected in sucrose and snap frozen in liquid nitrogen before ultrathin sectioning is carried out. The dehydration and embedding of the alternative low-temperature-resin methods and their inherent problems are therefore avoided. In our opinion this method is the most sensitive method of localizing antigens at subcellular level.

Acknowledgments

We would like to thank all those who contributed material used in this chapter in particular Tom Chappell, Graham Warren, Paul Nurse, Nigel Goode, Peter Parker, David Shima, Claewen James, and J. Hymes.

References

1. Nurse, P. (1990) Universal control mechanism regulating onset of M-phase *Nature* **344,** 503–508.
2. Yanagida, M. (1989) Gene products required for chromosome separation. *J Cell Sci. Suppl.* **12,** 213–29.
3. Simanis, V. (1995) The control of septum formation and cytokinesis in fissionyeast. *Semin. Cell Biol.* **6,** 79–87.
4. Armstrong, J., Pidoux, A., Bowden, S., Craighead, M., Bone, N., and Robinson, E. (1994) The ypt proteins of *Schizosaccharomyces pombe. Biochem. Soc. Trans.* **22,** 460–463.
5. Arellano, M., Duran, A., and Perez, P. (1996) Rho1 GTPase activates the (1–3) beta-D-glucan synthase and is involved in Schizosaccharomyces pombe morphogenesis. *EMBO J.* **13,** 4584–4591.
6. Egel, R. (1989) Mating-type genes, meiosis, and sporulation. in *Molecular Biology of the Fission Yeast.* Nasim, A., Young, P., and Johnson, B. F., eds. Academic Press, San Diego, pp. 31–73.
7. Hughes, D. A. (1995) Control of signal transduction and morphogenesis by Ras. *Semin. Cell Biol.* **6,** 89–94.
8. Wilkinson, M. G., Samuels, M., Takeda, T., Toone, W. M., Shieh, J. C., Toda, T., Millar, J. B., and Jones. N. (1996) The Atf1 transcription factor is a target for the Sty1 stress-xactivated MAP kinase pathway in fission yeast. *Genes Dev.* **10,** 2289–2301.
9. Sturm, S., and Okayama, H. (1996) Domains determining the functional distinction of the fission yeast cell cycle "start" molecules Res1 and Res2. *Mol. Biol. Cell* **7,** 1967–1976.
10. Zhu, Y., Takeda, T., Whitehall, S., Peat, N., and Jones, N. (1997) Functional characterization of the fission yeast Start-specific transcription factor Res2. *EMBO J.* **16,** 1023–1034.
11. Streiblova, E., and Wolf, A. (1972) Cell wall growth during the cell cycle of Schizosaccharomyces pombe. *Zeitschrift f. Allg. Mikrobiologie* **12,** 673–684.
12. Mitchison, J. M., and Nurse, P. (1985) Growth in cell length in the fission yeast Schizosaccharomyces pombe. *J. Cell Sci.* **75,** 357–76.
13. Verde, F., Mata, J., and Nurse, P. (1995) Fission yeast cell morphogenesis: identification of new genes and analysis of their roles during the cell cycle. *J. of Cell Biology* **131,** 1529–38.
14. Snell, V., and Nurse, P. (1993) Investigations into the control of cell form and polarity: the use of morphological mutants in fission yeast. *Dev. Suppl.* 289–299.
15. Marks, J., Hagan, I. M., and Hyams, J. S. (1986) Growth polarity and cytokinesis in fission yeast: the role of the cytoskeleton. *J. Cell Sci. Suppl.* **5,** 229–41.
16. Hagan, I. M., and Hyams, J. S., (1988) The use of cell division cycle mutants to investigate the control of microtubule distribution in the fission yeast Schizosaccharomyces pombe. *J. Cell Sci.* **89,** 343–357.
17. Robinow, C. F., and Hyams, J. S. (1989) General cytology of fission yeasts. in *Molecular Biology of the Fission Yeast.* Nasim, A., Young, P., and Johnson, B. F., eds. Academic Press, San Diego, pp. 273–330.

18. Chang, F., and Nurse, P. (1996) How fission yeast fission in the middle. *Cell* **84,** 191–194.
19 Tanaka, K., and Kanbe, T. (1986) Mitosis in the fission yeast S. pombe as revealed by freeze–substitution electron microscopy. *J. Cell Sci.* **80,** 253–268.
20. Yamamoto, M. (1996) The molecular control mechanisms of meiosis in fission yeast. *Trends Biochem. Sci.* **21,** 18–22.
21. Nielsen, O., and Davey, J. (1995) Pheromone communication in the fission yeast Schizosaccharomyces pombe. *Semin. Cell Biol.* **6,** 95–104.
22. Stern, B., and Nurse, P. (1997) Fission yeast pheromone blocks S-phase progression by inhibiting the G1 cyclin B-p34cdc2 kinase. *EMBO J.* **16,** 534–544.
23. Fukui, Y., Kaziro, Y., and Yamamoto, M. (1986) Mating-pheromone-like diffusible factor released by Schizosaccharomyces pombe. *EMBO J.* **5,** 1991–1993.
24. Leupold U. (1987) Sex appeal in fission yeast. *Curr. Genet.* **12,** 543–545.
25. Chikashige, Y., Ding, D. Q., Funabiki, H., Haraguchi, Y., Mashiko, S., Yanagida, M., and Hiraoka, Y. (1994) Telomere-led premeiotic chromosome movement in fission yeast. *Science* **264,** 270–273.
26. Hayles, J., and Nurse, P. (1992) Genetics of the fission yeast Schizosaccharomyces pombe. *Ann. Rev. Genet.* **26,** 373–402.
27. Hoheisel, J. D., Maier, E., Mott, R., McCarthy, L., Grigoriev, A. V., Schalkwyk, L. C., Nizetic, D., Francis, F., and Lehrach, H. (1993) High-resolution cosmid and P1-maps spanning the 14MB genome of the fission yeast S. pombe. *Cell* **73,** 109–120.
28. Grimm, C., Kohli, J., Murray, J., and Maundrell, K. (1988) Genetic engineering in Schizosaccharomyces pombe a system for gene disruption and replacement using the ura4 gene as a selectable marker. *Mol. Gen. Genet.* **215,** 81–86.
29. Maundrell, K. (1993) Thiamine-repressible expression vectors pREP and pRIP for fission yeast. *Gene.* **123,** 127–130.
30. Sipiczki, M. (1989) Taxonomy and phylogenesis, in *Molecular Biology of the Fission Yeast.* Nasim, A., Young, P., and Johnson, B. F., eds., Academic Press, San Diego, pp. 431–452.
31. Kaufer, N. F., Simanis, V., and Nurse, P. (1985) Fission yeast Schizosaccharomyces pombe correctly excises a mammalian RNA transcript intervening sequence. *Nature* **318,** 78–80.
32. Dubey, D. D., Kim, S. M., Todorov, I. T., and Huberman, J. A. (1996) Large, complex modular structure of a fission yeast DNA replication origin. *Curr. Biol.* **6,** 467–473.
33. Nakaseko, Y., Adachi, Y., Funakashi, S., Niwa, O., and Yanagida, M. (1986) Chromosome walking shows highly homologous repetitive sequence present in all the centromere regions of fission yeast. *EMBO J.* **5,** 1011–1021.
34. Moreno, S., Klar, A., and Nurse, P. (1991) Molecular genetic analysis of fission yeast Schizosaccharomyces pombe. *Meth. Enzymol.* **194,** 795–83.
35. Ayscough, K., Hajibagheri, N. M., Watson, R., and Warren, G. (1993) Stacking of Golgi cisternae in Schizosaccharomyces pombe requires intact microtubules. *J. Cell Sci.* **106,** 1227–1237.

36. Ruhrberg,C., Hajibagheri, M. A. N., Simon, M., Dooley, T., and Watt, F. M. (1996). Envoplakin, a novel precursor of the cornified envelope that has homology to desmoplakin. *J. Cell Biol.* **134(3)**, 715–29, 1996.

37. Ruhrberg,C., Hajibagheri, M. A. N., Parry, D. M., and Watt, F. M. (1997) Periplakin, a novel component cornified envelopes and desmosomes that belongs to the plakin family and forms complexes with envoplakin. J. Cell Biol. **139**, 1835–1849.

38. Chappell, T. J., Hajibagheri, M. A., Ayscough, K., Pierce, M., and Warren, G. (1994) Localization of an 1, 2 galactosyltransferase activity to the Golgi apparatus of Schizosaccharomyces pombe. *Mol. Biol. Cell* **5**, 519–528.

39. Goode, N. T, Hajibagheri, M. A. N., Warren, G. B., and Parker, P. J. (1994) Expression of mammalian protein kinase C in Schizosaccharomyces pombe: isotype specific induction of growth arrest, vesicle formation and endocytosis. *Mol. Biol. Cell* **5**, 907–920.

40. Goode, N. T., Hajibagheri, M. A. N., and Parker, P. J. (1995) Protein Kinase C (PKC)-induced PKC Down. *J. Biol. Chem.* **27**, 2669–2673.

41. Dittie, A., Hajibagheri, M. A. N., and Tooze, S. (1996) The Ap-1 adaptor complex binds to immature secretory granules from PC12 cells and is regulated by ADP-ribosylation factor. J. Cell Biol. **132**, 523–536.

42. Whitehouse, C., Burchell, J., Gschmeissner, S., Brockhausen, I., Lloyd, K. O., Taylor–Papadimitriou, J. (1997). A transfected sialyltransferase that is elevated in breast cancer and localizes to the medial/trans Golgi apparatus inhibits the development of core-2-based O-glycans. *J. Cell Biol.* **137(6)**, 1229–41.

4 3. James, C., Gschmeissner, S., Fraser, A., Evan, G. I. (1997) CED-4 induces chromatin condensation in Schizosaccharomyces pombe and is inhibited by direct physical association with CED-9. *Curr. Biol.* **7(4)**, 246–52.

13

Preparation of Double/Single-Stranded DNA and RNA Molecules for Electron Microscopy

M.A. Nasser Hajibagheri

1. Introduction

Electron microscopy (EM) has proved to be an increasingly powerful tool in the study of nucleic acids. It provides qualitative and quantitative data on the size and structure of native and experimentally manipulated DNA and RNA molecules. The electron microscope, as a tool for analyzing molecular structure, interactions, and processes, is one that can be used with increasing confidence, particularly in conjunction with biochemical and biophysical studies *(1–4)*. In order to prepare nucleic acids for electron microscopy, there are two important requirements. Since nucleic acids are thread-like molecules which form three-dimensional flexible random coils in aqueous solution, they must converted to two-dimensional unaggregated molecules. These requirements are met by spreading them together with a surface-active substance which traps the nucleic acid molecules in a monolayer floating on a hypophase of water *(1-8)*. This monolayer then is absorbed to a thin carbon film on a carbon-coated grid. To enhance contrast, the nucleic acid molecule is stained either with uranyl acetate (in particular, for dark-field microscopy) or by low-angle shadowing with evaporated heavy metal, for example, Pt/Pd alloy (*see* **Fig. 1**).

1.1. Dark-Field Electron Microscopy (EM)

The dark-field technique, well known and much used by the material scientists, has up to the present been little used by biological microscopists *(1–2* and *9–15)*. Dark-field is a technique used in transmission EM when it is desired to enhance the visibility of a second phase that is present in small amount. The electron beam having passed through the specimen is divided into a transmitted

From: *Methods in Molecular Biology,* vol. 117: *Electron Microscopy Methods and Protocols*
Edited by: N. Hajibagheri © Humana Press Inc., Totowa, NJ

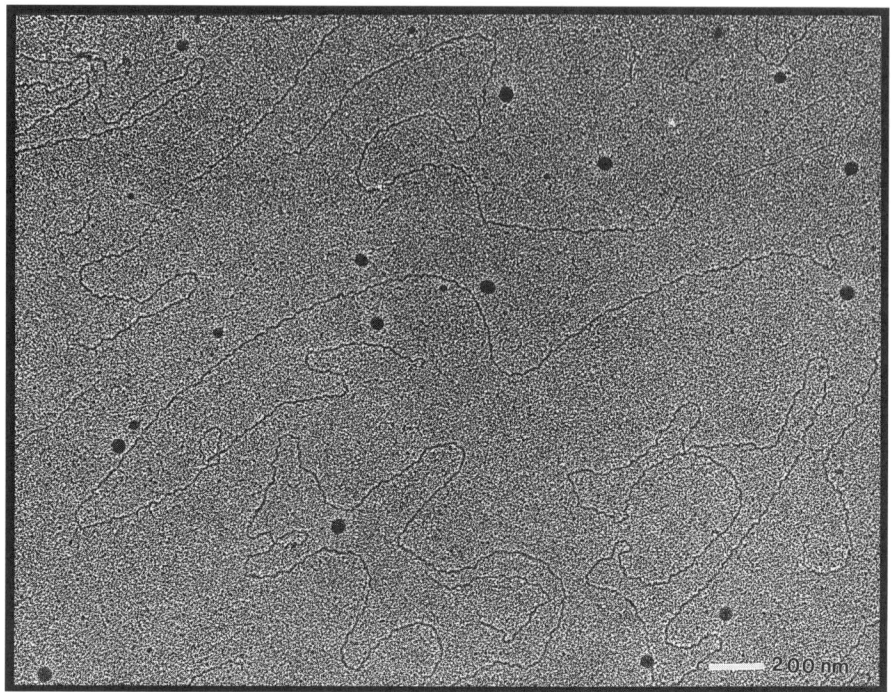

Fig. 1. Bright-field visualization of the linearized DNA (pXIr101A plasmid), spread in ammonium acetate and BAC and mounted on very thin carbon-coated grids (10 nm thickness), and shadowed with Pt/Pd alloy.

electron beam and a scattered (diffracted) electron beam. Usually, the image formed by the former is called the bright field image, while that formed by the latter is called the dark-field image. The dark field image is characterised by high contrasts although it usually has insufficient brightness (*see* **Note 1**). The dark-field technique is advantageous because the transmission image of a molecule is formed directly according to its incinerated mass rather than indirectley by the deposition and permeation of heavy metal salts as in conventional bright-field microscopy.

EM of nucleic acids has been improved by dark field observation of unshadowed, stained specimens, absorbed onto support film on the grids without a basic protein film *(1, 13–15)* because this technique permits more precise measurements of molecular size and shape than the more classical EM methods. We have used dark-field electron microscopy to visualize DNA molecules for a number of projects that will be described in this chapter without the use of denaturing agents or metal shadowing and we have found that it is an excellent method for the study of macromolecules, the low contrast of biological macromolecules being overcome relatively easily by dark-field visualization.

2. Materials

2.1. Equipment

1. Vacuum coating unit with rotary shadowing.
2. A Teflon-coated or translucent plastic dish, to hold hypophase solution.
3. Acid-clean glass slides.
4. Talcum (talc) or graphite powder.
5. Flat Teflon-coated surface or Parafilm.
6. Fine stainless steel forceps.
7. EM grids (200–400 mesh) coated with carbon.
8. Automatic pipet and disposable tips for measuring volumes of 2–100 μL.
9. Filter paper.
10. Platinum-palladium (Pt-Pd) wire.
11. Transmission electron microscope with dark-field facility.

2.2. Regents

1. 0.5 M ammonium acetate.
2. Deionized formamide.
3. 0.2% BAC (benzyldimethylalkylammonium chloride). Stock solution in deionized formamide.
4. BAC 1:50, stock diluted in 0.5 M ammonium acetate.
5. 0.1% cytochrome-c in triple distilled water.
6. 2% uranyl acetate in 95% ethanol.
7. 2% uranyl acetate in aqueous solution.
8. 100–150 μg/mL ethidium bromide in aqueous solution.
9. 30%, 90 and 100% ethanol.
10. 0.5 M EDTA, 1 M Tris-HCl, pH 7.4.
11. 0.1 M Phosphate buffer, pH 7.3.
12. Acetone.
13. Clean nucleic acid preparations of about 10 ng/ μL.
14 Streptavidin conjugated to colloidal gold (Sigma, 10 nm size).
15. 5 mM-MgCl2.
16. 2 mM ATP.
17. 10 μL of each of dATP, dGTP, dCTP, and bio-11-dUPT (Sigma).
18. 3 M urea.

2.3. Preparation of Thin Carbon Support Films (Thickness <5 nm)

1. To get a clean and very flat surface onto which the carbon is evaporated to form a thin film, a piece of mica is cleaved with a forceps or razor blade (care should be taken that all instruments used throughout the preparation (forceps, troughs, etc.) are cleaned from any grease by rising them with acetone).
2. The mica is then placed in a vacuum coating unit with the freshly opened surface oriented towards carbon gun.
3. Carbon deposition can be easier be followed by a little piece of paper under the mica so that it will be coated with carbon from the gun.

4. The carbon gun must be cleaned before use. The flattened tip of the carbon rod is to be placed in the middle of the coil that represents the electron source for the electron beam evaporation.
5. The vacuum coating unit is evacuated to about 1/100, 000 Torr.
6. The electron source is heated. A shutter is then opened for the time needed to produce the film thickness desired by indirect coating (*see* **Note 2**). The thickness of the carbon film can be estimated from the frequency change of the quartz crystal (*see* **Note 3**).
7. Then the coating unit is ventilated and the mica is placed into a heated moist chamber with carbon film uppermost.
8. A cleaned trough is filled with triple distilled water.
9. A wire mesh covered with a little piece of filter paper which carries the nickel grids is submerged.
10. Now the mica with carbon fill uppermost is pushed slowly at angle of 30° to 45° into the water.
11. The floating film is lowered onto the grids by slowly draining the hypophase from the trough with a water pump and then the grids, together with filter paper, are dried.

2.4. Dark-Field Imaging

A brief account of the dark-field imaging technique required to visualize the DNA molecules is as follows.

1. The electron microscope is initially set up and aligned for normal bright field imaging.
2. Use 50 μm objective aperture because it gives the best compromise between resolution and brightness in the dark field.
3. The diffraction mode is then selected using a camera length of around 100–200 cm.
4. After focusing the bright spot and aperture image using the brightness and diffraction focus controls, the objective aperture image must be centered around the bright spot.
5. The dark-field switch is then pressed in order to tilt the beam.
6. The defector X and Y controls are used to adjust the beam tilt so that the required diffraction ring or spot is in the center of the aperture image, and it is often sufficient to move the bright spot out of the center of the aperture in order to obtain a dark-field image.
7. It is not necessary to move the objective aperture at all on Jeol electron microscope, but you may need to do it in other type of the electron microscope.
8. As a result, high-resolution images are easily achieved in dark mode. When the conventional magnification mode is selected, the dark-field image is viewed as in bright field, and astigmatism, beam shift, etc., can be adjusted as normal.
9. Fully independent alignment memories for bright and dark field ensures that the TEM can now be freely switched between the two modes with no further adjustment (*see* **Note 4**).

2.5. Spreading Solution for DNA and RNA Molecules

2.5.1. Spreading Solution for Double-Stranded DNA (dsDNA)

For double-stranded DNA spreading mixture contains: 0.3 to 0.6 μL/mL dsDNA, 4 μL BAC 0.002% (*see* **Note 5**) and 4 μL 0.5 *M* ammonium acetate. Spread on water at room temperature.

2.5.2. Spreading Solution for Single-Stranded DNA (ssDNA)

One to 10 μL/mL ssDNA, 4 μL Cytochrome-c 0.01% and 4 μL 0.5 *M* ammonium acetate. Spread on water at room temperature.

2.5.3. Spreading Solution for ssRNA

For single-stranded RNA for secondary structural analysis the spreading mixture contains: RNA 2 to 5 μL/mL in 0.1 *M* Tris Buffer-HCl, 0.1 m*M* EDTA, pH 7.0, 3 *M* urea and 0.002% BAC. Spread on to water at 40° to 70°C (*see* **Note 6**).

2.5.4. Spreading RNA and Single-Stranded DNA Molecules (2, 18, 19)

RNA molecules (or single-stranded DNA) can be spread for electron microscopy using four different denaturing conditions:

1. 80% formamide spreading.
2. Urea-formamide (4 *M* urea-80% formamide) spreading.
3. 80% formamide spreading with glyoxal pretreatment.
4. Urea-formamide (4 *M* urea-80% formamide) spreading with a glyoxal pretreatment. Each of these denaturing agents has its own advantages and disadvantages as described in detail by Coggins (*3*).

The following procedure can be recommended for spreading RNA molecules for EM observation.

The standard formamide cytochrome-c spreading method can be used (*2*), normally with 50% formamide, 0.06 *M* Tris (pH 8.2), 6 m*M* EDTA and 0.05 m*M* NaCl in the spreading solution, and in the hypophase. The spreading mixture consists of the following:

1. 10 μL of 0.06 *M* Tris-HCl, pH 8.2, 6 m*M* EDTA, 0.05 m*M* NaCl.
2. 40 μL of RNA molecules (0.5 ng/mL).
3. 10 μL of cytochrome c (at 1 mg/mL).
4. 40 μL of formamide (50%).
5. Spread above RNA mixture over a hypophase (15% formamide, 0.01 M Tris-HCl, pH 8.2, 1 mM EDTA) as described in **Subheading 3.1.1.** for dark-field microscopy or stain with uranyl acetate and shadow with platinum for bright-field observation.

3. Methods

3.1. Pretreatment of Carbon-Coated Grids

The are several methods are available to make carbon-coated grids hydrophobic, positively charged so that nucleic acids are absorbed to the surface (*6-8*). Grids can be treated either by glow-discharge (physical method) or by treatment with ethidium bromide (chemical method) and amount of DNA absorption to the grids depend upon of concentration of DNA in solution and to the time allowed for absorption.

3.1.1. Pre-Treatment of Carbon-Coated Grids by Ethidium Bromide

To facilitate a quantitative absorption of nucleic acids to the mounting base, a pretreatment of the carbon film is carried out as follows.

1. The carbon-coated grids are placed on a drop of 100–150 µL/mL ethidium bromide for 5 min.
2. Then the grids are washed by briefly touching them to the surface of a drop of triple distilled water.
3. The treated grids are dried on filter paper and they should be used within half an hour.

3.1.2 Pretreatment of Carbon-Coated Grids by Glow Discharge Method

Dubochet (16) developed a method of absorbing DNA to a carbon film by first exposing it to a plasma discharge in a partial atmosphere of pure amylamine. This treatment apparently creates upon the film a number of amylamine polymers with positively charged groups that firmly bind the nucleic acids strands and prevent their aggregation upon drying.

Prior to the spreading procedure, carbon coated grids need to be made hydrophobic by glow discharge for 15 s to facilitate absorption of DNA molecules to the carbon films. The method is as follows.

1. Place carbon-coated grids on a piece of filter paper in the chamber of a vacuum Unit containing glow discharge equipment.
2. Switch on high voltage supply. A purple glow is produced in the chamber by admitting a small amount of air and is maximised by adjusting the current and the vacuum conditions.
3. Glow discharge the grids for about 15 s at 70–80 mtorr (9–11 Pa) air pressure in a Harrick plasma cleaner.
4. Switch off high voltage, remove the grids, and turn off the vacuum evaporator and use the grids within 30 min.

3.2. Spreading Nucleic Acids for EM Visualization

3.2.1. Spreading of DNA Molecules by BAC Method

A protein-free nucleic acid preparation method for EM is described here (*1*). The basic procedure is very similar to the classical protein monolayer spreading techniques. The carrier protein (usually cytochrome-c) is replaced by benzyldimethylalkylammonium chloride (BAC), a molecule whose weight is only about 350 Daltons. Both the hypophase method and the microdiffusion or droplet method can be applied with this compound.

All solutions for EM preparation should be filtered prior to use (0.1 or 0.22 µm pore size).

1. An acetone-cleaned Teflon trough is filled with triple-distilled water as the hypophase.
2. A Teflon-coated bar on the rim of the through allows the acid cleaned microscope slide to be arranged in the dish at the angle of about 30° to the hypophase.
3. A little talc powder is dusted on the hypophase in order to visualize and compress the film.
4. When the movement of the hypophase has stopped, 10 µL of spreading solution [0.3 to 0.6 µL/mL dsDNA, 4 µL BAC 0.002% and 4 µL 0.5 *M* ammonium acetate (*see* **Note 7**) are applied onto the glass ramp (above 1 cm above the level of the hypophase), the solution runs down the slide, pushing the talc away and spreads the DNA molecules in a monolayer floating on a hypophase of water. A very pure water is critical to good results (*see* **Note 8**).
5. After the spreading process has occurred, a pretreated carbon-coated grid (*see* **Subheading 3.1.**) is briefly touched to the monolayer.
6. Staining of the nucleic acid is achieved by transferring the grid with carbon film (*see* **Note 3**) onto a drop of 0.02% uranyl acetate (uranyl acetate stock solution diluted with 30% ethanol) for about 2 min (*see* **Note 9**).
7. The grid is then placed with carbon-coated side uppermost on filter paper in order to dry.
8. After drying, the grid can be examined in dark-field mode in an electron microscope (*see* **Figs. 2** and **3**).

3.2.2. Spreading DNA by Cytochrome-C Method without Using Formamide

Nucleic acids can be incorporated into a monolayer film of denatured cytochrome-c for subsequent transfer to a specimen support film *(1, 2)*. The effect of embedment of the nucleic acids in the relatively rigid monofilm is to produce a well-spread display of DNA strands. There are two methods of using cytochrome c available which are as follows.

1. The droplet method *(20)* is a simple modification of the original Kleinschmidt *(1)* technique, and is a one-phase system not requiring any spreading accessories. In the droplet method (*see* **Note 10**), the typical spreading m ixture contains DNA (0.3–0.6 µg/mL), ammonium acetate (0.5 *M*) and cytochrome-c (0.1%). The application is carried out by allowing 10 µL of spreading mixture to sit on the activated carbon-coated grid for approximately 1 min, followed by dropwise washing with triple-distilled water, and then stained with uranyl

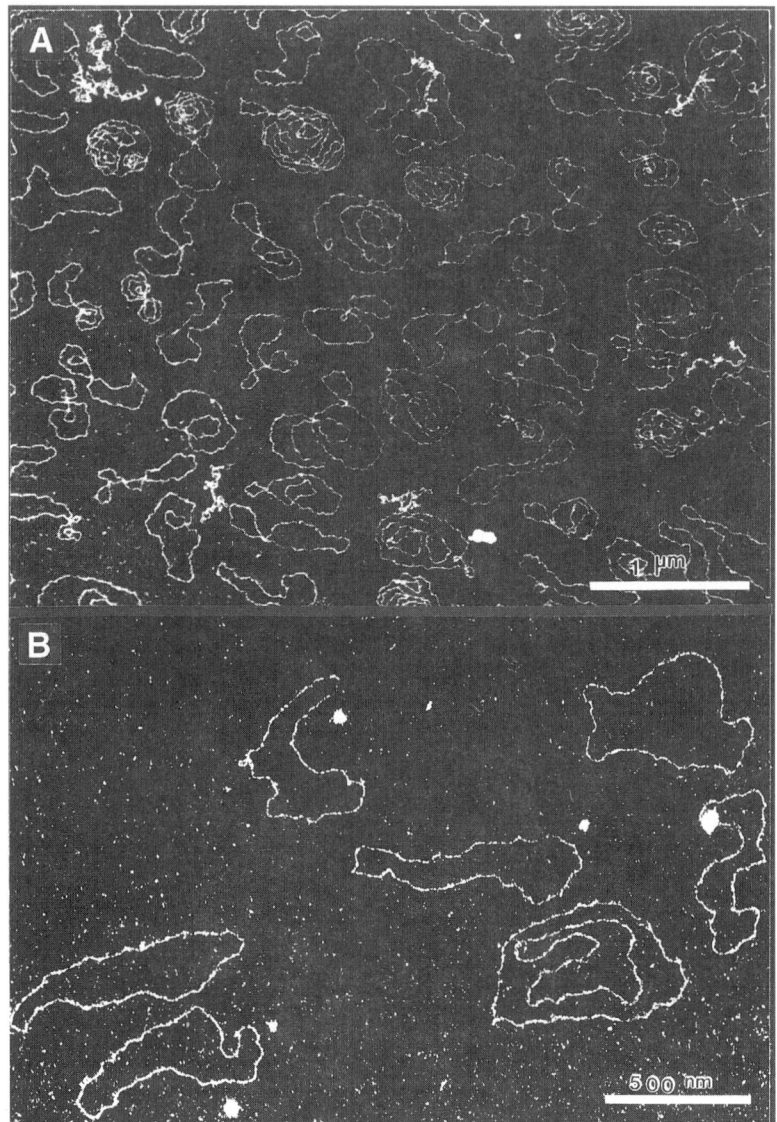

Fig. 2. Low (**A**) and high magnification (**B**) of dark-field images of DNA (pBS, 3000 base pairs) spread as described in **Subheading 3.3.**, provide qualitative and quantitative data on the size and structure of native and experimentally manipulated DNA molecules.

acetate for either dark-field observation (as described in **Subheading 3.3.1.**) or bright-field examination after metal shadowing.

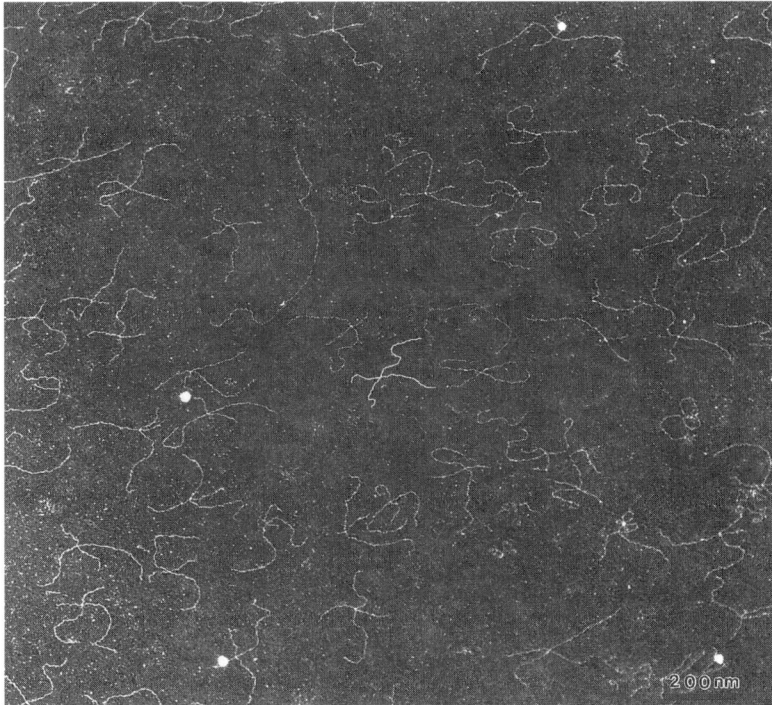

Fig. 3. Dark-field EM of X-DNA, indicating the lengths of each arm flanking the Holiday junction.

2. The second method of using cytochrome-c is carried out by spreading DNA molecules onto a monolayer of basic protein (cytochrome-c), followed by staining and shadowing with electron dense metal to give a sufficient contrast. This method is quick and accurate, and can be used to visualise the fate of the DNA molecules under investigation (*see* **Note 11**).

3.2.3. Diffusion Method (3)

1. Fill a round Teflon-coated trough with solution containing 20 ng/mL DNA, 0.2 *M* ammonium acetate.
2. A little talc powder can be dusted on the surface of solution to help to visualize formation of the protein films.
3. Dip a clean needle in cytochrome-c powder and touch it to the surface of the solution and leave 5–40 min to allow the DNA to diffuse into the surface (*see* **Note 12**).
4. Pick up a sample by touching with a carbon-coated grid (the grid needs to be activated first by glow discharge) and dehydrate by dipping in 95% ethanol and then either stain for dark-field microscopy as described in **Subheading 3.3.1.** or shadow with platinum/palladium for bright-field observation.

3.2.4. Watch Glass Technique for Spreading DNA Molecules (21)

This technique has been used successfully by Banerjee and Iyer *(21)* for the analysis of replicating branch points in large number of DNA plasmid (pCU 918). No denaturing agent is needed to visualise replication bubbles.

In principle, the method involves monolayer formation by spreading a hyperphase over a hypophase *(22)*. The usual spreading accessories are replaced by a watch glass. The hyperphase containing the DNA sample is allowed to slide down the side and then allowed to form a monolayer for 10 to 20 min. Formvare-coated grids can be used to lift the DNA followed by staining and rotary shadowing.

3.3. Spreading of DNA-Bounded Protein Complex

1. For spreading 2 μL of DNA/protein reaction mixture (fixed in 0.2% glutaraldehyde for 15 min at 37°C or unfixed) is diluted in 2–5 m*M* magnesium acetate. After few seconds, a glow discharged grid (*see* **Subheading 3.1.2.**) is touched to the droplet placed on Parafilm.
2. After removing excess solution, grids need to be washed in 2–5 m*M* magnesium acetate for 15 min.
3. Put a grid on 50 μL of 2% aqueous solution of uranyl acetate for 5 min for negative staining (*see* Chapter 2 for details).
4. The grid is then blotted on filter paper and air dried.

3.4. Immunogold Labeling
of DNA Repair Patches Following UV-Irradiation

The following protocol has been developed *(23)* to investigated the size and distribution of repair patches induced by human cell extracts in ultraviolet (UV) light-irradiated plasmid DNA [*see (23)*]. Repair synthesis was carried out in a buffer substituting biotinylated dUTP for dTTP, to allow repair patches to be detected by EM after streptavidin/colloidal gold labeling. Here is the following protocol.

1. Plasmid DNA (irradiated and nonirradiated) is spread onto carbon-coated nickel grids in solution containing 4 μL of BAC (diluted 1:50 in 0.5 *M* ammonium acetate; *see* **Note 13**), 4 μL of ammonium acetate (0.5 *M*) and 2 μL of DNA solution (20–30 ng/ μL) as described in **Subheading 3.3.1.** up to **Step 6** (before staining).
2. To detect biotinylated dUTP in repair patches by EM, the grids need to be labeled with streptavidin conjugated to colloidal gold (Sigma, 10 nm size) by soaking a grid for 2 h at room temperature in 1:100 dilution of stock with 0.1 *M* phosphate-buffered saline, pH 7.3.
3. The grids are then washed three times with phosphate-buffered saline, pH 7.3 and two times with distilled water.
4. The DNA molecules need to be stained lightly by incubating the grids on a drop of a 0.02% (w/v) uranyl acetate solution in 30% (v/V) ethanol for 2 min.

5 . Air dry the grids and examine by dark-field or normal bright-field imaging in a electron microscope.

3.5. Examples

The following examples are described to demonstrate the usefulness of the above techniques for EM visualization (*see* **Figs. 1**, **2** and **3**).

3.5.1. Electron Microscopic Analysis of DNA Replication

EM is particularly suitable to study the replication process because it allows observation of individual replicate intermediates at various consecutive stages *(17)*.

The idea behind the project is to look for origins of replication that cease expansion after several hundred base pairs of DNA have been synthesized. In collaboration with Hiro Mahbubani and Julian Blow, the author has tried to test: a) whether microbubbles really exit under spreading conditions that are not nondenaturing and b) whether their appearance is confined to a part of *S*-phase. The spreading method was performed in the absence of formamide or any other agent that might artefactually denature the DNA. He has used the method described in **Subheading 3.3.** to examine directly the initiation of DNA replication on plasmid DNA molecules microinjected into Xenopus eggs. Most plasmid molecules were simply linear, 0.5% of DNA molecules contained a bubble (*see* **Note 14**) consistent with their being a replicative intermediate. Only a single replication bubble is *seen* on each molecule. The maximum bubble size per DNA molecule observed by EM. was over 1 kb (*see* **Figs. 4** and **5**). These results are consistent with the "Jesuit" model of DNA initiation.

3.5.2. Electron Microscopy of DNA Excision Repair Patches Produced by Human Cell Extract

To investigate the possibility of a human DNA repair process on a defined substrate in vitro, in collaboration with David Szymkowski and Rick Wood [*see* *(15)*], we have directly examined the size and distribution of individual repair patches in UV-irradiated plasmids by using EM to detect repair synthesis produced by human cell free extracts. To examine individual repair patches formed in human cell extracts, we have modified this cell-free system to detect repair synthesis in damaged plasmid DNA using EM.

In this type of the work, it is possible to show that bio-11-dUTP is incorporated into UV-irradiated plasmid DNA by repair-proficient human cell extracts, and that individual repair event can be detected by staining with streptavidin/colloidal gold.

To characterize the process by which the mammalian nucleotide excision repair complex interact with DNA to recognize and repair lesions, we have investigated the size and distribution of repair patches induced by human cell

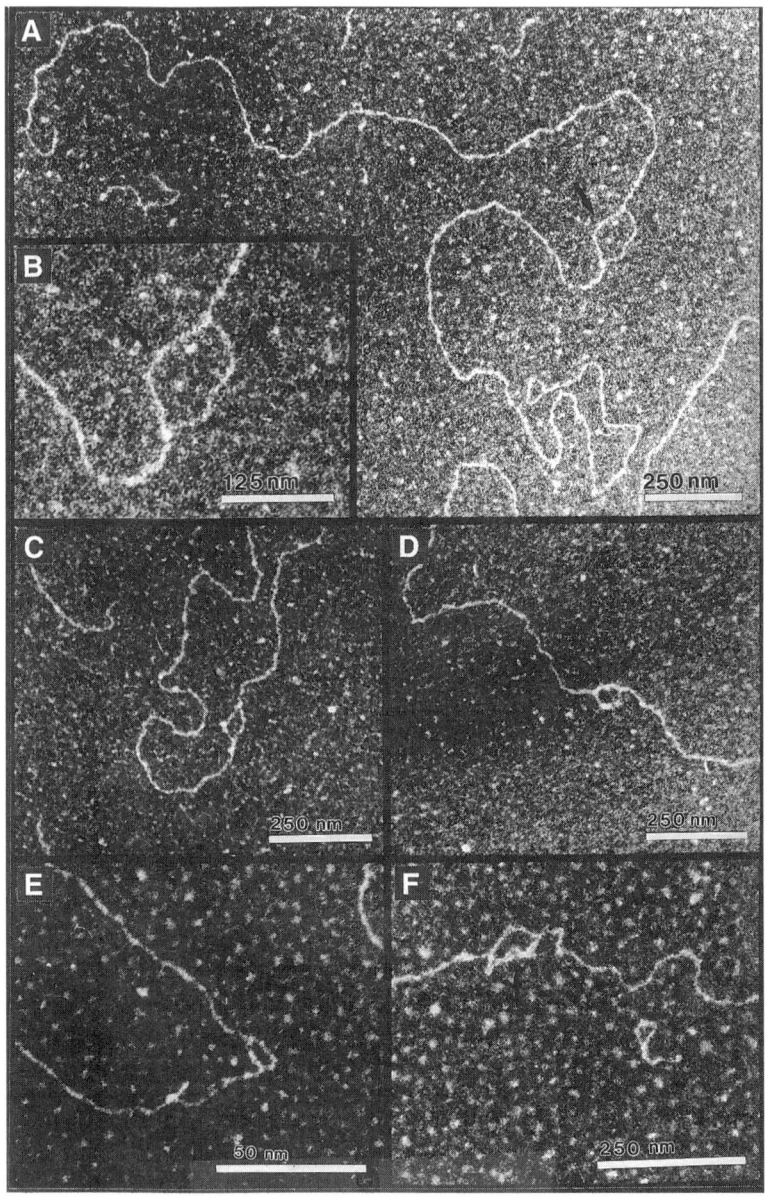

Fig. 4. EM of pXl01A molecules replicating in Xenopus egg extract. pXl01A
replicating synchronously Xenopus extract was isolated, linearized, and prepared
for EM as described in **Subheading 3.3.** (**A**). A complete molecule of pXIr101A
containing bubble structure. (**B**) Magnification of bubble structure in (**A**). (**C–F**)
Details of different bubble structures.

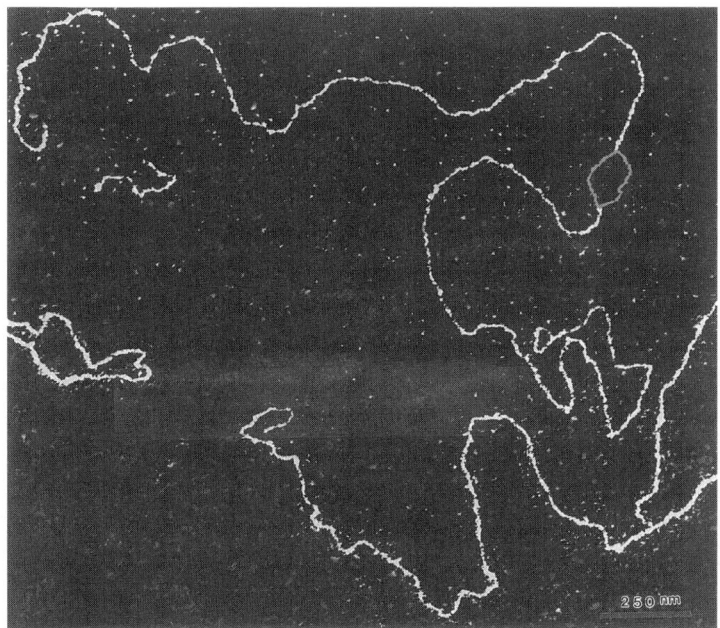

Fig. 5. Purified DNA replication in a cell free extract from Xenopus eggs. pXIrl01A, 1 15 kb plasmid, containing a single copy of the Xenopus ribosomal DNA gene repeat, was incubated in Xenopus egg extract DNA. DNA added to this extract is replicated under normal cell cycle control. Midway through the replication reaction, DNA was isolated, linearised and prepared for EM as described in **Subheading 3.3.1.** Two-dimensional gel analysis suggests that each plasmid molecule is replicated by a pair of replication forks initiated at a single site on the DNA. (This is expected to give rise to a single replication bubble on each replicating molecule.)

extracts in UV light-irradiated plasmid DNA. Repair synthesis was carried out in a buffer substituting biotinylated dUTP for dTTP, to allow repair patches to be detected by EM after streptavidin/colloidal gold labeling. Individual repair events on circular plasmids that had undergone repair synthesis in cell extracts were scored as gold particles bound specifically to irradiated molecules (*see* **Fig. 6**). Samples of over 2000 irradiated and unirradiated plasmids were counted. Repair synthesis at UV light photoproducts typically replaced about 30 nucleotides, because 69% of patches contained only one particle of 10 nm gold and 24% of patches contained two gold particles (each covering approx 29 nucleotides). In addition, the ordering of repair events among damaged plasmids closely fitted a Poisson distribution indicating that repair of lesions is achieved via a nonprocessive, random diffusion mechanism. This suggests that the repair complex is not intrinsically processive.

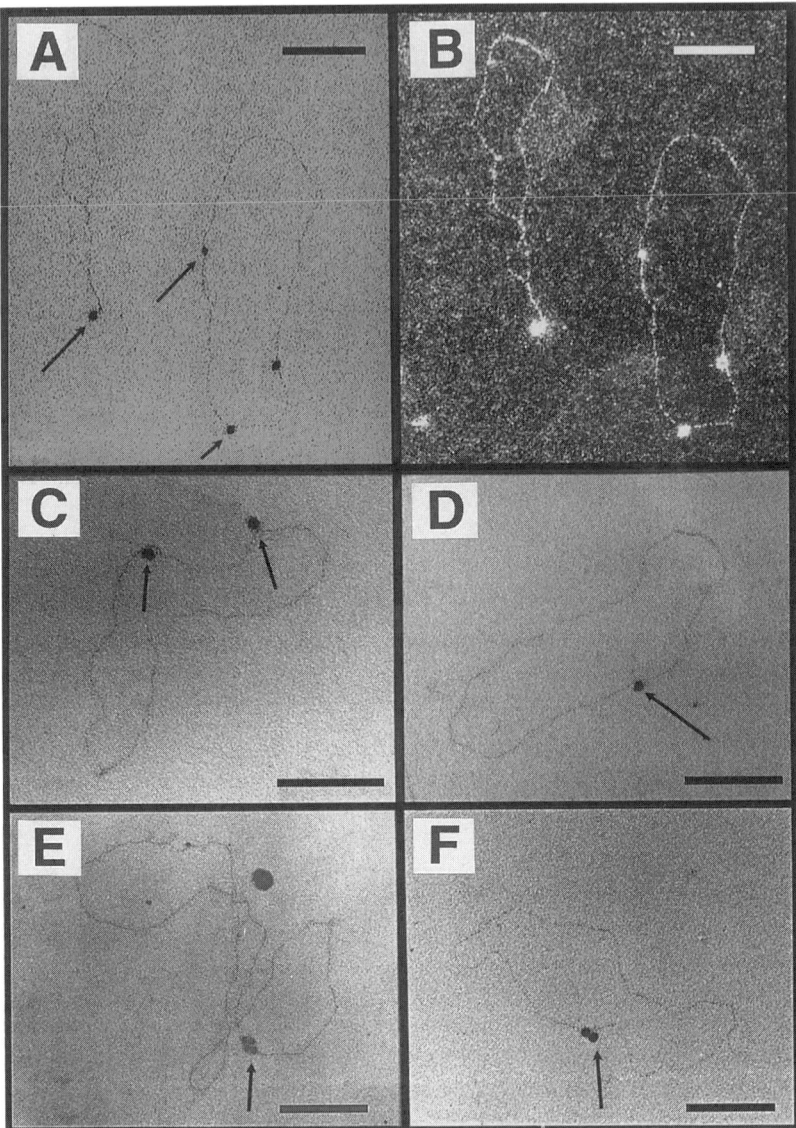

Fig. 6. EM is used to identify repair patches in irradiated pBluescript KS+. The UV-irradiated plasmid was repaired in the presence of bio-11-dUTP, extensively purified and prepared for EM as described in **Subheading 3.5.2.** Bright-field (**A**) and dark-field (**B**) views of an irradiated plasmid DNA containing one repair patch. (**C**) plasmid with two repair events. (**D**) Plasmid containing one repair patch incorporating one gold particle. (**E**) Unusual plasmid containing a repair patch binding four gold particles. (**F**) Plasmid with one repair patch containing two gold particles. Scale bar = 100 nm.

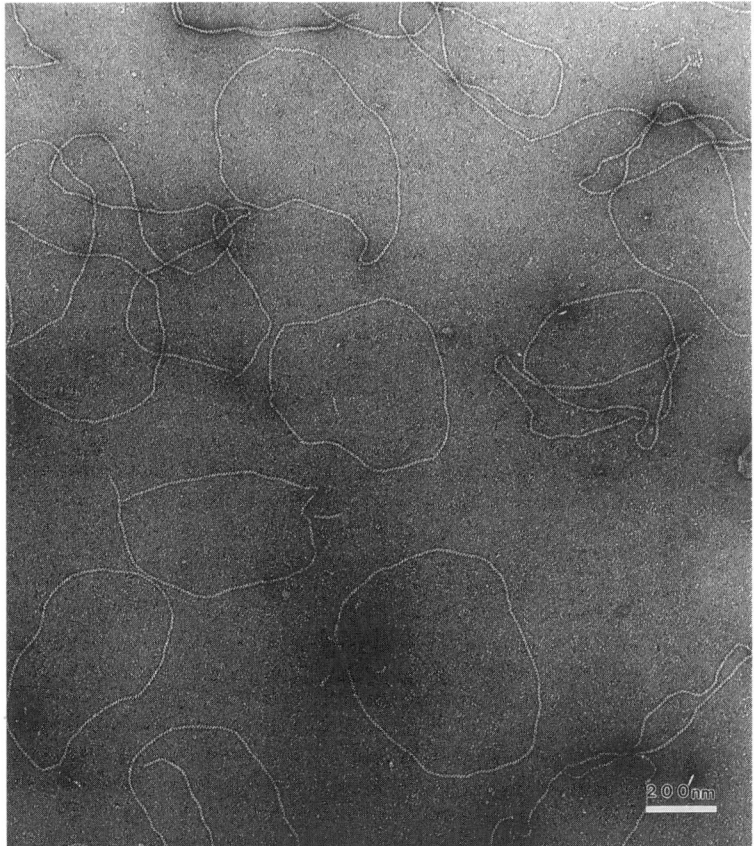

Fig. 7. Visualization of recombination reactions. RecA-DNA complexes formed in the presence of ATP.

3.5.3. EM Visualization of Binding Protein to Nucleic Acid

EM is one of the several powerful methods for studying recombination reactions, as has been shown in the analysis of in vitro recombination reactions promoted by the *E. coli* RecA protein *(23, 24)*. This method allows one to study the low-resolution molecular structure of DNA-protein complexes involved in DNA recombination *(23–26)*. We can observe different states of DNA and protein components [*see (24* and *26)* for details] and electron micrographs taken during sequential stages of recombination reaction allow for an understanding of the multistage processes involved [*see (23)* for review) and will contribute to the eventual molecular models of DNA recombination which will emerge in the future.

Figs. 7 and **8** show RecA-dsDNA complexes formed in the present of ATP and prepared for EM visualization as described in **Subheading 3.4.**

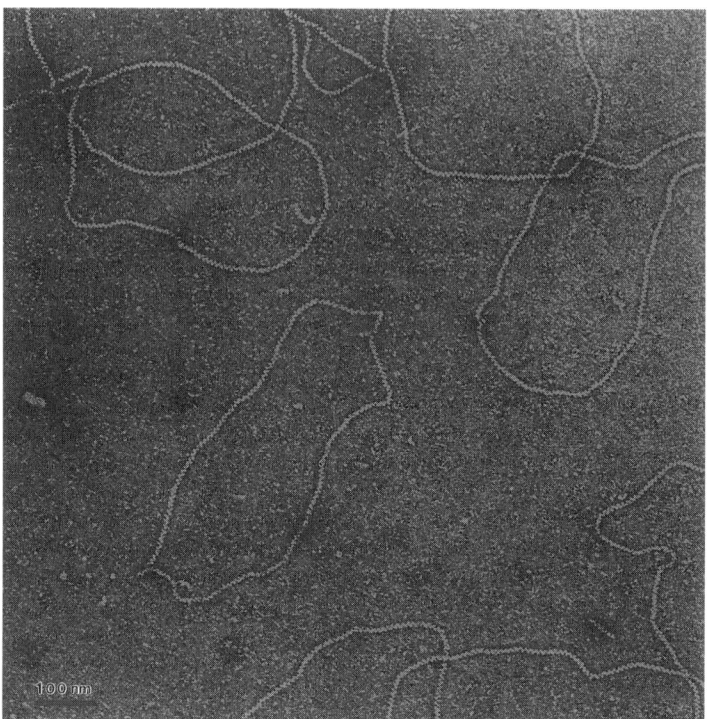

Fig. 8. High magnification of RecA-DNA complex on negatively stained preparation as described in **Subheading 3.5.**, illustrating a zigzag-shaped substructure, which is generally believed to be because of a helical distribution of RecA protein on DNA molecule.

The complexes reveal a zigzag-shaped substructure that is generally believed to be because of a helical distribution of RecA protein.

4. Notes

1. Insufficient brightness in dark field can be easily overcome by using high accelerating voltage in the electron microscope; unlike bright field, higher contrast can be achieved with 100–120 KV of an accelerating voltage.
2. Indirect carbon coating on the grid is recommended for visualization of DNA molecules by EM because the indirect deposition of carbon yields a film that is composed of particles that are more stable in the beam of the electron microscope.
3. It is necessary to have very thin carbon film because the grids carrying a thin carbon support film (less than 5 nm thickness) give lower background grain and better contrast.
4. For photography in dark field, it is necessary to have longer exposure (32 s) in comparison with bright-field (2 s exposure time) microscopy in order to get good quality pictures.

5. In comparison with cytochrome-c method, benzyldimethylalkylammonium chloride does not lead to any apparent thickening of the nucleic acid strands. Partially denatured DNA spread with this reagent shows a loosened structure with a foamy appearance in the regions previously considered to be "unmelted," which open up locally into melted loops of different size. Specifically bound proteins, such as RNA polymerase on bacteriophage T7 DNA, can be detected unambiguously *(7)* by this component.

6. A few examples of spreading mixtures are presented in **Subheading 3.2.** Many alterations of the protocol are possible. For example, it is possible to add a step before spreading in which DNA is partially denatured (either by an alkaline treatment or by rising in temperature). The partial denaturation can be done irreversibly; then spreading is carried out at room temperature.

7. Increasing the ammonium acetate concentration increases contrast, but if the concentration in the hypophase is above 0.5 M, little or no DNA sticks to the films.

8. Water should be purified until the required quality is obtained, e.g., by triple distillation in a quartz still or by reverse osmosis. You can test the quality of water by shaking it in a glass bottle. Surface bubbles should disappear within a few seconds.

9. The more concentrated stock solution of uranyl acetate generally gives a more highly contrasted DNA, but also more uranyl oxide precipitate in the background. The contrast can be increased by rotary-shadowing as described in *(12)* with heavy metal, but is not recommended when single-stranded DNA are present. Excellent contrast can be obtained with the aqueous technique by staining alone.

10. It has been reported in *(21)* when the droplet method has been used to visualize the linearized plasmid DNA (pCU 918), a significant fraction of molecules (30–40%) had been induced to circularized artificially. Lang and Mitani *(20)* found that because of the small surface droplet, spreading of the protein becomes exceedingly difficult to control and it may serve to promote such artificial circularization. Therefore, they recommended that when the droplet method is used, formaldeyhde (0.004 to 0.11 M) is essential to prevent the undesirable orientation of DNA molecules.

11. Cytochrome-c method has been proven enormously useful and popular for observation of nucleic acids, but it cannot be satisfactorily used for association of nucleic acids with protein molecules, such as enzyme, because of the obscuring effect of the bound cytochrom-c on the protein.

12. Diffusion method is the best method when dealing with very long DNA molecules or with nucleic acid preparation with low DNA concentration in spreading solution.

13. The BAC technique has the advantage over other protein-free spreading techniques in that single- and double-stranded nucleic acids and nucleic acid-protein complexes can be visualized. In certain laboratories, the BAC technique has been found to be difficult to apply. However, some of the these difficulties are because of unfavorable surface properties of the carbon-support films, which do not bind the DNA-BAC complex. As described in this chapter a simple way to overcome this problem is to pretreat the carbon-support film with ethidium bromide prior to absorption of the DNA as described in **Subheading 3.3.1.**

14. In DNA replication studies, it is essential to be careful when you are looking at a replication bubble, the size of the two arms of the replicate bubble should be the same. If the arm of the replicate bubble are not the same it could be artefact (for example, DNA is simply twisted).

Acknowledgments

The author would like to thank all those who contributed material used in this chapter; in particular, Rick Wood, Steve West, Julian Blow, David Szymkowski, Hiro Mahbubani, and Peter Baumann from ICRF. The author thank Professor Kleinschmidt for teaching this method and passing on new ideas for improvements in this field. The author also like to thank Ken Blight and Michael Gorn for preparing carbon-coated grids, and Carol Upton for critically reading this manuscript.

References

1. Kleinschmidt, A. K. (1968) Monolayer technique in electron microscopy of nucleic acid molecules in *Method in Enzymology* (Grossman, L. and Moldave, K., eds.), vol. 12B, Academic, New York, pp. 361–376.
2. Davis, R. W., Simon, M., and Davidson, N. (1971) Electron microscope heteroduplex methods for mapping regions of base sequence homology in nucleic acids, in *Methods in Enzymology*, vol. 21, (Grossman, L. and Moldave, K., eds.), Academic, New York, pp. 413–428.
3. Coggins, L. W. (1987) Preparation of nucleic acids for electron microscopy, in *Electron Microscopy in Molecular Biology* (Sommerville, J. and Scheer, U., eds.), IRL Press, Oxford, England, pp. 1–29.
4. Poon, N. H. and Seligy, V. L. (1998) Fine structure of mononucleosomes derived by bright and dark field electron microscopy. Experimental dark field electron microscopy. *Experiment. Cell Res.* **125,** 313–331.
5. Murcia, F., Bilen., J. J., Ittel, M. E., Mandel, P., and Delain, E. (1983) Poly (ADP-ribose) polymerase auto-modification and interaction with DNA: electron microscopic visualization. *The EMBO J.* **2,** 543–548.
6. Grifith, J. and Christiansen, G. (1978) Electron microscopic visualisation of chromatin and other DNA-protein complexes. *Annu. Res. Biophys. Biophs2.* **7,** 19–35.
7. Vollenweider, H., Sogo, J., and Koller, T. (1975) A routine method for protein-free spreading of double and single-strand nucleic acid molecules. *Proc. Natl. Acad. Sci. USA.* **72,** 83–87.
8. Grifith, J. (1978) DNA structure: evidence from electron microscopy. *Science* **201,** 525–527.
9. Wessel, R., Muller, H., and Hoffmann-Berling, H. (1990) Electron microscopy of DNA helicase I complexes in the act of strand separation. *Eur.J.Biochem.* **189,** 277–285.
10. Ruben, G., Spielman, P., Tu., C. D., Jay, E. Siegel, B., and Wu, R. (1977) Relaxed circular SV40 DNA as cleavage intermediate of two restriction endonucleases. *Nuclei Acids Res.* **4 ,** 1803–1813.

11. Bazett-Jones, D. P. and Ottensmeyer, A. (1982) DNA organization in nucleosomes. *Can. J. Biochem. V* **60**, 364–370.
12. Wessel, R., Ramsperger, U., Stahl, H., and Kinippers, R. (1992) The interaction of SV40 Large T antigen with unspecific double-stranded DNA: an electron microscopy study. *Virology* **189**, 293–303.
13. Hajibagheri, M. A. N, Mahbubani, H. M., Blow, J. J., Szymkowski, D., Blight, J. K., and Wood, R. D. (1994) Dark-field microscopic visualization of DNA/protein interaction, DNA replication and detection of DNA repairs patches induced by human cell extract in UV irradiated plasmid DNA. *ICEM13* **V3**, 4–6.
14. Revert, B., Malinge, J. M., LeBret, M., and Leng., M. (1984) Electron microscopic measurement of chain flexibility of poly (dG-dC), poly 9dG-dC0 modified by cis–diamminedichloroplatinum (II). *Nucleic Acids Res.* **12**, 8349–8362.
15. Szymkowski, D. E., Hajibagheri, M. A., and Wood R. D. (1993). Electron microscopy of DNA excision repair patches produced by human cell extracts. *J. Molec. Biol.* **231(2)**, 251–60.
16. Dubochet, J., Ducommun, M., Zollinger M., and Kellenberger, E. (1971) A new preparation method for dark-field electron microscopy of biomacromolecules. *J. Ultrastruct. Res.* **35(1)**, 147–67.
17. Preiser, P. R., Wilson, R. J., Moore, P. W., McCready, S., Hajibagheri, M. A., Blight, K. J. (1996) Recombination associated with replication of malarial mitochondrial DNA. *EMBO J.* **15(3)**, 684–693.
18. Martinez, J., Hewlett, J., Pettersson, R. F., and Baltimore, D. (1977) Circular forms of Uukuniemi Virion RNA: an electron microscopy study. *J. Virol.* **21**, 1085–1093.
19. Glass, J. and Wertz, G. W. (1980). Different base per unit length ratios exist in single-stranded RNA and single stranded DNA. *Nucleic Acids Res.* **23**, 5739–5751.
20. Lang, D. and Mitani, M. (1970) Simplified quantitative electron microscopy of biopolymers. *Biopolymers* **9**, 373–379.
21. Banerjee, S. K. and Iyer, N. V. (1995) Quick and easy spreading techniques for electron microscopy of DNA. *Biotechniques* **18**, 946–947.
22. Burkardt, H. J. and Puhler, A. (1988) Electron microscopy of plasmid DNA. *Methods Microbiol.* **21**, 155–177.
23. Stasiak, A. and Egelman, E. H. (1988) Visualization of recombination reactions, in *Genetic Recombination.* (Kucherlapati, R. and Smith, G. R., eds.), American Society for Microbiology, Washington, pp. 265–308.
24. Parsons, C. A., Stasiak, A., Bennett, R. J., and West, S. C., (1995) Structure of a multisubunit complex that promotes DNA branch migration. *Nature* **374(6520)**, 375–378.
25. Stasiak, A., Egelman, E. H. (1994) Structure and function of RecA-DNA complexes. *Experientia* **50(3)**, 192–203.
26. Dyck, E. V., Hajibagheri, N. M. A., Stasiak, A., and West, S. C. (1998) Electron microscopic visualization of human Rad52 protein and its complexed with Rad51 and DNA. *J. Molec. Biol.* (in press).

14

Applications of Electron Microscopy for Studying Protein-DNA Complexes

Maria Schnos and Ross B. Inman

1. Introduction

Electron microscopic (EM) studies of DNA-protein complexes have led to many insights into biological reactions. There are three general methods that can be used to great advantage in such studies, each gives a different type of information, but each has its own disadvantages. In this chapter, we will discuss these methods and show examples of the usefulness of each.

The first method, cytochrome-c spreading, allows DNA to be visualized after it is adsorbed to a monolayer of unfolded and partially denatured cytochrome-c protein. One of the main advantages of this procedure is that one can accurately *see* what type of DNA structures are present in a deproteinized reaction mixture. By its very nature, this method will tend to denature proteins and therefore double-stranded and single-stranded DNA can be visualized quite well even in relatively crude mixtures containing unreacted proteins. Additionally, DNA-protein complexes can be studied if the protein is first stabilized by crosslinking. The method is often used to compare DNA structures present both before and after removal of bound proteins. If the procedure is carried out as described in **Subheading 1.1.**, negatively supercoiled DNA circles (which normally are more difficult to *see* because of the coiling) are clearly visualized. This results from the conversion of the supercoils into small unwound or denatured areas within the DNA; consequently, the DNA is relaxed and unusual DNA structures such as branch points and bound proteins, can be more easily observed.

A second very general method simply consists of the direct adsorption of DNA-protein complexes onto a thin carbon film. The film can be activated in a variety of ways to increase the efficiency of adsorption. This method has the great advantage that there is generally very much less disruptive force applied

From: *Methods in Molecular Biology, vol. 117: Electron Microscopy Methods and Protocols*
Edited by: N. Hajibagheri © Humana Press Inc., Totowa, NJ

to the complexes during adsorption and drying compared to the cytochrome-c procedure. For instance, the sample can be adsorbed and dried in buffers containing up to 20% glycerin, which can greatly stabilize delicate structures. The main disadvantage of direct adsorption is that nonvolatile buffer residues are often difficult to reduce to acceptable levels, and can therefore result in rough sample background. This can become a problem if preservation of protein structures requires high salt concentrations. A further disadvantage of this method is that single-stranded DNA is not as clearly resolved compared with the cytochrome-c method; one reason for this arises from the more "natural" collapsed single-stranded DNA configuration that results from direct adsorption. As with the cytochrome-c method, complexes can be further stabilized by crosslinking.

The final method is an extension of either of the above methods to allow proteins to be specifically labeled with antibodies conjugated to gold particles. This method has the obvious advantage that it allows the unambiguous identification of the proteins involved in a DNA complex. Furthermore, if the protein of interest is too small for direct visualization, the gold spheres show approximate positions. The disadvantage of the gold labeling technique is that resolution of DNA-protein complexes is usually inferior to the above two methods.

The electron micrographs shown in this chapter, along with many more, can be viewed at a web page entitled DNA, DNA-protein complexes and virus at: http://phage.bocklabs.wisc.edu.

1.1. Cytochrome-c Spreading Method

The principle of this method is that a DNA solution, mixed with cytochrome-c, is slowly introduced onto a liquid surface. Because of the air-water interfacial tension, cytochrome-c tends to unfold or denature and remains at the interface. DNA that is bound to the protein therefore also remains at the interface. Ideally, this process occurs at protein concentrations that result in molecules well separated on the water surface; such a final surface concentration gives the most efficient cytochrome-c unfolding. The surface is now compressed to a point where the unfolded protein molecules just touch each other and form a continuous monolayer. This monolayer, along with the bound DNA, is then picked up on a thin carbon or plastic film and shadowed with a heavy metal for electron microscopy. This method, as originally introduced in 1962 (*1*), yields excellent samples, but requires specialized apparatus and large quantities of material. The procedure has been modified by several investigators and one such method, developed in this laboratory, is described below. The main difference from the ideal spreading procedure described above is that the monolayer is formed on a small drop of water rather than on a large Langmuir trough. The concentration of added cytochrome-c and the exposed area of the drop are arranged so that the final surface approximates a monolayer of cytochrome-c. Complete details are given in *(2)*.

2. Materials
2.1. Materials Needed for Cytochrome-c Spreading

1. Spreading block: White teflon, about 6 in × 6 in × 1/8 in is screwed to an aluminium block and several shallow wells are milled into the teflon, each approx 0.7 in diam. × 0.1 in deep. After each use the teflon surface should be scrubbed with detergent followed by distilled water, ethyl alcohol, and distilled water rinses.
2. Spreading rods: Stainless steel rods (5 in × 0.25 in diam.) with a 0.6 in tapered end are polished smooth (avoid polishing compounds that contain silicone). These rods should be cleaned with detergent, rinsed and stored in water after each use.
3. Carbon films: Freshly cleaved mica sheets are placed on clean paper, cleaved side down, and fixed along two edges with adhesive tape. Many disks (about 0.3 in diam.) are now punched out from these sheets using a punch and dye machine. After each disk has been blown clean, the freshly cleaved side is vertically shadowed with carbon to form a layer easily visible to the naked eye and thick enough to resist the various manipulations described below (this requires the evaporation of about 0.25 in of 0.04 in diam. carbon at a distance of 5 in). These carbon coated mica disks can be stored for years in a desiccator containing silica gel. Before used disks are prepared by scratching away a very small ring of carbon from the periphery. Mica dust should then be removed with a burst of air. The freshly exposed hydrophilic mica ring helps keep the liquid drop in place (*see* **Subheading 3.**).

3. Methods
3.1. Cytochrome-c Spreading Procedure

1. 1 mL of quartz distilled water is placed in a well on the spreading block.
2. The following buffer is prepared: 32 µL 1 M Na_2CO_3, 40 µL 0.126 M Na_2EDTA, 200 µL distilled H_2O, 200 µL 37% HCHO, 292 µL 100% formamide and 20 µL 1 M HCl.
3. To 5 µL of the above buffer is added 7 µL of DNA or DNA-protein complex (DNA concentration 4 mg/mL).
4. Add 2 µL of 0.1% cytochrome-c in water.
5. 3 µL of this solution is now taken up in an Eppendorf pipet. A spreading rod is primed with about 0.5 µL of the above solution. This priming extends along a line from the pointed tip to about 3/4 length of the tapered section. The rod is now inserted into the periphery of the water drop at an angle of about 45°, until about one-half of the tapered section is submerged. The primed region must be kept at the top most part of the rod. The remaining portion of the liquid is now expelled at the top of the primed region and allowed to run down the rod onto the drop. The rod is then gently withdrawn from the drop. After 1–5 min a sample is picked up

by touching a mica disk (carbon side down) to the drop. The disk is removed, along with a pendant drop of liquid and immediately submerged in ethyl alcohol, washed with further ethyl alcohol and then dried under a stream of warm nitrogen gas. The disk should be kept horizontal until it is submerged in the alcohol and should then be held at an angle of 45° (carbon side up) during drying. The mica disk, containing the sample, is then shadowed with platinum, and finally, the carbon film is floated off at a clean air/water interface and picked up from below on 200 mesh copper EM grids. This method has the following advantages. As little as 6 ng DNA is needed for a single spreading. The carbon film is held flat, by the mica, thus resulting in a constant shadow-height angle during shadowing. The spreading buffer can be adjusted so that double stranded DNA remains native or is partially or fully denatured. As a rough guide, if 20 μL of 5 M NaOH is added, instead of the HCl indicated above, then low and high degrees of partial denaturation occur when the spreading mixture (before adding cytochrome-c) is heated at 50° for 10 or 40 m, respectively. The exact conditions will depend on the A+T content of the DNA and often will need to be determined by trial and error.

3.2. Direct Adsorption Method

DNA-protein complexes can be directly adsorbed to carbon films attached to 400 mesh copper EM grids as described below. In some cases it is advantageous to "activate" the carbon film by pretreatment with alcian blue or various other chemicals. Alcian blue (0.2% in 3% acetic acid) is diluted 100-fold with water just before use. The grid is simply floated on a small drop of the Alcian blue (0.2% in 3% acetic acid) dye for 5–10 m, touched to a water drop and floated on a second water drop for 5–10 m followed by withdrawal of the liquid with a capillary pipet and drying under a heat lamp.

3.3. Materials Needed for Direct Adsorption

1. Alcian blue (0.2% in 3% acetic acid) is diluted 100-fold with water just before use.
2. 100 mM ammonium acetate.
3. 5% uranyl acetate. To prepare the stain, a mixture of 5% uranyl acetate in water is left for several days, most of the liquid is then decanted and clarified through a 0.22-μm Millipore filter and stored in a dark bottle.

3.4. Direct Adsorption Procedure

1. About 7 μL of the DNA-protein complex (DNA concentration 0.4 μg/mL) is placed on the carbon film for 1–3 m (the grid is held with locking forceps).
2. Grids are then floated on water or 100 mM ammonium acetate or any necessary buffer for 1–5 m and stained with uranyl acetate.
3. The grids are touched to a drop of the 5% uranyl acetate and floated on another for 20–30 s, touched briefly to 5 drops of water or 100 mM ammonium acetate and finally liquid is withdrawn from the grids (held at about 45°) with a capillary pipet.

4. The samples can then be shadowed with platinum if further contrast enhancement is needed.

There is a reasonable degree of flexibility in the buffers that can be used. Most parts of the procedure will work even in the presence of 15% glycerol. However, the uranyl acetate staining step can only tolerate 5% glycerol and even this will result in precipitation unless the stain is used within a few hours.

3.5. Gold Labeling of Specific Proteins in DNA-Protein Complexes

The method previously employed to gold-label proteins at the molecular level uses the following strategy. A reaction mixture containing the DNA-protein complex is prepared and antibody to the protein (raised in rabbit) is added followed by protein A-gold. The final result is that gold spheres are now coupled with the protein and due to the high contrast of the gold, their location is unmistakable. The experiment can also be performed in other ways; for instance, the protein A-gold conjugate could be reacted with the antibody, which in turn could be mixed with the protein before the DNA complex is formed. Whichever way is chosen, the procedure as outlined above requires compromises in the optimal concentrations of the protein of interest, the antibodies, and the protein A-gold because of the persistent aggregating tendency of the antibodies in these reactions. Although concentrations can usually be found that work around this problem, it is a time consuming process to find these conditions and, in our hands, different proteins require different compromises in these concentrations.

As described above, the conventional gold tagging reaction is performed in true solution and as a result there is always a likelihood of aggregation. The important modification, described below, avoids this problem by simply performing the reactions on the protein-DNA complex while it is immobilized on the carbon film used to support the molecules during visualization in the electron microscope. In brief, the reaction mixture is prepared in exactly the same way as if the sample is to be stained or shadowed for EM, but once the sample is adsorbed onto a carbon film it is first gold tagged before staining or shadowing by simply floating the grids on drops of the various reagents. Because the DNA-protein complex is immobilized on a surface before the addition of the antibody, the aggregation problem is completely avoided.

3.6. Gold Labeling of Specific Proteins in DNA-Protein Complexes

The procedure can be carried out in exactly the same way as described above in **Subheadings 3.1.** up to the step preceding shadowing, or in **Subheading**

3.4., up to the step preceding staining. These samples can be treated directly, as described below, or the samples can be dried and stored in a desiccator for use at a later time. In the following steps, the samples are transferred between the drops as quickly as possible to minimize drying out until the final wash.

Step 1: Reaction of the protein in the DNA-protein complex with antibody. Float the grid containing the sample on a 50 μL drop of 0.1% BSA in PBS (150 mM NaCl, 8.3 mM Na$_2$HPO$_4$, 1.85 mM NaH$_2$PO$_4$) for 5 min. Place the grid on a 20 μL drop of a purified antibody solution (diluted 1:5000 in 0.1% BSA, obviously, this dilution will depend on the actual antibody concentration) and shake (*see* **Notes**) for 20 min. The polyclonal antibody should be raised in rabbit and normally contains about 2 mg/mL total antibody; with probably only a small proportion of this being directed against the protein of interest.

Step 2: Attach the protein A-gold conjugate to the bound antibody. Float the grid on a 50 μL drop of 0.1% BSA in PBS for 3 min as before. Float and shake (*see* **Notes**) on a 20 μL drop of protein A-gold diluted 1:20 in 0.1% BSA in PBS for 20 min. Float again on a 50 μL drop of 0.1% BSA in PBS for 3 min.

Step 3: Staining (and/or shadowing). Wash the grids on two successive 100 μL water drops for 5 m each. Touch to a 20 μL drop of 5% uranyl acetate for 1 s. Float on a second drop of uranyl acetate for 25 s. Touch the grid to six successive 100 μL drops of water, each for 1 s. Withdraw water from the grid using a capillary pipet. If needed, the sample can now be shadowed in the conventional way.

3.7. Examples of the Use of These Methods

(A) Exploration of the proximity of DNA and protein within Virus or Bacteriophage. A very simple example will first be discussed that enables one to deduce which DNA end is first injected through the tail of a bacteriophage upon infection. The phage is first subjected to very gentle crosslinking conditions and then spread for EM by the cytochrome-c procedure. This denatures the phage head proteins with high efficiency, but leaves the tail predominately intact. Because the DNA end to be first injected is already within, or in close proximity to the proximal opening within the tail, one can expect that this contact will be preserved to some extent and be made more permanent by the crosslinking. EM spreading show the DNA to be released from the phage head and when carried out under the correct crosslinking conditions, a high proportion are found to be connected to the proximal end of the phage tail. An example is shown for bacteriophage Mu in **Fig. 1**. The DNA binding is known to occur at the proximal end of the tail, because in a low proportion of the complexes, remnant head proteins can still be seen, and in these cases, the DNA emerges from the tail at the same end as the remaining head proteins. If the spreading conditions are arranged so that the DNA is partially denatured, as shown in **Fig. 1**, the DNA end attached to the tail can also be deduced from the resulting denaturation pattern. In a variety of phages (λ, Mu, P2, P4, and 186), it is found that the attached ends are always unique and can be expected to represent the DNA end to first enter the bacteria upon infection. Such experiments show that the genetic right ends of λ and P4 are injected first whereas with P2 and 186 the left ends are the first to enter the bacteria *(3)*. In the case of phage Mu, the end known as the "variable end" is injected first *(4)*.

The principle of this type of experiment can be extended to more complex situations. For instance, Reo virus contains 10 double-stranded RNA molecules/infective particle. When experiments, exactly similar to that described above, are performed on this virus, spider-like structures are observed (**Fig. 2**). Partial denaturation allows one to map the ends that are fixed to the residual viral proteins. This method is sufficiently accurate to map 8 of the complement of 10 RNAs and in all cases the attached ends are unique. Other evidence shows that it is this end to which the viral polymerase binds. It appears therefore that within the intact virus, the viral polymerase could already be in place and bound to the correct RNA end to commence production of the viral RNA strands once infection takes place. The two RNA fragments that could not be accurately mapped had in one case a denaturation pattern that was almost symmetrical and in the other there were too few denatured sites for accurate mapping. In both these cases, therefore, it was not possible to determine if an unique end was again bound to the residual viral proteins.

Another complex example involves a polydnavirus from the Australian wasp Microplitus demolitor. Isolated virus contains a variety of head volumes (**Fig. 3**) and the isolated DNA consists of a collection of circles of various sizes. **Fig. 4A** shows a histogram of the circle lengths. About 13 species can be identified between 6.5 and 35.2 Kbp. The most frequent circle size is 13.8 Kbp which accounts for almost 15% of the numbers of circles. There may also be a low proportion of even larger circles. Because of the variable head and DNA size, it is possible that each virus particle contains just one DNA species. If this virus is treated in such a way as to allow the DNA to be released but still held to remnant viral proteins, it should be possible to determine if there is just a single DNA species/virus particle. If the virus is subjected to the same treatment as in the above two examples, the result shown in **Fig. 5** is obtained. **Fig. 4B** shows the same data as **Fig. 4A**, but DNA lengths have been converted to DNA volumes (this calculation is simply the volume expected of a cylinder of solid DNA of 20A diam and of measured length). Similar DNA volumes measured for DNA released from individual virus, such as the three examples shown in **Fig. 5**, are given in **Fig. 4C**, both profiles are roughly similar and support the notion that each viral particle contains just a single DNA species. Finally, **Fig. 4D** shows the calculated volume of the variable DNA-containing cylinder

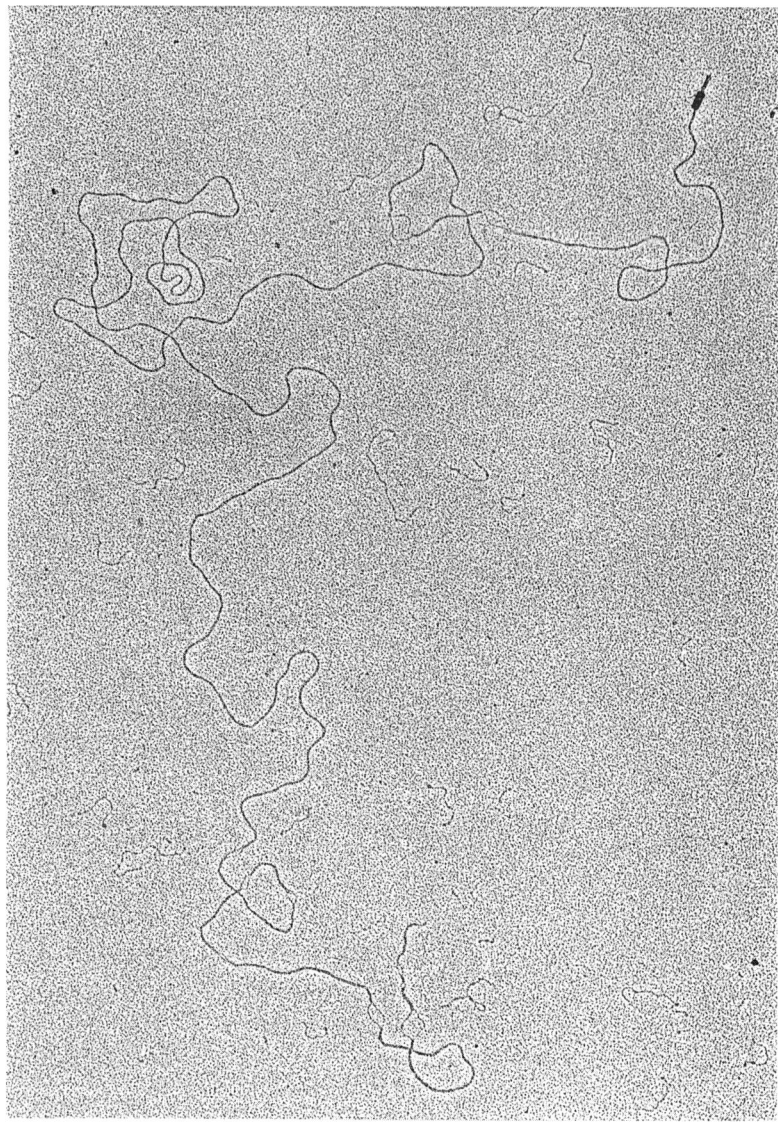

Fig. 1. Demonstration of close proximity of a DNA end and the proximal end of the tail in bacteriophage Mu. The spreading conditions were arranged so that DNA was released from the head and was partially denatured (to allow identification of DNA ends). At the top right in this micrograph can be seen the phage tail bound to a DNA end. Bacteriophage Mu contains about 39 Kbp of DNA. Prepared by the cytochrome-c method.

within intact virus. It should be emphasized that in this particular type of experiment there are the following possibilities of error. A significant fraction

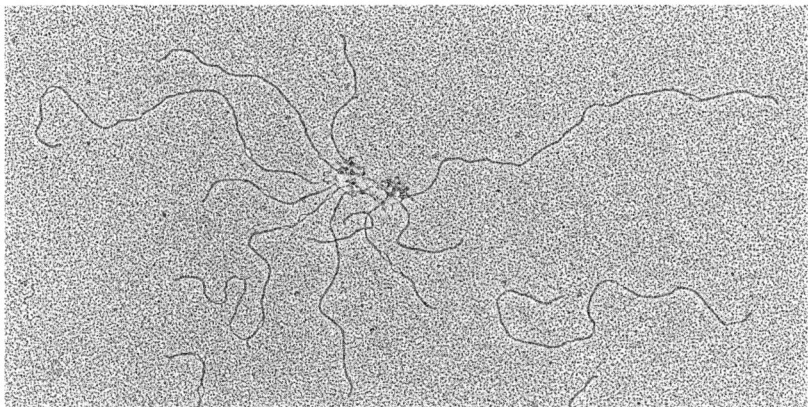

Fig. 2. Reo virus core particles were subjected to similar conditions to that shown for phage Mu in **Fig. 1**. In at least 8 of the 10 double-stranded RNA units contained within this virus, the ends joined to the viral protein remnants are unique as judged by partial denaturation mapping. The shortest and longest Reo RNA species are known to be 1.196 and 3.854 Kbp, respectively. Prepared by the cytochrome-c method.

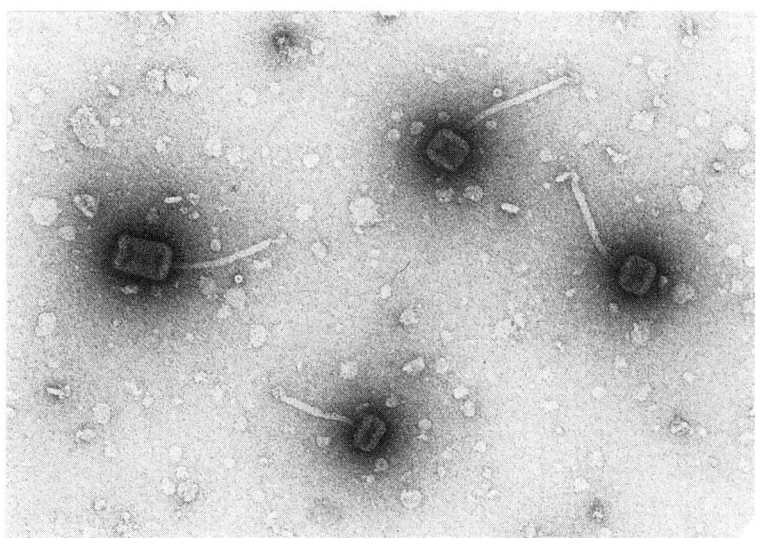

Fig. 3. Wasp virus particles isolated from calyx fluid of Micropilitis demolitor. Individual viral volumes vary widely. The outside diameter is constant at about 480A with an average protein shell width of about 33A and the head height varies from 300 to 790A with a protein shell width of about 115A. Prepared by negative staining.

of such complexes were not measurable because DNA release was incomplete, leading to unresolved DNA, and therefore, a possibility of bias in the results.

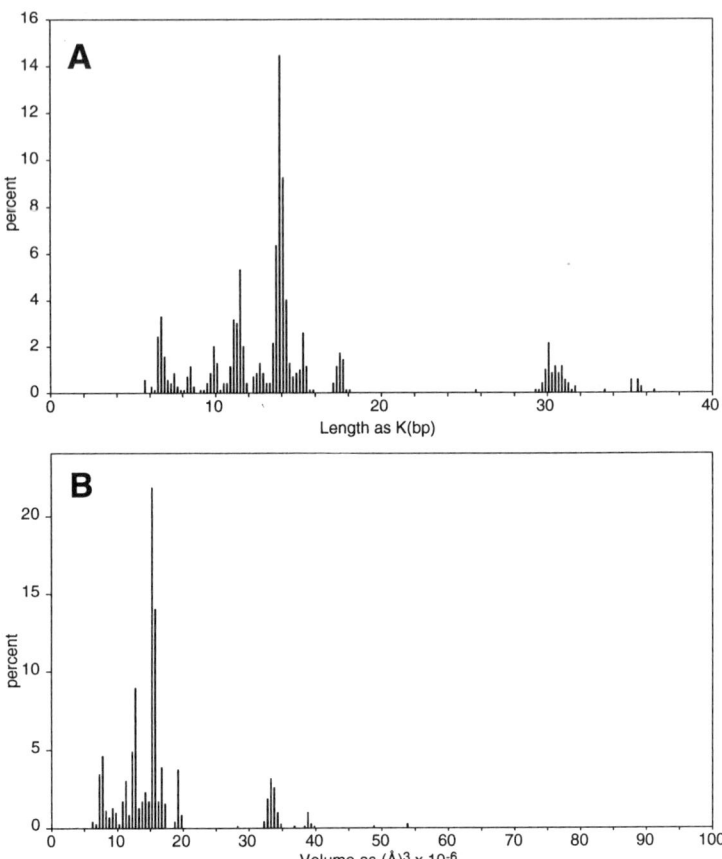

Fig. 4. Length and volume histograms of Wasp virus DNA and virus. (**A**) Length of DNA isolated from virus showing many different length classes of double-stranded circles. (**B**) Same data, but now expressed as DNA volume.

Additionally, it could be argued that some of the released DNA was not retained by the viral proteins and was not therefore measured as part of the viral compliment.

(B) Effect of RecO and RecR on RecA-DNA Filaments *(5)*. RecA protein readily forms filaments on single stranded DNA in the presence of single stranded binding protein (SSB) and an ATP regeneration system. The process apparently initiates at one or more points and the filament propagates in a cooperative manner in a 5' to 3' direction along the DNA. In the case of single-stranded circles, the filamentation is complete and only a very small proportion of molecules are found to contain minor imperfections. **Fig. 6A** shows an

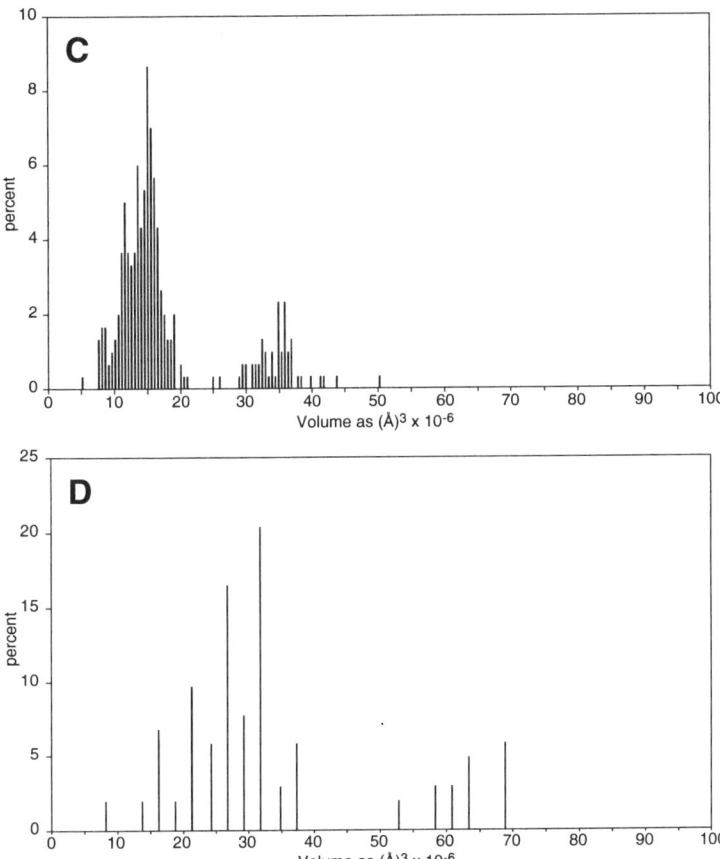

Fig. 4. Length and volume histograms of Wasp virus DNA and virus. (**C**) DNA volumes associated with disrupted virus such as those shown in **Fig. 5**. (**D**) Viral volumes of the intact virus as shown in **Fig. 3**.

example of a completely filamented single-stranded circle. When linear single-stranded molecules are reacted with RecA, filamentation again occurs, but one end of each molecule is generally not coated. As the incubation is continued, the end regions lacking RecA increase in length, **Fig. 6B**. Studies show that the 5' to 3' filament propagation is also accompanied by dissociation of RecA from 5' ends, thus explaining the lack of RecA at 5' ends. **Fig. 6C** shows that the gold labeling technique can be used, with SSB antibody, to show that SSB is now bound to the ends not coated with RecA. When similar experiments are carried out in the presence of RecO and RecR, RecA no longer dissociates from 5' ends. In fact, addition of RecO and RecR after an incubation that would other-

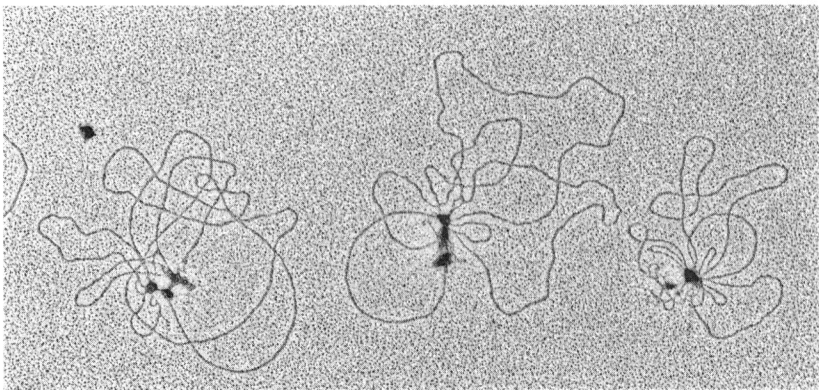

Fig. 5. Wasp virus treated to allow disruption of head protein during spreading. The idea is to try and preserve the DNA compliment of each viral particle in order to determine if each particle contains just one of the many length species present in the total viral population. The DNA lengths are unique and, as shown in **Fig. 4A**, span the range 6.5–35.2Kbp. This example shows three disrupted virus. Prepared by the cytochrome-c method.

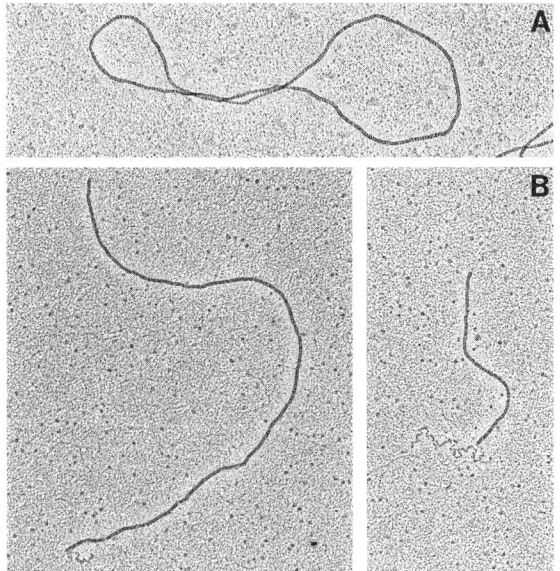

Fig. 6. RecA filaments are stabilized by RecO and RecR. (**A**) Single-stranded DNA circle completely filamented by RecA. (**B**) Under similar conditions, linear single strands are not completely filamented, but rather have one end complexed with SSB. The size of this unfilamented region increases as the reaction proceeds (compare the left and right panels of [**B**]).

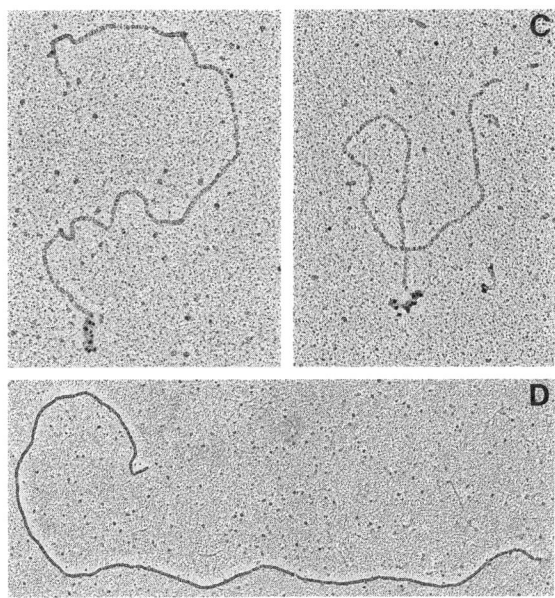

Fig. 6. **(C)** Gold labeling shows that the nonfilamented region contains SSB (dense black spheres at the lower ends of each molecule). **(D)** In the presence of RecO and RecR, complete RecA filamentation is observed. The single-stranded circle (8.266 Kb) used in these experiments, and shown in Fig. 6A, becomes very extended when filamented with RecA; however, the DNA becomes greatly forshortened when bound by SSB. Prepared by the direct adsorption method, stained and shadowed.

wise cause significant RecA dissociation leads to full RecA filaments. **Fig. 6D** shows an example of the full RecA filamentation that occurs in the presence of RecO and RecR.

(C) Effect of RecF and RecR on RecA-DNA filaments *(6)*. RecA quickly filaments single-stranded DNA, but does not readily bind to double-stranded DNA unless the pH is reduced. However, at neutral pH a double-stranded circle containing a single-stranded gap can become fully filamented as shown in **Fig. 7**(A). This occurs because filamentation initiates within the single-stranded region and is then easily able to progate into the emaining double-stranded DNA. In the absence of RecA, the protein RecF binds non-cooperatively to double circles, resulting in a broad distribution of bound protein. As the concentration of RecF is increased, the DNA becomes more density packed with protein (*see* **Fig. 7B**).

In the presense of RecF and RecR, full RecA propagation around gapped circles is prevented and as the concentrations of RecF and RecR are increased the length of the RecA filament is porportionally decreased. The effect is shown in

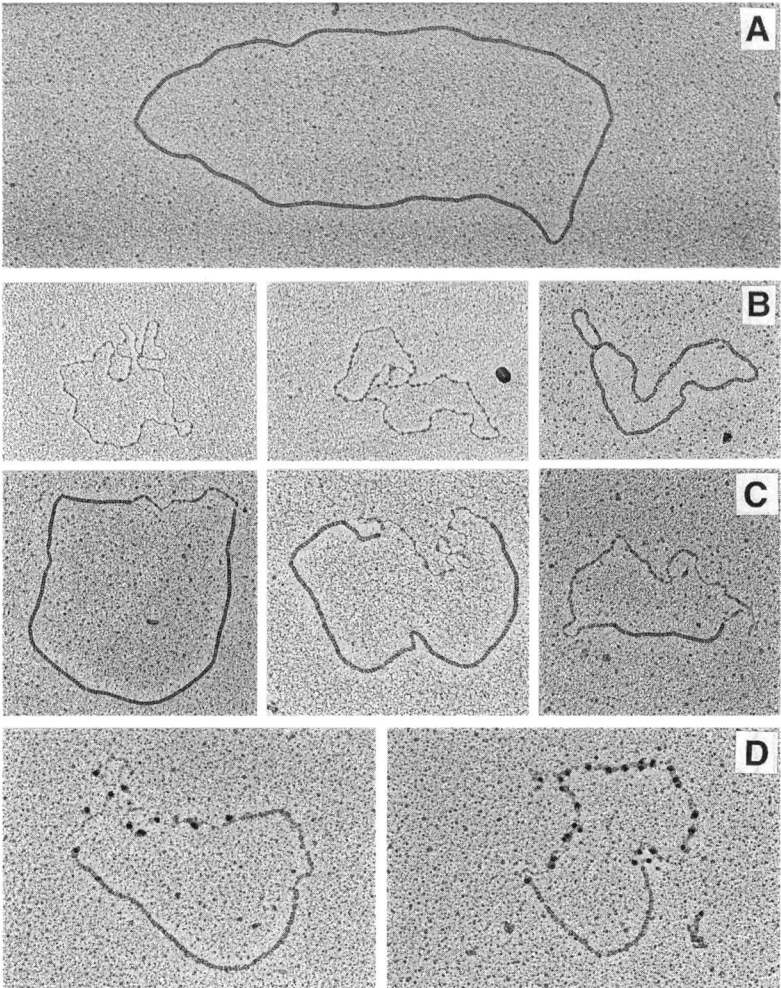

Fig. 7. RecA filamentation is prevented by RecF and RecR. (**A**) Gapped DNA circles are completely filamented by RecA. (**B**) RecF binds to double-stranded circles in a noncooperative manner and at high enough concentrations the DNA is fully covered with an amorphous RecF coating (compare the three panels in [**B**] which show increasing concentrations of RecF protein from left to right). (**C**) The progression of RecA filamentation is prevented by the presence of bound RecF and RecR. In this panel, the concentrations of RecF and RecR increase from left to right, but the range of RecF concentrations does not extend to the same high level as shown in (**B**) above. (**D**) Gold labeling with antibody to RecF (left panel) and RecR (right panel) shows that both proteins are bound in the nonfilamented region. The gapped DNA circles used in these experiments contained a 1.037 Kb single-stranded region and 7.229 Kbp of double-stranded DNA. Filamentation of the double stranded region causes a very large extension (about 1.6-fold) of the double helical DNA length. Prepared by the direct adsorption method, stained, and shadowed.

Fig. 7C. These results are consistent with the prevention of RecA filamentation by the first bound RecF, or RecF-RecR complex encountered by the leading end of the propagating filament. Such an effect may play an important role in the recombination reaction. **Fig. 7D** shows similar reactions to that in **Fig. 7C** except that the structures have been gold labeled, using RecF and RecR antibodies, to show the presence of RecF and RecR respectively on the nonfilamented regions of the complex.

4. Notes

1. Shaking: During the incubation with antibody and with the protein A-gold, we try to have good movement of the liquid across the immobilized DNA-protein complexes. The Teflon block, containing the drop and EM grid, is held against a vibrator which is set to a level which just causes a visible low-frequency vibration of the EM grid with respect to the drop.
2. Crosslinking: The protein-DNA complex can be initially crosslinked with glutaraldehyde or HCHO. However, depending on the particular protein involved, crosslinking may not be needed to stabilize the structure.
3. Antibody: The serum was partially purified by adsorption to a protein A - Sepharose 5B column followed by elution with 100 mM glycine at pH3; 200 µL fractions were collected into tubes containing 20 µL of 1 M Tris (pH 8.0).
4. Protein-A gold: This was purchased from Sigma and contained 0.031 mg protein A/mL and the gold diameter was 10 nm.

References

1. Kleinschmidt, A., Lang, D., Jacherts, D., and Zahn, R. (1962) Darstellung und Längenmessungen des gesamten Desoxyribonucleinsäure-inhaltes von T$_2$-bakteriophagen. *Biochim. Biophys. Acta* **61,** 857–864.
2. Inman, R. B. and Schnos, M. (1970) Partial denaturation of thymine- and 5-bromouracil-containing l DNA in alkali. *J. Mol. Biol.* **49,** 93–98.
3. Chattoraj, D. and Inman, R. B. (1974) Location of DNA ends in P2, 186, P2, and lambda bacteriophage heads. *J. Mol. Biol.* **87,** 11–22.
4. Inman, R. B., Schnos, M., and Howe, M. (1976) Location of the "variable end" of Mu DNA within the bacteriophage particle. *Virology* **72,** 393–401.
5. Shan, Q., Bork, J. M., Webb, B. L., Inman, R. B., and Cox, M. M. (1997) RecA protein filaments: end-dependent dissociation from ssDNA and stabilization by RecO and RecR protein. *J. Mol. Biol.* **265,** 519–540.
6. Webb, B. L., Cox, M. M., and Inman, R. B. (1997) Recombinational DNA repair: the RecF and RecR proteins limit the extension of RecA filaments beyond single-strand DNA gaps. *Gene* **91,** 1–20.

15

X-Ray Microanalysis Techniques

A. John Morgan, Carole Winters, and Stephen Stürzenbaum

1. Introduction

X-ray production is often the inevitable consequence of the electron irradiation of atoms, whether the atoms reside in biological or materials specimens that are either thin or infinitely thick. As long as the incident electrons are invested with sufficient kinetic energy to ionize individual atoms, X-rays will emanate from them. The technique of biological X-ray microanalysis, henceforth generically referred to as EPXMA (electron probe X-ray microanalysis), has been evolving for over 25 years, and its principles and practicalities have been described in detail on a number of occasions over that period (*see [1–5]* for primary sources). It is a technique that detects and measures elements within structurally defined "compartments"; it cannot distinguish "bound" from "free" elemental pools; it cannot differentiate between isotopic or redox states. The spatial resolution of EPXMA is unsurpassed by any other analytical technique that combines simultaneous compositional and morphological observation. The sensitivity of EPXMA seems impressive (10^{-18} to 10^{-19} g of an element like Ca under favorable conditions), until this is converted into a concentration value for the local analyzed compartment (about 2 mmoles/kg weight, or 200 µg/g, for Ca), which is several orders of magnitude poorer than, for example, "bulk" analytical techniques such as atomic absorption spectrophotometry and induction coupled plasma analysis. It is worth bearing these basic facts in mind before embarking on a tortuous EPXMA study, whilst also acknowledging that trace elements may not necessarily be homogeneously distributed through a biological system: they may be localized within a certain cohort of specialized cells and/or sequestered within discrete subcellular compartments. The limiting factor for EPXMA detection is local not bulk concentration.

From: *Methods in Molecular Medicine Biology*, vol. 117: *Electron Microscopy Methods and Protocols*
Edited by: N. Hajibagheri © Humana Press Inc., Totowa, NJ

In common with other physical probe techniques such as PIXE (proton induced X-ray emission), SIMS (secondary ion mass spectrometry), and LAMMA (laser microprobe mass analysis), the interlinked chain of practical steps that delivers EPXMA data for biological samples has two "domains" that are the main determinants of the success or otherwise of the venture. Of course, the ultimate objective of all of these techniques is to derive interpretable and meaningful data. The signal generated within, and locally released from, the specimen must be captured, qualitatively displayed, and (very often) mathematically manipulated to yield concentration values. This represents the second or output domain alluded to above. Modern EPXMA hardware is supported by software that facilitate data display and reduction according to well-validated theoretical models *(6–9)*. Whereas the advent of such commercial software packages, based in most thin-film biological cases on the Hall Continuum Normalization Method *(9)*, has made quantitative EPXMA more widely accessible, the operator still plays a central role in setting analytical conditions, calibrating the system, experimental design, sampling strategy, statistical analysis, and data interpretation. However, "domain two" is not the major challenge that it was for many EPXMA operators ten or more years ago.

We now switch to what we call "domain one" in the EPXMA protocol chain. State-of-the-art EPXMA equipment will yield data from a biological specimen at the time of its analysis, which reflects its elemental composition, providing standard operational guidelines are followed. One objective of this article is to furnish these guidelines. However, note the fundamentally important phrase "specimen at the time of its analysis." This specimen is the object being analyzed; it may or may not be far removed from the object that the operator wanted to analyze or, indeed, thought was being analyzed! No matter how sophisticated the EPXMA and accompanying software, compositional damage deliberately, or inadvertently caused to a specimen prior to analysis is irreversible. Minimizing damage to a biological specimen's fidelity during its sampling and subsequent preparation is, therefore, paramount. It is a technical challenge exacerbated by the hydrated and dynamic nature of cells and tissues, together with the incompatible high-vacuum environment extant within the operational electron microscope. Specimen fidelity is inevitably compromised to some extent prior to analysis. The trick is to keep the damage within bounds compatible with the biological objectives of the EPXMA exercise being undertaken.

The main aim of this chapter is to describe protocols for the preparation of thin biological specimens for quantitative EPXMA. (Note: this does not mean that the protocols are restricted to specimens that are intrinsically thin, like bacteria or cultured cell monolayers. It refers to techniques that render soft biological tissues thin enough for X-ray analysis in the transmission electron microscope).

The reader may wish to note a few other general points.

1. The selected protocols that we describe have worked on a more-or-less routine basis for us and/or others. They are not by any means foolproof. Success will normally accompany vigilance, which is itself, fueled by an understanding of basic principles. For example, an appreciation that cryomicrotomy is a good deal easier with well-frozen rather than severely ice-damaged specimens, and that certain freezing practices are significantly better than others, is a reasonable starting point. (The reader is referred to six sources for indepth treatment of the crucially important low-temperature methods for electron microscopy (EM) *see [10–15]*).

2. Given that the predominant emphasis of this chapter is on specimen preparation for X-ray microanalysis, and that the low-temperature variants are collectively the methods of choice *(16)*, the information contents of a number of other Chapters (6,7,9,15) in the present book are also instructive.

3. The principles, and to a greater or lesser extent, the practice of specimen preparation for EPXMA apply equally well to other probe techniques. Chemical fixation, for example, leaches solutes from tissues just as effectively whatever the end-use of the specimen is to be. Thus, the protocols described may be applicable to other analytical situations. There is no virtue in reinventing wheels!

4. Performing X-ray microanalysis badly is easy, even with outstandingly good equipment. Obtaining biologically valid data is rather more difficult, with specimen preparation being the most formidable and unavoidable hurdle that must be negotiated.

5. Finally, before embarking on an EPXMA "run" it is worth asking whether you are using the most appropriate analytical tool to match your specific biological problem? Asking subsidiary questions may be helpful and here are some examples. Do you want to know whether or not the electrolyte concentration in an organ changes with developmental age, a life-cycle parameter, or drug treatment? If so, you may be able to obtain a good answer relatively quickly using, say, atomic absorption spectrophotometry. Do you want to measure the concentration and kinetics of Ca^{2+} released from intracellular stores by an external stimulus? If so, the best technique will almost certainly entail the use of one of the several commercially available Ca^{2+}-sensitive molecular probes. Do you want to determine whether Cd is compartmentalized within a given cell type, and to comment on the spatial distribution of other elements within the same cell? If so, then EPXMA is probably the technique of choice. Matching tools to jobs is a cardinal rule in research laboratories. Given the time and effort, not to mention the cost, required to deliver valid subcellular X-ray data, it is a rule particularly worth observing in the EPXMA laboratory.

2. Materials

This chapter describes a diverse family of techniques centered around the electron probe X-ray microanalyzer. The names and addresses of the major suppliers (restricted mainly to the UK) of hardware and software needed to accomplish these techniques are provided under **Subheading 2.1**. All reagents and solvents are commercially available. Thus, the apparatus and consumables required are standard in most well-founded electron microscope laboratories.

2.1. Names and Addresses of Some Major EPXMA and Associated Equipment Suppliers in the U.K.

2.1.1. Transmission Electron Microscopes

1. Hitachi Scientific Instruments, Nissei Sangyo Co. Ltd., 7 Ivanhoe Road, Haywood Lane Industrial Estate, Finchampstead, Wokingham, Berks. RG40 4QQ.
2. JEOL (UK) Ltd., Jeol House, Silver Court, Watchmead, Welwyn Garden City, Herts, AC7 1LT.
3. Philips Electron Optics, York Street, Cambridge, CB1 2QU.

2.1.2. Energy Dispersive Detectors, Analyzers and Software

1. Oxford Instruments Microanalysis Group, Halifax Road, High Wycombe, Bucks, HP12 3SE.
2. Princeton Gamma-Tech, 2 The Metro Centre, Welbeck Way, Woodston, Peterborough, PE2 0UH.

2.1.3. Cryoultramicrotomes

1. LEICA UK Ltd., Davy Avenue, Knowhill, Milton Keynes, MK5 8LB.

2.1.4. Transfer Devices and Cold Stages

1. Gatan Inc., 6678 Owens Drive, Pleasantion, CA 94588.
2. (KEVEX) L.O.T.-Oriel, 1 Mole Business Park, Leatherhead, Surrey, KT22 7AU.
3. Oxford Instruments Scientific Research Division, Old Station Way, Eynsham, Witney, Oxon, OX8 1TL.

2.1.5. Cryofixation, Freeze-Drying and Freeze-Substitution Equipment

1. Agar Scientific Ltd, 66A Cambridge Road, Stansted, Essex, CM24 8DA.
2. Balzers High Vacuum Ltd., Bradbourne Drive, Tilbrook, Milton Keynes, MK7 8AZ.
3. Edwards High Vacuum International, Manor Royal, Crawley, West Sussex, RH10 2LW.
4. LEICA UK Ltd (*see* above).
5. Princeton Gamma-Tech, 2 The Metro Centre, Welbeck Way, Woodston, Peterborough, PE2 0UH.

3. Methods

3.1. Specimen Preparation

1. This section, despite its unarguable importance, does not provide an exhaustive array of preparative protocols. It is, for a number of reasons, very selective. First,

it would be impossible to describe all available methodologies and their variants within the compass of an article of this prescribed length. Second, the value of such a raw compilation presented without extensive commentaries is questionable. Third, specific biological tissues and/or biological problems may necessitate minor, but critical, modifications; it would be prohibitively difficult to anticipate all of these. We have elected, therefore, to focus on certain core protocols, and we try to identify some of the steps within them where modifications could be made to optimize for particular purpose.

2. Preparation of specimens is a central issue in biological EPXMA because many biologically interesting elements are electrolytes existing almost entirely in free solution (e.g., Na^+, K^+, and Cl^-). These, and other solutes, are readily leached from, or translocated within, soft-tissues during chemical fixation. Aqueous dehydrating solutions, ambient-temperature resins, and contrasting agents will also extract residual solutes. Even if these chemical reagents did not interfere directly with the elemental composition of cells and tissues, they would still be precluded from the majority of EPXMA applications because their rates of penetration are far slower than the flux rates of most physiologically active electrolytes, and about 10^3 to 10^4 times slower than cryofixation *(14,16,17)*. Drastically reducing standard fixation time does not alleviate the imposed compositional artefacts to any useful degree.

3. Chemically fixed tissues may, exceptionally, yield useful data in the cases of highly insoluble, discrete, heavily mineralized compartments (**Fig. 1**). Even in these cases, a measure of circumspection is not misplaced. Certainly, it is demonstrably unsound to assume that so-called heavy metals sequestered by biological tissues are recalcitrant to fixative leaching effects. Whereas the losses of divalent cations may be reduced by the inclusion of various precipitant anions in the aqueous preparative media (e.g., pyroantimonate for Na^+ and Ca^{2+}; oxalate for Ca^{2+}; sulphide for "soft acid" metals like Hg and Cd), the efficacies of these agents fall well short of complete immobilization *(18)*.

4. Most successful soft-tissue preparative procedures for EPXMA share one key feature: they are anhydrous. Furthermore, the majority are cryoprocedures where fixation is achieved by quench freezing (or "snap freezing") fresh tissues.

3.2. Tissue smearing, air-drying [See: Note 1, Section 4]

1. Take a small "pinch" (~2 mm^3, say) of fresh tissue directly from an appropriately immobilized organism using fine curved forceps.

2. Immediately brush the surface of a coated EM grid (preferably 100 mesh- a compromise between good film support and minimal extraneous-background contribution) with the tissue. Brush gently to avoid rupturing the coating layer and support the grid on hardened filter paper. The newly formed smear should appear merely as a thin gloss to the naked eye, no solid tissue deposits should be apparent. (Note: it is important that the support films are of very high quality).

3. Transfer and store the smeared grids over silica gel in a dessicator.

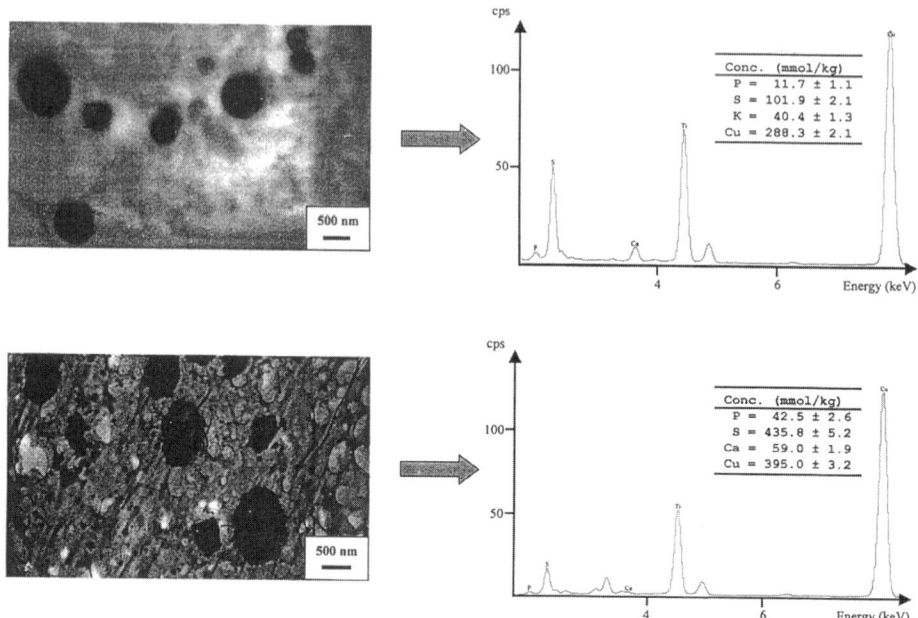

Fig. 1. Glutaraldehyde-fixed, resin embedded (Top Panel) and ultrathin, freeze-dried, cryosection (Bottom) of woodlouse, *Oniscus asellus,* and hepatopancreatic S-cell. The accompanying X-ray energy spectra (and element concentrations) are derived from the Cu-rich inclusions of these cells. Note that the element concentrations are systematically lower in the resin-embedded preparation, indicating that the values are more similar to wet-weight concentrations than dry-weight concentrations because the resin has replaced the extracted water. The lower values should not automatically be interpreted as evidence of element leaching.

4. Observe the grid in the transmission EM. The grid surface should contain a well-dispersed monolayer of granules (dense intracellular compartment) against a barely visible background (cytosolic film). Probe and analyze individual granules.

3.3. Air-Borne Microparticulates

1. Ambient airborne PM10s are collected in a high-volume air sampler (Negretti, UK) fitted with a PM10-selective inlet head. A flow rate of 30 L/min is suitable, and the particulates are trapped on polycarbonate filter (pore size = 0.67 μm; diameter = 55 mm; Millipore, U.K.).

2. A collection filter containing captured particulates is added to a volume of HPLC-grade water (Sigma, U.K.) in a sterile, wide-bottomed, plastic container. Contact with metal tools (tweezers, scalpel blades, etc.) is scrupulously avoided. The container is placed in an ultrasonic bath containing iced water, for 15 min. This pro-

cedure helps to "wet" the hydrophobic particles and shake the majority of them from the filter into suspension.

3. 2 μL samples of PM suspensions are aliquoted by micropipet onto pioloform-coated nickel grids (100 mesh; TAAB UK), and allowed to air dry at room temperature in covered Petri dishes.

4. Individual particles are analyzed by EPXMA. A nickel signal in the X-ray spectra does not interfere with the detection of other elements commonly found in PM10s (*see* **Note 2**, **Subheading 4.**).

3.4. Cryofixation for EPXMA

There are four general cryofixation techniques available (**Fig. 2**) whose common roles in conjunction with EPXMA are a) to rapidly arrest cellular activity so that ion movements are severely restricted, b) to solidify the specimen in its native medium so that it may be sectioned, if necessary, without infiltration by extraneous media (i.e., resins), and c) to achieve cooling rates that are sufficiently high to minimize ice damage in the superficial layers of the tissue specimen. When reflecting upon the above four techniques a number of practical points need to be noted (**Notes 3–14**).

Cryofixation for purposes other than EPXMA is dealt with elsewhere in this book (Chapter 9). EPXMA, however, makes special demands of cryofixation; demands that warrant practical responses that are modifications of the four themes outlined in **Fig. 2**. Some of the specific goals that certain EPXMA-associated cryopreparative protocols aim to meet are to minimize postmortem damage, to freeze tissues in defined functional states, to cryofix cultured cells, and to obtain specimens from human subjects.

3.4.1. Minimizing Postmortem Damage: In Situ Freezing

Many of the changes linked to dissection delays and the intrusive nature of sampling (described earlier) would be much reduced if specimens were cryofixed without prior removal from the animal, or indeed plant. Ideally, cryofixation and specimen retrieval should be simultaneous events. A number of cryofixation devices have been designed, and fairly widely used, that aim to realize this ideal. Most are based on the slam-freezing principle. A good, simple, example of the approach is represented by Hagler and Buja's (1984) "cryopliers" *(20)*. The device consists of smooth-jawed electrical pliers to which two highly polished copper blocks are silver-soldered to make a parallel gap when closed of 0.2 to 0.3 mm. Agar Scientific Ltd. (Essex, England) supply two types of cryopliers: one suitable for relatively thin specimens, the other for thicker (1 mm) specimens. (An improvement on the basic design is the inclusion of adjuster screws, so that the parallelism of the gap between what are facing metal mirrors can be retained through several cycles of cryofixa-

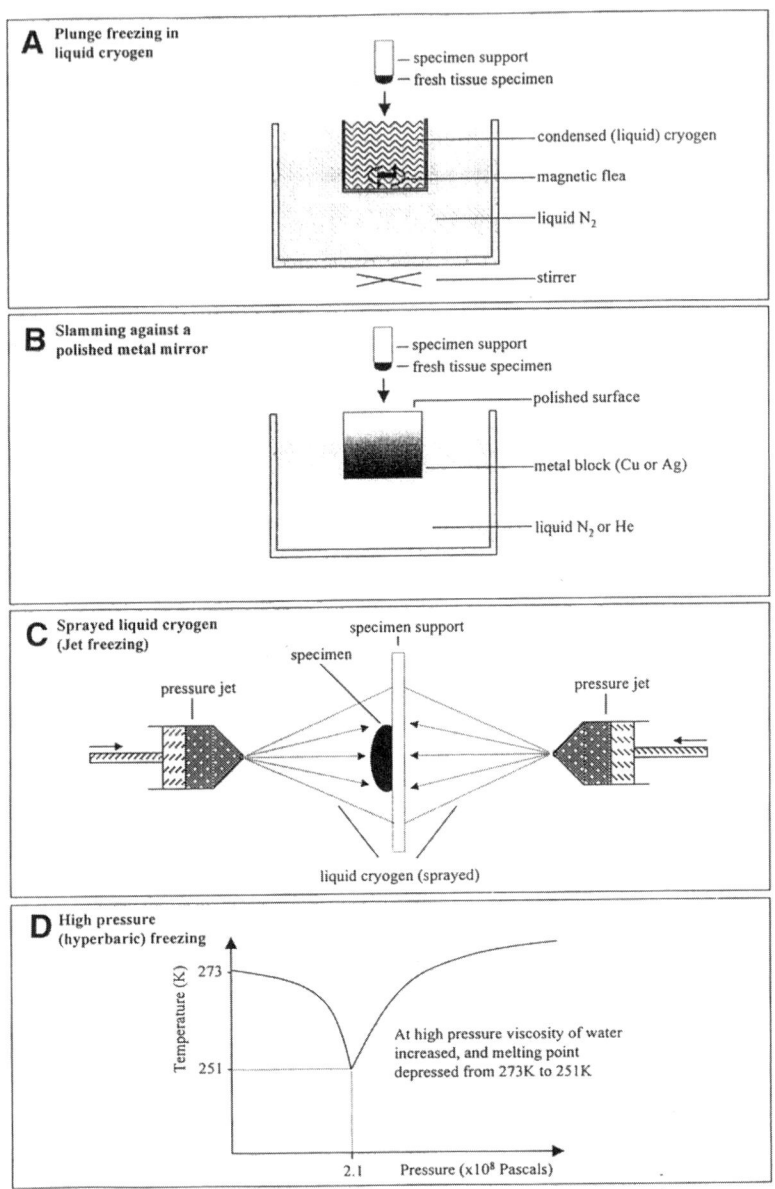

Fig. 2. Schematic representation of the four main cryofixation techniques (after **ref.** *37*).

tion—*see [13]*). The use of these cryopliers (*see* **Notes 15–22**) is schematically outlined in **Fig. 3**, and can be described in a series of practical steps.

1. Polish, dry, and clean the metal inserts before starting. We do this with Wenol (Agar Scientific Ltd.) metal polish, followed by ultrasonication in propan-2-ol.
2. Immerse the jaws in liquid N_2 in the closed position to prevent water condensation on the mirror surfaces.
3. Remove pliers from liquid N_2, still in the closed position. Quickly move towards the exposed tissue (typically supported in a deeply anaesthetized animal), open the jaws, and clamp them onto the tissue.
4. Twist the pliers sharply so that the clamped frozen tissue is immediately detached. Return the closed pliers with the excised tissue wafer to the liquid N_2 bath.
5. The wafer of tissue, if well frozen, should be fairly glossy in appearance; milky whiteness usually indicates the presence of severe ice damage. The wafer of tissue should possess two parallel well-frozen surfaces. Reject, by trimming with a cold scalpel, any region of the specimen that obviously protruded outside the confines of the jaw inserts.
6. Transfer the frozen tissue wafer with cold forceps to a copper block immersed under the liquid N_2 bath. (You may need a second cold forceps to gently tease the frozen specimen away from the first. Alternatively, ceramic tools may reduce static at cryogenic temperatures.) Cut the wafer into suitably sized pieces for mounting under liquid N_2 in the clamp-type jaws of a cryomicrotome specimen holder. Overtightening the jaws will crack the brittle specimen. If necessary, loosely clamp the specimen and stabilize it with a small droplet of cryoglue.

3.4.2. Time-Resolved Studies: Freezing in Defined Functional States (see [21])

Two good examples of devices designed to cryofix very different biological specimens in time-resolved and defined physiological states, respectively, are schematically illustrated in **Fig. 4** (*see* **Notes 23** and **24**). Brief descriptions of these elegant arrangements are warranted because they may serve a heuristic purpose.

3.4.2.1. TIME-RESOLVED FREEZING OF THE STINGING CAPSULES (NEMATOCYSTS) OF *HYDRA* (*19,24*)

Electrical stimulation can cause discharge of the harpoon-like tubule of *Hydra* nematocytes within 50 ms. The device shown in **Fig. 4A** is designed to plunge-freeze *Hydra* attached to a metal support at defined time intervals after stimulation by contact with an electric current conducted through a flexible copper foil.

3.4.2.2 CRYOFIXATION OF ISOLATED MYOCYTES FROM GUINEA PIG (*23*; **Fig. 4B**)

A patch electrode was used to transfer individual ventricular myocytes from a growth chamber onto a pioloform support across an aperture in a silver-foil holder. Resting potential was recorded to ensure that cells were not functionally damaged during transfer from the shallow growth chamber. The voltage

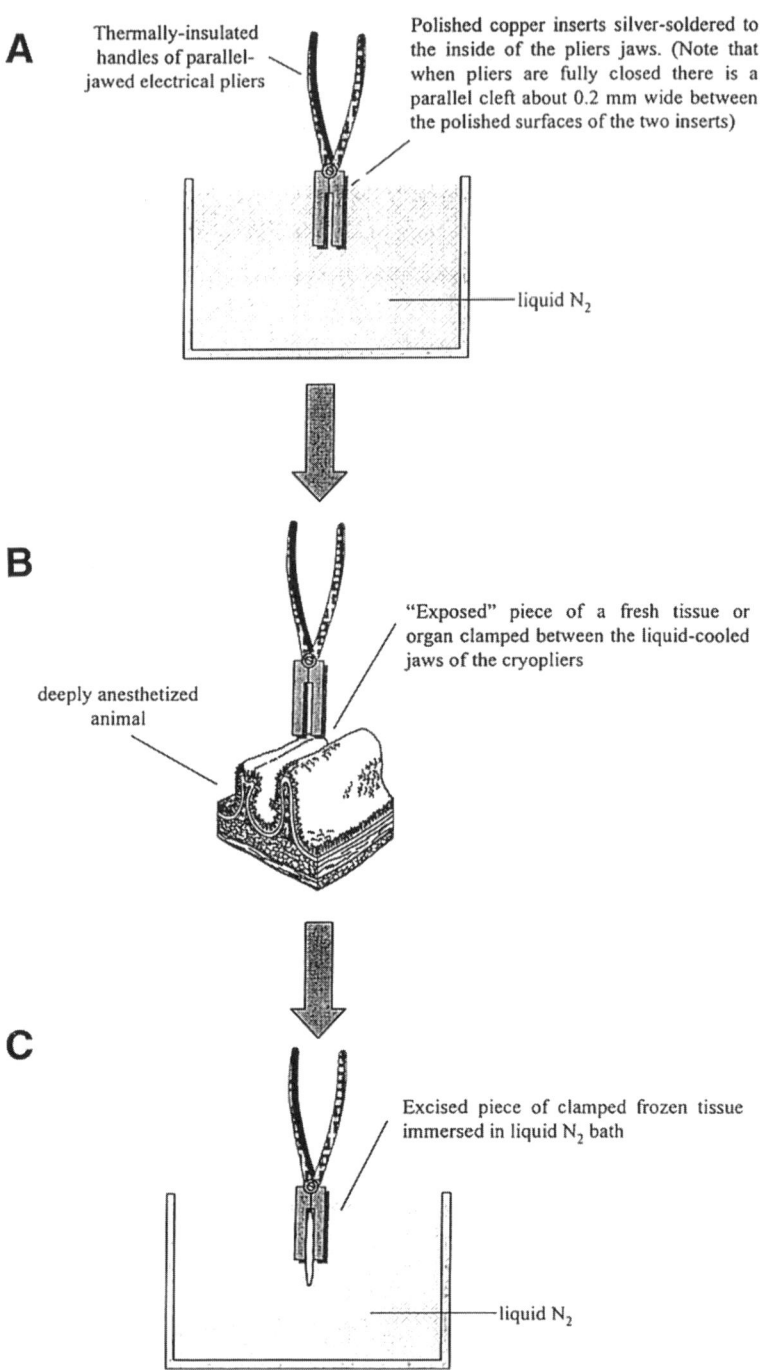

Fig. 3. Schematic diagram of *in situ* cryofixation with, for example, a home-made pair of cryopliers.

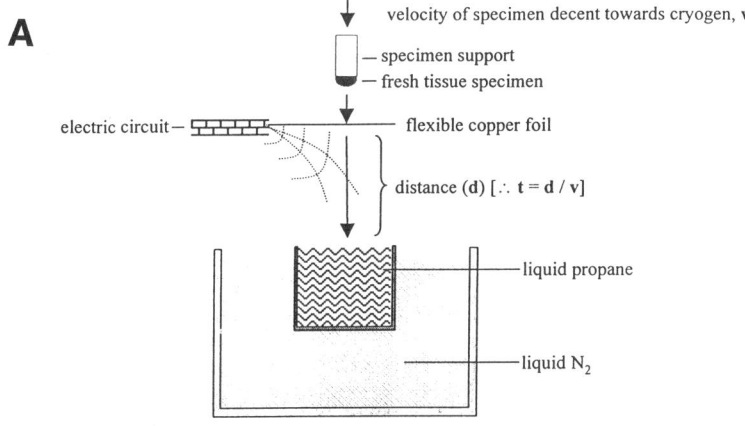

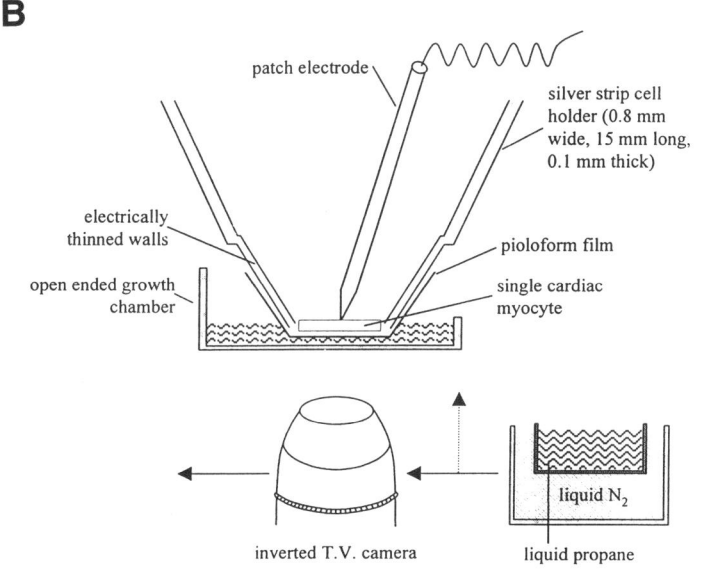

Fig. 4. Schematic diagrams of devices used for time-resolved (**A**; after **ref. 59**) and snap arrest of dynamic biological events in defined states (**B**; after **ref. 39**) by cryofixation.

clamp was switched on and the cells were stimulated. The entire assembly was then rearranged under automated control: the cell holder was held absolutely static; the microscope and open-ended growth chamber were driven pneumati-

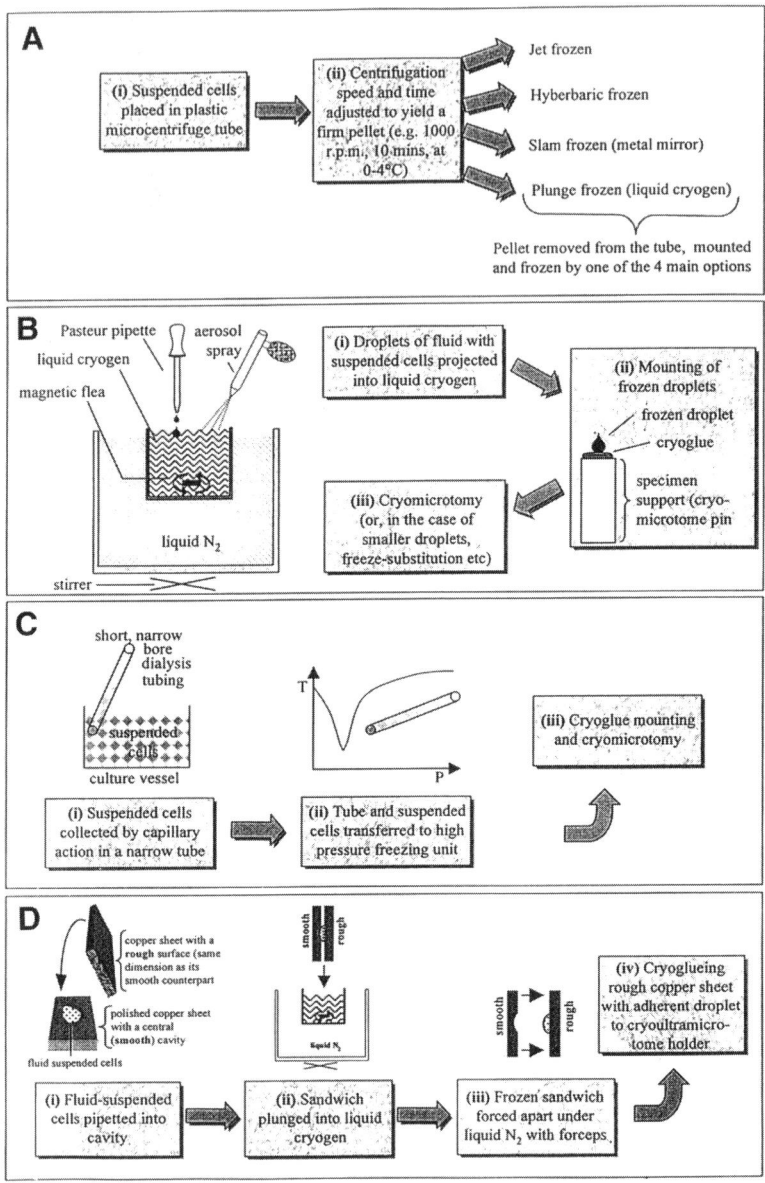

Fig. 5. Schematic diagrams of cryopreparative techniques designed for fluid—suspended cells (in vitro cell cultures or native cells).

cally to the side; a bath of liquid propane was driven under and upwards so that the specimen and holder were immersed; the detached holder with cryofixed myocyte was transferred to a cryoultramicrotome for thin sectioning, and eventually, EPXMA.

3.4.3. Cryofixation of Cells Cultured in Fluid Suspension

Cultured cells offer a number of significant advantages to the cell physiologist. They can, in some cases, be synchronised so that individual cells enter and exit the various phases of the cell-cycle in unison; they are easy to experimentally manipulate and monitor using a full array of cell biological, biochemical and physiological techniques; they reduce the costs and ethical problems associated with the housing and experimentation on mammals. In addition, they may be viewed as "small specimens" and, thus, ideal potential subjects for cryofixation. However, their smallness and dispersed nature presents manipulative difficulties. Several simple protocols have been described for the cryofixation and subsequent cryopreparation of cells growing in vitro in fluid suspension (*see* **Notes 25–29**) and cells that are substrate attached (*see* **Note 30**). The important variants of these protocols are outlined schematically in **Figs. 5** and **6**.

3.4.4. Cells Grown in vitro as Monolayers on Solid Substrates

A problem presented by cells growing on solid substrates in vitro is that they are covered by a very deep layer, in cryogenic terms, of aqueous growth medium. Furthermore, even thin-film growth supports form a formidable thermal impediment to a cryofixative front approaching from the cell-attachment side. **Figure 6** illustrates some of the practical solutions to these problems

3.4.5. Tissue Specimens from Human Patients

The protocols outlined in **Subheadings 3.4.1.–3.4.4.** have been more-or-less optimized to meet certain experimental aims and objectives. For a number of reasons, it is seldom possible to cryofix tissues from living or dead human subjects, even under approximately optimal conditions. Cryofixing such material, therefore, involves quite substantial procedural and analytical compromises. These have been discussed at length by Roomans *(25)*.

3.4.6. Postcryofixation Processing of Thin Sections

Having cryofixed the biological specimen, a number of preparative options are available. These are outlined in **Fig. 7**. Since detailed aspects of freeze-substitution (Chapters 7 and 9) and ultracryotomy (Chapter 6) are dealt with elsewhere in this book, we restrict ourselves to a few practical notes (*see* **Notes 33–43**) relating these techniques to EPXMA.

3.5. Operating the Analytical Transmission Electron Microscope

The acquisition of good X-ray data from thin biological specimens is facilitated if care is devoted to certain analytical hardware parameters.

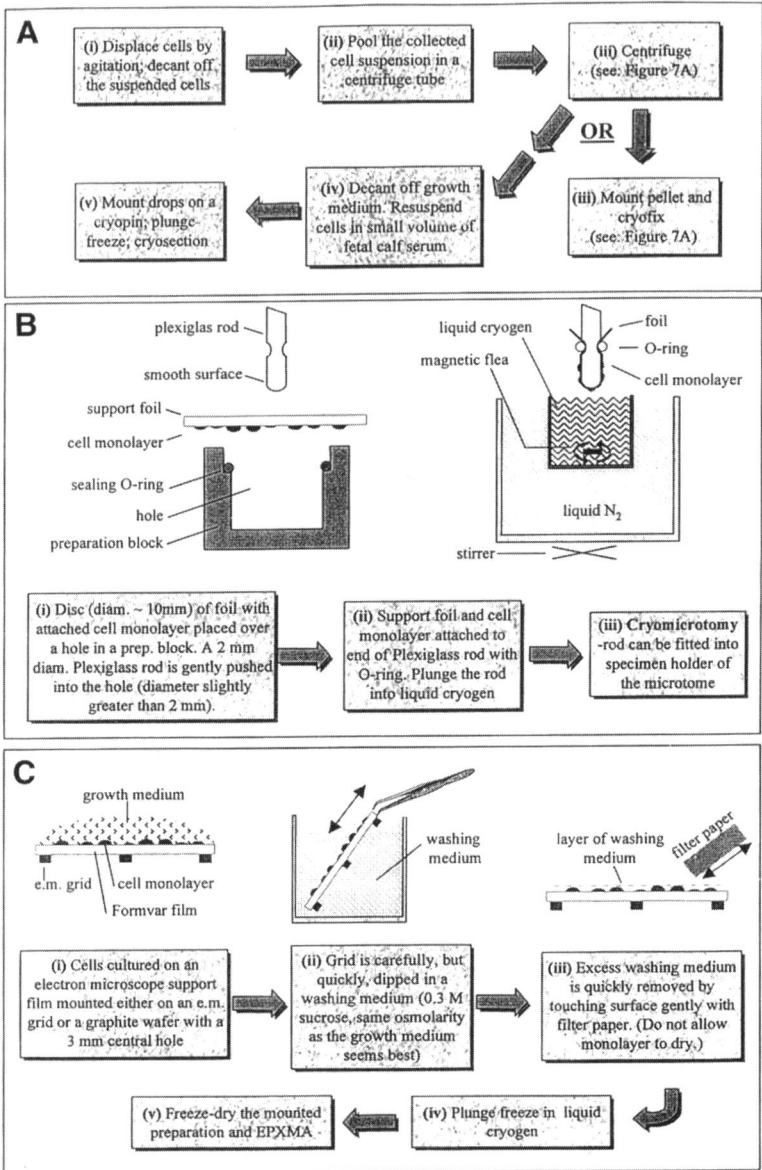

Fig. 6. Schematic diagrams of cryopreparative techniques designed for cell mono-layers grown in vitro on solid substrates. (Note that the drawings are not to scale.)

1. Select the most appropriate accelerating voltage (usually about 80 kV) that yields the best peak-to-background ratios and minimizes beam damage. Note that for

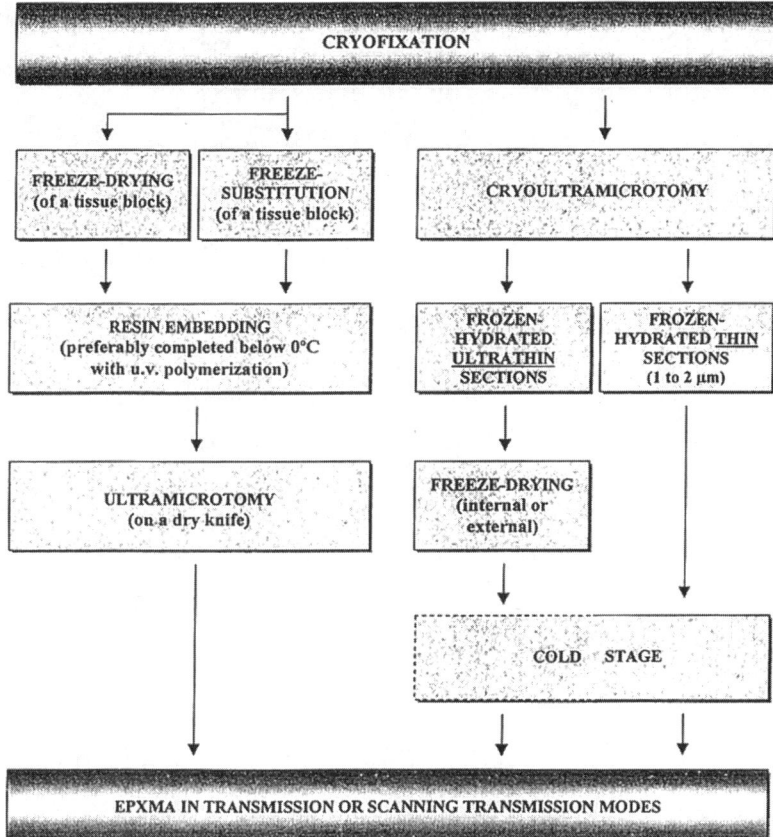

Fig. 7. Schematic representation of the post cryofixation pathways for the preparation of ultrathin (or analytically thin) sections for EPXMA.

thin specimens, higher energy produces least damage.

2. Optimize the detector-specimen geometry, by adjusting the detector-to-specimen distance (if possible, but be careful not to approach the specimen holder too closely), and by adjusting the specimen tilt angle.

3. Try to keep the measured X-ray count rates at about 800 to 1000 counts per second by adjusting the beam current. High beam current, although it improves the count rate, can cause organic mass loss from the specimen and specimen instability (a very important artefact in the case of quantitative X-ray mapping with total acquisition times of several hours).

4. Keep magnification as low as possible and the probe diameter as high as possible (within limitations dictated by the objectives of the analytical study). Always search the specimen at low magnification; try not to preexpose the precise areas to be analyzed by higher magnification morphological scrutiny.

5. Eliminate all obvious sources of extraneous (to the specimen itself) peaks and sources of continuum X-rays. Several steps may be taken to achieve these ends:-
   a. collimate the detector if necessary (but to be done at installation by a qualified engineer).
   b. use coarse mesh support grids (e.g., 50 mesh for cryosections).
   c. remove the objective aperture during analysis.
   d. use a small (100 mm) condenser aperature.
   e. make sure that the column is well aligned.
   f. maintain a clean vacuum.
   g. install an efficient anticontaminator (a **very** efficient one if a cold stage is used).
   h. operate at as low a specimen tilt angle as possible.
   i. coat metallic parts in the vicinity of the specimen (notably the pole-pieces and anticontamintor) with colloidal graphite; a graphite "top-hat-type" hard X-ray kit inserted into the pole-piece achieves a similar function (again, to be done by an engineer at installation).
   j. select, where possible, regions of the specimen for EPXMA that are well removed from grid bars.
   k. use thick apertures.
6. Make sure that the X-ray path from specimen to detector is not obstructed by the edge of the grid holder.

3.6. Converting Measured X-ray Intensities into Element Concentrations

Without delving into the intricacies of EPXMA quantitation *(6–9)* suffice to say here that the conversion of measured characteristic X-ray counts for given elements into concentration units requires thin-film standards of known composition—*concentration standards*. (Other standards may be required for *peak profiling*, so that overlapping peaks can be deconvoluted, and for support-film correction). All concentration standards must possess the following properties *(26)*:

1. A well characterized, or easily measurable by independent analytical procedures, bulk elemental composition;
2. A homogeneous distribution of all the EPXMA-detectable elements. (This should be systematically checked by probing at intervals across a thin-film standard);
3. A matrix that behaves under electron irradiation in a manner that mimics fairly well the behavior of an ultrathin tissue section (whether resin-embedded or freeze-dried cryosection).

We describe two different types of standards that meet the above criteria.

3.6.1. Cryosectioned Standards (see **Note 44**)

1. Make a mixture of one or more binary mineral salts (the concentration range expressed on a dry mass basis should, ideally, cover the range of element concentrations anticipated in the specimens to be analyzed, although this is not an abso-

lute requirement) in 20 to 30% gelatin or albumin with 5% glycerol added as a cryoprotectant. It is advisable to analyze the mixtures by (say) atomic absorption to obtain "actual" rather than nominal bulk concentrations.

2. A standard that has been used as an incubating medium and "peripheral standard" for insect tissues contains: $K = 20$ mM; $Na = 132$ mM; $Cl = 160$ mM; $Mg = 9$ mM; $H_2PO_4 = 4$ mM; $HCO_3 = 10$ mM; glucose = 30 mM; made up in an aqueous solution of 10 to 20% high molecular weight (M.W. - 237,000) Dextran. This, in fact, is a Dextran-Ringer solution, the Dextran acting as a cryoprotectant and organic matrix for the standard.

3. Place well-mixed (with a magnetic stirrer) small drops of standard solution on cryoultramicrotome pins. Cryofix by plunging in liquid cryogen (preferably propane, although liquid N_2 may be adequate if the solution is well cryoprotected).

4. Cut ultrathin cryosections of the standard, and freeze-dry plus analyze them under identical conditions to the tissue sections.

3.6.2. Aminoplastic Standards

These very versatile standards (into which a wide range of light and heavy elements may be incorporated) behave like thin, freeze-dried, tissue cryosections under electron irradiation *(27)*. They may be stored indefinitely in a desiccator, either as blocks or sections on grids.

1. Pre-weigh small (~10 mL capacity), clean, dry flat-bottomed glass vials to four decimal places.

2. Make aqueous solutions, in analytical quality water, of inorganic salts (concentration range from 0.125 M to 2 M, if possible). The salts should all be analytical grade. They may be dissolved at the higher concentrations by the addition of analytical grade HNO_3 to a final concentration of 10%. Acidifying the salt solutions does not interfere with the polymerization of the blocks, although high concentrations of certain salts can cause problems. (In the case of electrolytes, it may be possible to make aminoplastics containing a mixture of different salts—the only practical constraints are solubility and polymerization.)

3. Make Solution A by heating a mixture of 5 mL 50% glutaraldehyde (2 mL deionized H_2O + 5 mL 70% double-distilled glutaraldehyde yields a 50% solution) and 16 g of urea in a 50-mL conical flask. Stir the mixture continuously (by hand if the flask is suspended in a heated water bath, or with a magnetic flea). Cool the mixture with the addition of 1 mL cold deionized water and stir just before use.

4. Solution B is made directly in the individual, weighed, glass polymerization vials (*see* Step 1) by adding 0.5 mL of 70% glutaraldehyde and 0.2 mL of salt solution. Mix by swirling.

5. Add 0.7 mL of Solution A to Solution B in each vial. Mix thoroughly—we often use the heat-sealed ends of glass microcapillaries for this purpose. The mixtures should spontaneously polymerize, but preferably not for several minutes. In fact the polymerization time varies with the individual salt and its concentration:the higher the concentration the faster the polymerization seems to be. Discard any opaque or powdery blocks; only perfectly transparent blocks provide good ho-

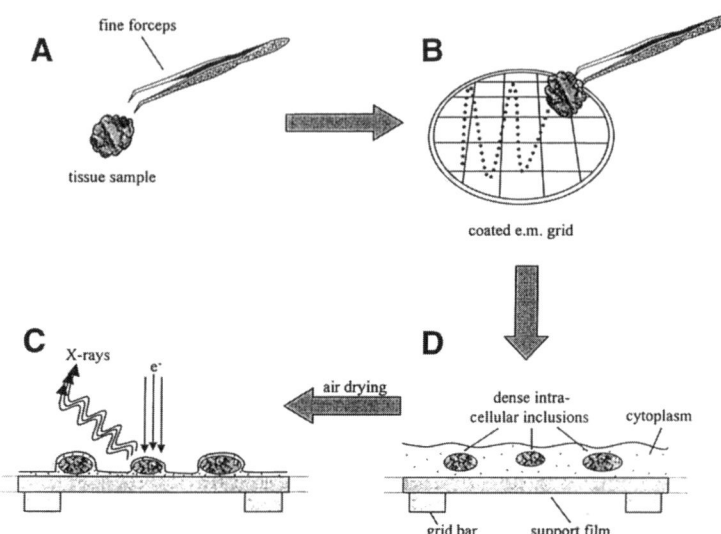

Fig. 8. Schematic diagram illustrating the principles of fresh tissue smears followed by air drying. **(A)** Fine forceps with freshly excised, fully-hydrated soft-tissue "biopsy" sample; **(B)** tissue sample is gently brushed over the entire support film/coat of an appropriate EM grid (grid material is selected to be compatible with the requirements of analysis, e.g., if Cu is subject of analysis then, obviously, Cu grids are to be avoided); **(C)** a very thin layer of smeared tissue is deposited on the support film, with the dense intracellular components of analytical interest suspended in the hydrated cytoplasm; **(D)** after air drying the intracellular dense granules are not significantly collapsed, but are smeared above and below with layers of dried cytoplasm. Individual "granules" are easily located in the thin monolayer.

mogeneous standards after sectioning. The cracking of blocks as polymerization progresses does not matter. Polymerization is deemed to be complete when the blocks have reached a constant weight at ambient (not elevated) temperature. Element concentration can be calculated or determined by atomic absorption spectrophotometry of acid-digested samples taken from individual blocks.

6. Cut sections (0.25–0.5 μm thick) on dry knives. Transfer the sections onto coated grids with an eyelash and flatten gently with a smooth-surfaced metal rod.

4. Notes

1. Tissue smearing, air drying is the simplest of all the anhydrous preparative methods, whose applicability is restricted (**Fig. 8**). It is especially useful in the cases of cells containing relatively large (0.5–2 μm diameter) inclusions whose densities and concentrations of certain elements of interest significantly exceed those of the surrounding cytosol. Metal-accumulating chloragocytes of earthworms are a good example of cells suitable for this mode of preparation for EPXMA (*28*;

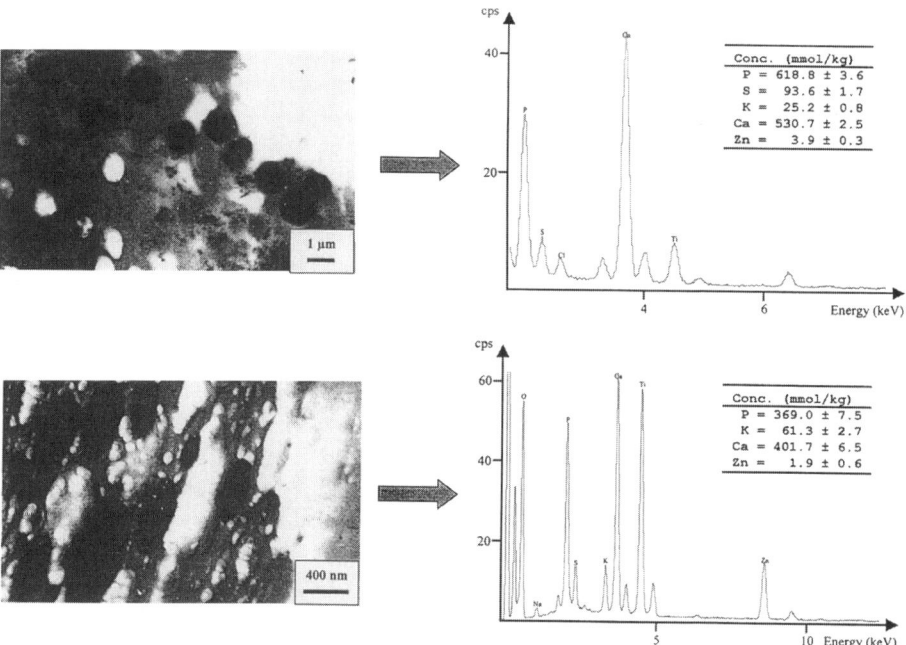

Fig. 9. Electron micrographs of an air-dried fresh tissue smear (Top Panel) and ultrathin, freeze-dried cryosection (Bottom) of earthworm chloragogenous tissue. The accompanying X-ray energy spectra (and element concentrations) are derived from the characteristic electron dense inclusions, the chloragosome granules. Note that the concentration values are comparable in both preparations.

Fig. 9). **Table 1** summarizes the advantages and disadvantages of the technique that yields final preparations, once graphically described by a colleague, as akin to "viewing crockery (the dense subcellular inclusions) on a table (the support film) through a flimsy tablecloth cast over them (the collapsed layer of dried cytoplasm)."

2. The preparation and analysis of air-borne particulates (e.g., asbestos fibers; particulates with a mean aerodynamic diameter <10μm, PM10s, derived from vehicle exhausts, industrial processes, volcanic eruptions, etc.) is, strictly, outside the scope of biological EPXMA. However, these particulates engender widespread biomedical concern because they accumulate in, and compromise the function of, certain biological tissues, notably in the respiratory tract. For this reason we provide a protocol *(29)* for sampling them from air, and for depositing them on coated EM grids for quantitative EPXMA.

3. Given that it is usual not to chemically fix or cryoprotect biological tissues prior to cryofixation for EPXMA (in contrast with specimens destined for immunogold cytochemistry by the so-called "Tokuyasu method": Chapter 4, this book), a de-

Table 1
Advantages and Disadvantages of the Tissue-Smearing/Air Drying Preparative Procedure

Advantages	Disadvantages
1. Simple, rapid and cheap.	1. Limited applicability (i.e., it is a minority technique).
2. High sampling yield. Very easy to meet statistical requirements in terms of number of compartments per organism and number of organisms per experimental group (although it may not be possible to quote the number of cells from which compartments were derived).	2. Morphological characterization is very poor. Cannot identify cell boundaries or organelles that are not large and dense.
3. Anhydrous—no chemicals of any kind involved. Elemental loss or intrusion involving mineralized or metal-sequestering compartments is probably low.	3. Concentration values are erroneous. Since the dense "granules" selected for analysis contain, by definition, much higher concentrations of certain elements than the surrounding cytosol, the "contaminating" cytosolic layers above and below the analyzed compartments contribute a relatively small mass to the analyzed volume that, nevertheless, reduces (by "dilution") the measured concentration of the major constituents in the putative compartment.
4. Limited comparative studies involving cryomicrotomy/freeze-drying indicate that biologically meaningful data can be obtained from appropriate specimens.	4. X-ray absorption within large, dense, air-dried compartments may be very high. It must always be carefully monitored.
5. Unbiased sampling. Compartments can only be selected for analysis on the basis of size and density.	

gree of ice damage is inevitable in snap-frozen fresh tissue specimens. Fortunately, low-resolution EPXMA is acceptable even in fairly severely ice-damaged tissue regions providing that the dimension of the electron probe exceeds by a considerable margin the area of ice-crystal "ghosts" that it encloses (*16*). Limiting ice damage because it ensures more accessible analytical areas and limited compositional disruption, must still be a major EPXMA priority. Ice damage is a function of cooling rates within the tissue. The thermal conductivity of ice is relatively poor, so that cooling rates deteriorate sharply as the cooling front penetrates into tissue away from the specimen/cryogen interface. Plunge freezing, jet freezing, and slam freezing (**Fig. 2**) provide tissue blocks with an ice damage-free superficial rind that is restricted to a depth of the order of 5 to 30 μm from the surface (i.e., one or two cell diameters). Hyperbaric freezing increases this well-frozen depth by at least an order of magnitude (300–500 μm; *30*).

4. High-pressure freezing offers significant advantages for cryofixing morphologically complex specimens. This advantage is overwhelming if the specific cellular

targets of analysis are distributed at considerable depth from the surface of a heterogeneous organ or organism. However, commercial high-cost hyperbaric freezing units are available in, or to, but a limited number of EPXMA laboratories. The potential of this technique has not been fully explored.

5. Plunge freezing, jet freezing, and slam freezing units can be bought from commercial sources. Unlike hyperbaric freezing units, very good devices can be fairly easily built with basic workshop facilities and a careful appreciation of physical and safety matters *(31)*.

6. Liquid propane (melting point = 84 K, boiling point = 231 K) is the commonest cryogen used for plunge freezing. Liquid propane, like liquid ethane (melting point = 90 K, boiling point = 184 K,), is highly explosive. "Homemade" liquid propane freezing units should incorporate the following features: a) the entire unit should be installed inside a fume-hood to minimize the possibility of propane gas build-up, b) the unit should be isolated from all possible sources of electric sparks or naked flames, c) propane gas should be delivered through a liquid N_2-cooled, coiled, Cu pipe of appropriate diameter (internal bore of 3 mm is suitable), so that it is partially liquefied prior to deposition in the liquid N_2-cooled condensation chamber, d) liquid propane should not be allowed to overflow the condensation chamber into the surrounding insulated liquid N_2 reservoir, e) the unit should be mounted on a magnetic stirrer so that the liquid propane can continuously be stirred with a rotating "flea," thus ensuring that the coolant is supercooled at liquid N_2 temperature, f) liquid propane should not be allowed to contact bare skin or eyes of the operator—it is, after all, a far superior cryogen than liquid N_2, and g) after completing the cryofixation run, liquid propane should be carefully and immediately evaporated inside a fume hood while stirring or, preferably, by burning in a proprietary device (*see 11*).

7. Liquid ethane is even more efficient than propane *(15)*, but its higher cost and safety requirements restrict its cryogenic use particularly to the "bare-grid" preparation of vitrified aqueous films for high-resolution, low-electron dose, structural studies *(12)*.

8. The entry velocity of the specimen, and the depth of travel through the liquid coolant, are both important parameters that determine the maintenance of the highest possible thermal gradient between the cryogen and specimen. Ryan *(32)* provides an indepth description of these practicalities *(32)*.

9. Every effort must be made to ensure that the tissue specimen does not travel through a significant cold gas layer prior either to entry into a condensed liquid cryogen, or prior to being slammed against a cooled metal mirror *(33)*.

10. The efficiency of plunge freezing is determined by the physical properties of the cryogen used, its depth, presence or absence of a gas layer, as well as the entry velocity of the mounted specimens. However, specimen size and shape, as well as the size and mass of its support, are especially important considerations. Ideally, the specimen should be as small as possible, and fairly streamlined to allow nonturbulent coolant flow over its surface. For the same reason it is advantageous if the supporting metal pin is also streamlined.

11. Small tissue specimens require low-mass metal supports, so that cooling rates are not impeded by a large transfer of heat from support to specimen. Thermally insulated aluminium foil arches glued to cryopins have been described for this purpose (*33,34*), although not widely used to our knowledge. Such foil supports are fairly flimsy, but can be made more mechanically stable for cryomicrotomy purposes by carefully infilling the arch with a cryoglue (*see* below) after snap-freezing the tissue specimen.

12. Mounting tissue specimens on their supports and freezing them, by whatever the chosen method, must be achieved with minimal delay. Dissecting as well as trimming the specimen to size and shape can cause major postmortem compositional change. Trimming alone can cause severe physical damage to cells in the very superficial layers of the tissue block that are going to be optimally cryofixed. Ice damage can be avoided if a small tissue block is deliberately or accidentally allowed to air dry before freezing, but the cost in terms of solute fluxes within the constituent cells may well be disastrous. Herein lies a paradox that underscores the difficulty of specimen preparation for EPXMA:small streamlined specimens promote good freezing, but obtaining such specimens from bulk tissue can cause unacceptable damage prior to cryofixation. Thus, and in the context of cryofixation, "speed," "simplicity," "efficiency," and "reproducibility" are worthwhile bywords.

13. Metal mirror freezing (also known as "impact freezing," slam freezing," or "metal block freezing") is most suitable for large pieces of soft tissue. It is most important that the specimen does not reverberate away from the cold mirror upon impact. Commercially available slamming devices employ magnets, pneumatic pressure, or some shock-absorbing strategy to ensure specimen-mirror contact. Another important practical point is that metal mirrors must always be well polished and scratch free, again to ensure maximal specimen-mirror contact. Copper appears to form the most efficient metal mirror at liquid N_2 temperature (*34*).

14. Although it is a very inefficient cryofixative, liquid N_2 is superb for the indefinite storage of cryofixed tissues. In addition, cryofixed tissues can be trimmed and mounted in, for example, the jaw-type specimen holders of cryomicrotomes being held under liquid N_2. It should be noted that amorphous ice can transform to cubic and hexagonal ice at elevated temperatures; furthermore, hexagonal ice crystals can grow at subzero temperatures (*13*). For these reasons, it is most important that cryofixed specimens are always handled with tools that have been precooled to liquid N_2 temperature.

15. Other devices, have been designed and fabricated in research laboratories, many of them relatively cheaply, for *in situ* cryofixation. These include the various liquid N_2-cooled tubular metal "corers" of various degrees of sophistication:the "cryogun" (*35*), "cryoballistic needle" (*36*), the manual "cryobioptic needle" (*37*), and the pneumatic "cryopunch" (*38*). *In situ* cryofixation has also been achieved with a jet of liquid cryogen (*39*) and with falling cryoblocks (*40*).

16. Since the cryopliers are operated by hand, the reproducibility of cryofixation is often poor compared with other techniques. Practice, as always, improves perfor-

mance. Moreover, whereas the technique can be done by a single person, it is advisable to have a second person holding the exposed tissue/organ to be frozen so that the first can concentrate on smooth, swift, movements, and accurate aiming.

17. The cryopliers can be used for freezing fairly delicate excised organs (e.g., small tubular elements) mounted across a hole punched in a Ringer-saturated filter paper. The potential advantage of this approach over immersion freezing and conventional slam freezing is the fact that the pliers presents two cold surfaces to the specimen.

18. The cryopliers technique, in common with the more robust and conventional slamming techniques *(41)*, causes tissue deformations on a macro scale, but remarkably little surface damage at a microscopic scale.

19. After each cryofixation "run" the cryopliers should be dried immediately. A hair dryer is particularly useful if the pliers are to be quickly recycled for a second "run."

20. A commercial version of the cryopliers, called the "Cryosnapper" (Gatan, Model 669) has been available for some time, and offers improved reproducibility although it has not been widely validated *(42)*.

21. Cryomicrotomy requires a frozen tissue specimen attached to a suitable support that can then be inserted into the arm of the microtome. Some good cryofixation techniques yield "free" specimens that are not mounted on such supports. All the *in situ* cryofixation methods fall into this category (*see* description of the cryopliers method); so does high-pressure freezing, and techniques that involve the spraying of fluid-suspended cells as droplets into liquid cryogens. Low temperature glues must, therefore, be used to attach the frozen specimens onto supports. Cryoglues must have a melting point lower than the temperature at which the specimen will be mounted (typically <193 K), with a significantly higher boiling point and low vapor pressure. Materials with melting points of about 170 K (−100°C) are potentially useful *(11)*.

22. Steinbrecht and Zierold (1984; ref. *43*) described the use on n-heptane (melting point = 182 K; boiling point = 372 K; vapor pressure = 5.33×10^{-1} Pa at 173 K), which can be used to "stick" cryofixed specimens to metal supports at temperatures warmer than 183 K (−90°C); it then solidifies at temperatures below the melting point, and permits successful cryomicrotomy in the useful range from 158 K (−115°C) to 143 K (−130°C). Care must be exercised to ensure that the application of the cryoglue does not occur at too high a temperature for too long a time.

23. Even the simplest tissues and organs consist of heterogeneous cellular populations: cohorts of cells, and their constituent organelles, at different stages in their life-cycles and undergoing different functional phases within the normal physiological range to, in some cases, include pathological decline. It is a basic tenet of cell biology that changes in functional status accompanies changes in ion content and distribution. A major advantage of EPXMA is the ability to investigate compositional heterogeneity. However, in the case of solid tissues, and particu-

larly in the absence of corroborative morphological data (not unusual in cryopreparations designated primarily for EPXMA), it may be difficult to interpret observed heterogeneity within a cellular population/community in physiological terms. For this reason, it is sometimes possible to resolve the conundrum by arresting the cell or cell population in defined physiological states and then determining their ion composition. Surprisingly, a number of devices have been described that achieve this objective, and they all capatalize on the rapidity of cryofixation, which presumably overtakes the diffusional migration of ions over analytically resolvable distances in intact cells. (Note the admonitory calculations, based on simple aqueous "models," made by Zierold *[19]*).

24. It is technically instructive to note that Zierold and Schafer *(44)* and Zierold *(45)* used a modified propane-jet to cryofix streaming *Amoeba* cells while being observed under a binocular microscope. The cryofixed protozoa were then examined by scanning electron microscopy combined with X-ray microanalysis.

25. Centrifuged pellets (**Fig. 5A**) may be treated like solid tissue samples. It is not clear whether centrifugation introduces compositional changes. It is, therefore, prudent to minimize both the period and speed of centrifugation.

26. Squirting, or spraying suspended cells into liquid cryogens (**Fig. 5B**) has been extensively used *(46,47)* in cell biology applications. In one study on the unicell, *Paramecium*, the technique afforded time-resolved fixation of exocytosis *(47)*. Aerosol spraying is better suited to cells suspended in aqueous media. Squirting larger droplets, which may subsequently be individually mounted on cryomicrotome pins with a cryoglue like n-hexane, is best done with rather viscous fluids.

27. Hyperbaric freezing of short lengths of dialysis tubing filled with fluid suspended cells (**Fig. 5C**) has been done by Hohenberg et al. *(48)*. (Members of the same laboratory have recently reported the successful hyperbaric freezing within a few seconds of fresh tissue cores sampled by biopsy needle *[49]*).

28. The droplet sandwiching technique (**Fig. 5D**) was developed and described by Zierold *(19,50)*. It is an improvement over straight-forward mounting of "exposed" droplets on metal supports prior to freezing because it reduces the risk of evaporative drying, and attendant osmotic changes, that can occur in small fluid droplets.

29. Cells "freed" from solid tissues by controlled enzyme digestion (e.g., hepatocytes—*[51]*; renal tubular cell aggregates—*[45,52]*) may be suspended in growth medium and cryofixed by any of the methods described in **Fig. 5**.

30. **Figure 6A** describes two variants of a method whereby cell monolayers are taken into fluid suspension and cryofixed as described in **Fig. 5A**. (Do note that some of the other protocols outlined in **Fig. 5** may also be applicable.) Chinese hamster ovary cells *(53)* in synchronous culture have been processed for EPXMA by this method. It is also instructive to note that pelletted isolated lymphocyte-nuclei have also been cryofixed for EPXMA by an analogous protocol *(54)*.

31. The film of growth medium "contaminating" the surface of cell monolayers, grown for example on a gas-permeable flexible plastic foil like Petriperm® *(51)*,

must somehow be removed, otherwise its constituent ions will render meaningless EPXMA measurements supposedly made on the cells. Cryomicrotomy circumvents the problem (**Fig. 6B**, *55*). However, the growth medium film on the surface of the cells must be "thinned" sufficiently prior to cryofixation so that it does not impede good freezing in the monolayer. A quick touch with filter paper near the O-ring should suffice.

32. The protocol illustrated in **Fig. 6C** *(56)* represents a very direct approach:cell monolayers are grown on substrates that may be subsequently inserted into an analytical electron microscope; the adhering layer of culture medium is removed prior to plunge freezing by dipping in a washing medium. Several washing media have been investigated (0.15 *M* ammonium acetate, 0.3 *M* sucrose, even ice-cold distilled H$_2$O *[57]*). Sucrose appears to be the best for most purposes *(56)* except: a) it should be removed completely immediately prior to plunging in the cryogen, otherwise, it will impede freezing and crystallize on the surface of the freeze-dried section, thus interfering with analysis and b) the validity of any washing procedure should be determined, where possible, by EPXMA comparison against unwashed cryosectioned cells. Only low-resolution analysis (cytoplasm/nucleus) can be performed on thin, washed, "whole cell mount" preparations).

33. Freeze-substitution is a technique that is especially associated with the preparation of botanical specimens containing large air spaces and/or fluid-filled vacuoles *(58)*. Such specimens can be difficult to prepare by cryoultramicrotomy, although we have experienced a measure of success with a cultured tobacco cell-line cryofixed at high pressure.

34. Diethyl ether is probably the favored solvent for freeze-substituting in the context of EPXMA, although acetone is a good alternative. It is absolutely essential that substituting and embedding media are rigorously dehydrated, otherwise ion retention in the tissue preparation is severely compromised. This is achieved by the inclusion of silica gel or activated molecular sieve in the media, and by maintaining a low-humidity atmosphere in the substitution chamber.

35. Debate still continues about the degree of solute retention in freeze-substituted and in freeze-dried/resin-embedded preparations *(59–61)*. Our view is that these techniques should be avoided if cryoultramicrotomy is an option. Resin embedding does yield specimens that are easily stored, reorientated and sectioned on a conventional ultramicrotome, but it does not resolve the challenges presented by cryofixation, and may cause significant solute displacement. However, the prospect of combining the techniques of ion localization by EPXMA and antigen detection by immunogold cytochemistry, both in cryosubstituted preparations, is an intriguing experimental prospect worthy of consideration.

36. Cryoultramicrotomy is best considered as conventional ultramicrotomy performed at a low temperature *(13)*. Many of the problems encountered with cryofixed specimens can be resolved if the operator thinks in terms of resin-embedded blocks. For instance, if the block is too brittle, soften it by increasing the temperature in 5°C steps; if these adjustments fail then it is likely that the specimen is poorly frozen (analogous to an inadequately polymerized plastic block).

37. Cryoultramicrotomy demands special care and some special tools. All tools used for specimen handling, tightening the knife, adjusting the knife, etc., must be clean and cooled in liquid N_2. Wash snow away from the knife edge and other critical instrument parts by dribbling liquid N_2 over them. Mount the ultrathin cryosections on coated grids using an eyelash (or hair from a Dalmatian dog is much prized!) glued to a straw or end of an artists' paint brush.

38. The frozen specimen must not thaw, or even be maintained at any temperature above either that at which it is to be cryosectioned (usually about 153 K) or freeze-dried (about 178 K), at any stage.

39. Freeze-drying of ultrathin cryosections may be achieved either externally (to the microscope) or internally (on a cold-stage under the microscope vacuum). External freeze-drying may be done in commercial units, or in simple home-made liquid N_2-cooled transfer pots introduced into the vacuum chamber of a carbon-coating unit. External drying provides a high specimen throughput, but great care must be exercised to ensure that the dried specimen does not rehydrate. Thin, freeze-dried, tissue sections will absorb moisture from a humid atmosphere almost as rapidly as paper tissue in a toilet pan—a slight, but useful, exaggeration! Internal freeze-drying uses a microscope furnished with a temperature-adjustable cold-stage as a large (and expensive) freeze-drier; specimen throughput is low, of course, but it provides good reproducible results.

40. Ultrathin cryosections must be thoroughly freeze-dried. This can be assured by completing the drying, internal as well as external, at an elevated temperature (– 30°C, say). Residual water in a section appears to exacerbate the problem of radiation damage under electron irradiation.

41. Mass loss in an electron beam of a microprobe is excessive, even near liquid N_2 temperature, in ultrathin frozen-hydrated sections *(22)*. Since mass loss is a surface-etching phenomenon, the EPXMA analysis of fully hydrated sections may only be done with sections 1–2 μm thick, but which still conform to the analytical definition of thinness because of the limited internal absorption of excited X-rays *(62)*.

42. It is advantageous to analyze freeze-dried sections at low temperatures. A cold-stage is a useful accessory even if it is not used for internal freeze-drying. A combination of low temperature and carbon coating minimizes the inevitable loss of organic mass that occurs at the high electron doses demanded by EPXMA *(13,22)*.

43. From the above, it may be seen that the essential truth of Zierold's statement (that) "the appropriate freezing technique has to be developed with respect to the experimental aim" *(19)* may be extended to the entire cryopreparative pathway. Despite the difficulties, biologically useful data may be obtained from ultrathin cryosections (**Fig. 10**).

44. Cryostandards have been widely used (*see [23]*) because they are simple to produce and are intuitively considered to resemble biological soft tissues in their radiation stability. They are especially useful for low atomic number elements; heavier elements, especially at higher concentrations, may be subject to heterogeneous distribution due to matrix binding and/or precipitation effects. EPXMA

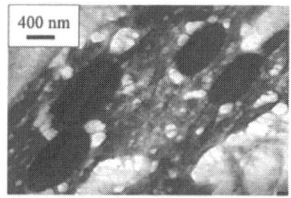

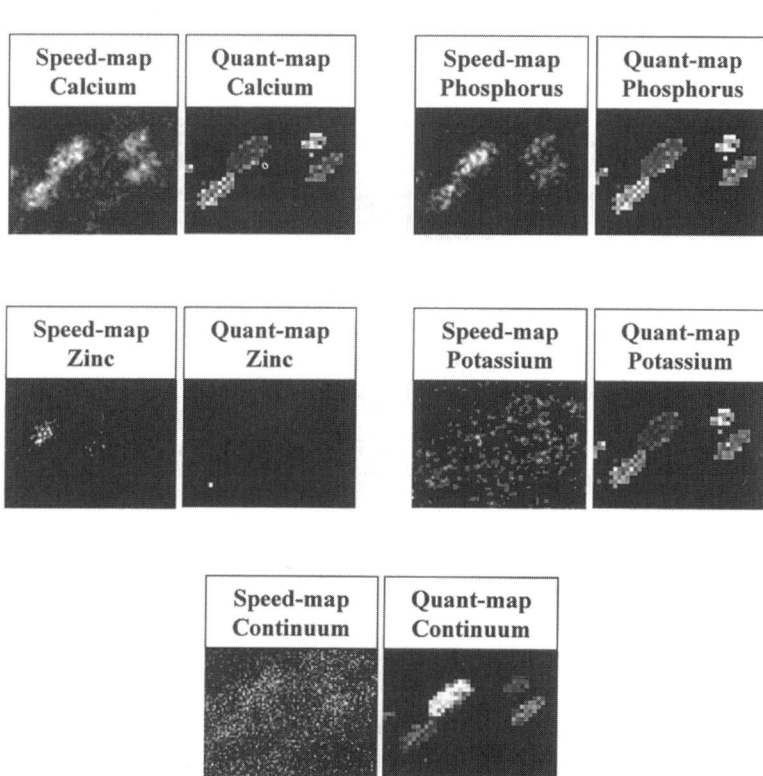

Fig. 10. X-ray distribution maps obtained from a defined region of an ultrathin, freeze dried, cryosection (top) of earthworm, *Lumbricus rubellus,* and chloragogenous tissue. The element maps are paired, with a qualitative "speed map" on the left and a quantitative map on the right. In the quantitative maps, each pixel corresponds to an energy spectrum that is processsed to yield dry mass concentrations; these data can be recovered and subjected to, for example, multivariate statistical analysis to derive information about element correlations with specified subcellular compartments.

has been used for a wide range of biological studies. The scope of these studies is far too broad to even summarize adequately. However, the biomedical observations that EPXMA has yielded may be very crudely divided into "qualitative" and "quantitative" categories. In recent years these two aspects have been elegantly combined with the advent of quantitative (elemental) mapping *(63)*, a technique that provides more comprehensive high-resolution images of compositional heterogeneities within cells (**Fig. 10**).

Acknowledgments

The authors have borrowed heavily on the published works of friends and colleagues in the cryo-EPXMA field during the preparation of the present article. They thank them collectively, and apologize for misrepresenting their efforts in one important respect: we have selected examples to illustrate our text that represent parochial rather than more generalized research interests. Mrs. Nicola Bassett deserves our special thanks for her invaluable help in the preparation of the manuscript. This article was written while A. J. Morgan was a recipient, with Dr. P. Kille, of two research grants (GR/311225, GST/02/1782) awarded by the Natural Environment Research Council.

References

1. Erasmus, D. A., ed. (1978) *Electron Probe Microanalysis in Biology*, Chapman and Hall, London, England.
2. Chandler, J. A. (1977) *X-Ray Microanalysis in the Electron Microscope*, North-Holland, Amsterdam, Holland.
3. Hayat, M. A., ed. (1980) *X-Ray Microanalysis in Biology*, University Park Press, Baltimore, MD.
4. Morgan, A. J. (1985) *X-Ray Microanalysis in Electron Microscopy for Biologists*, Oxford University Press, Royal Microscopical Society, Oxford, England.
5. Warley, A. (1997) *Practical Methods in Electron Microscopy*, vol. 16 (Glauert, A. M., series ed.), Portland Press, London and Miami, p. 276.
6. Hall, T. A. and Gupta, B. L. (1983) The localization and assay of chemical elements by microprobe methods. *Q. Rev. Biophys.* **16,** 279–339.
7. Roomans, G. M. (1988) Quantitative X-ray microanalysis of biological specimens. *J. Electron Microsc. Tech.* **9,** 19–43.
8. Hall, T. A. (1989) Quantitative electron probe X-ray microanalysis in biology. *Scanning Microsc.* **3,** 461–466.
9. Roomans, G. M. (1990) The Hall Method in the quantitative X-ray microanalysis of biological specimens: a review. *Scanning Microsc.* **4,** 1055–1063.
10. Steinbrecht, R. A. and Zierold, K., eds. (1987) *Cryotechniques in Biological Electron Microscopy*, Springer-Verlag, Berlin, Germany.
11. Robards, A. W. and Sleytr, U. B. (1985) *Low Temperature Methods in Biological Electron Microscopy*, Practical Methods in Electron Microscopy, vol. 10. (Glauert, A. M., ed.). Elsevier, Amsterdam, New York, Oxford.
12. Dubochet, J., Adrian, M., Chang, J.-J., Homo, J.-C., Lepault, J., McDowall, A. W. and Schultz, P. (1988) Cryo-electron microscopy of vitrified specimens. *Q. Rev. Biophys.* **21,** 129–228.

13. Roos, N. and Morgan, A. J. (1990) *Cryopreparation of Thin Biological Specimens for Electron Microscopy: Methods and Applications*, Oxford University Press, Royal Microscopical Society, Oxford, England.

14. Echlin, P. (1992) *Low-Temperature Microscopy and Analysis*, Plenum, New York.

15. Quintana, C. (1994) Cryofixation, cryosubstitution, cryoembedding for ultrastructural, immunocytochemical and microanalytical studies. *Micron.* **25,** 63–99.

16. Elder, H. Y., Nicholson, W. A. P., Mackenzie, M., and Johnson, A. D. (1997) A comparative look at microanalytical thin sample preparation problems in biology and materials science. *Proc. Roy. Microsc. Soc.* **31,** 14–19.

17. Elder, H. Y. and Bovell, D. L. (1988) Biological cryofixation: why and how?, in *EUREM 88: Proc. 9th Europ. Congr. Electron Microscopy* (Goodhew, P. J., ed.), Institute of Physics, Bristol, pp. 13–22.

18. Morgan, A. J. (1980) Preparation of specimens. Changes in chemical integrity, in *X-Ray Microanalysis in Biology* (Hayat, M. A., ed.), University Park Press, Baltimore, MD, pp. 65–165.

19. Zierold, K. (1993) Rapid freezing techniques for biological electron probe microanalysis, in *X-Ray Microanalysis in Biology: Experimental Techniques and Applications* (Sigee, D. C., Morgan, A. J., Sumner, A. T., and Warley, A., eds.), Cambridge University Press, Cambridge, England, pp. 101–116.

20. Hagler, H. K. and Buja, L. M. (1984) New techniques for the preparation of thin freeze-dried cryosections for X-ray microanalysis, in *The Science of Biological Specimen Preparation* (Revel, J.-P., Barnard, T., Haggis, G. H., eds.), Scanning Electron Microscopy, AMF O'Hare, IL, pp. 161–166.

21. Ryan, K. P. and Knoll, G. (1994) Time-resolved cryofixation methods for the study of dynamic cellular events by electron microscopy: a review. *Scanning Microsc.* **8,** 259–288.

22. Zierold, K. (1988) X-ray microanalysis of freeze-dried and frozen-hydrated cryosections. *J. Electron Microsc. Tech.* **9,** 65–82.

23. Wendt-Gallitelli, M. F. and Isenberg, G. (1989) X-ray microanalysis of single cardiac myocytes frozen under voltage-clamp conditions. *Am. J. Physiol.* **256,** H574–H583.

24. Zierold, K., Gerke, I., and Schmitz, M. (1989) X-ray microanalysis of fast exocytotic processes, in *Electron Probe Microanalysis Applications in Biology and Medicine* (Zierold, K. and Hagler, H. K., eds.), Springer-Verlag, Berlin, Germany, pp. 281–292.

25. Roomans, G. M. (1993) Applications of X-ray microanalysis in biomedicine: an overview, in *X-Ray Microanalysis in Biology: Experimental Techniques and Applications* (Sigee, D. C., Morgan, A. J., Sumner, A. T., and Warley, A., eds.), Cambridge University Press, Cambridge, England, pp. 297–315.

26. Warley, A. (1990) Standards for the application of X-ray microanalysis to biological specimens. *J. Microsc.* **157,** 135–147.

27. Morgan, A. J., Roos, N., Morgan, J. E., and Winters, C. (1989) The subcellular accumulation of toxic heavy metals: qualitative and quantitative X-ray microanalysis, in *Electron Probe Microanalysis. Applications in Biology and Medicine* (Zierold, K. and Hagler, H. K., eds.), Springer-Verlag, Berlin, Germany, pp. 59–72.

28. Morgan, A. J. and Winters, C. (1991) Diapause in the earthworm, Aporrectodea longa: morphological and quantitative X-ray microanalysis of cryosectioned chloragogenous tissue. *Scanning Microsc.* **5,** 219–227.
29. BéruBé, K. A., Jones, T. P., and Williamson, B. J. (1997) Electron microscopy of urban airborne particulate matter. *Microsc. Analysis* **61,** 11–13.
30. Sartori, N., Richter, K., and Dubochet, J. (1993) Vitrification depth can be increased more than 10-fold by high-pressure freezing. *J. Microsc.* **172,** 55–61.
31. Ryan, K. P. and Liddicoat, M. I. (1987) Safety considerations regarding the use of propane and other liquified gases as coolants for rapid freezing purposes. *J. Microsc.* **147,** 337–340.
32. Ryan, K. P. (1992) Cryofixation of tissues for electron microscopy: a review of plunge cooling methods. *Scanning Microsc.* **6,** 715–743.
33. Ryan, K. P. and Purse, D. H. (1984) Rapid freezing: specimen supports and cold gas layers. *J. Microsc.* **136,** RP5–RP6.
34. Bald, W. B. (1983) Optimizing the cooling block for the quick freeze method. *J. Microsc.* **131,** 11–23.
35. Chang, S. H., Mergner, W. J., Pendergrass, R. E., Bulger, R. E., Berezesky, I. K., and Trump, B. F. (1980) A rapid method of cryofixation of tissues in situ for ultracryotomy. *J. Histochem. Cytochem.* **28,** 47–51.
38. Zierold, K. (1993) The cryopuncher: a pneumatic cryofixation device for X-ray microanalysis of tissue specimens. *J. Microsc.* **171,** 267–272.
39. Greene, W. B. and Walsh, L. G. (1994) Cryo-jet preservation of calcium in the rat's spinal cord. *Scanning Microsc.* **8,** 587–600.
40. Severs, N. J., Gourdrie, R. G., Harfst, E., Peters, N. S., and Green, C. R. (1993) Intercellular junctions and the application of microscopical techniques: the cardiac gap junction as a model. *J. Microsc.* **169,** 299–328.
41. Sitte, H. (1996) Advanced instrumentation and methodology related to cryoultramicrotomy: a review. *Scanning Microsc. Suppl.* **10,** 387–466.
42. Hagler, H. K., Morris, A. C., and Buja, L. M. (1989) X-ray microanalysis and free calcium measurements in cultured neonatal rat ventricular myocytes, in Electron Probe Microanalysis Applications in Biology and Medicine (Zierold, K. and Hagler, H., eds.), Springer-Verlag, Berlin, Germany, pp. 181–197.
43. Steinbrecht, R. A. and Zierold, K. (1984) A cryoembedding method for cutting ultting ultrathin cryosections from small specimens. *J. Microsc.* **136,** 69–75.
44. Zierold, K. and Schafer, D. (1988) Preparation of cultured and isolated cells for X-ray microanalysis. *Scanning Microsc.* **2,** 1775–1790.
45. Zierold, K. (1991) Cryofixation methods for ion localization in cells by electron probe microanalysis: a review. *J. Microsc.* **161,** 367–366.
46. Bachmann, L (1987) Freeze-etching of dispersions, emulsions and macromolecular solutions of biological interest, in *Cryotechniques in Biological Electron Microscopy* (Steinbrecht, R. A. and Zierold, K., eds.), Springer-Verlag, Berlin, Germany, pp. 192–204.
47. Knoll, G., Braun, C., and Plattner, H. (1991) Quenched flow analysis of exocytosis in Paramecium cells: time course, changes in membrane structure, and calcium

requirements revealed after rapid mixing and rapid freezing of intact cells. *J. Cell Biol.* **113**, 1295–1304.

48. Hohenberg, H., Mannweiler, K., and Müller, M. (1994) High pressure freezing of cell suspensions in cellulose capillary tubes. *J. Microsc.* **175**, 34–43.
49. Hohenberg, H., Tobler, M., and Müller, M. (1996) High-pressure freezing of tissue obtained by fine-needle biopsy. *J. Microsc.* **183**, 133–139.
50. Zierold, K. (1988) Electron probe microanalysis of cryosections from cell suspensions, in *Methods in Microbiology*, vol. 20, *Electron Microscopy in Microbiology* (Mayer, F., ed.), Academic, London, England, pp. 91–111.
51. Petzinger, E. and Frimmer, M. (1988) Comparative investigations of the uptake of phallotoxins, bile acids, bovine lactoperoxidase and horseradish peroxidase into rat hepatocytes in suspension and in cell cultures. *Biochim. Biophys. Acta* **937**, 135–144.
52. Pavenstädt-Grupp, I., Grupp, C., and Kinne, R. K. H. (1989) Measurement of element content in isolated papillary collecting duct cells by electron probe microanalysis. *Pflüg Archiv E. J. Physiol.* **413**, 378–384.
53. Edwards, P. G., Kendall, M. D., and Morris, I. W. (1991) Effect of a platinum chemotherapy drug on intracellular elements during the cell cycle, using X-ray microanalysis. *Scanning Microsc.* **5**, 797–810.
54. Zglinicki von, T., Ziervogel, H., and Bimmler, M. (1989) Binding of ions to nuclear chromatin. *Scanning Microsc.* **3**, 1231–1239.
55. Zierold, K. (1989) Cryotechniques for biological microanalysis, in *Microbeam Analysis 1989* (Russel, P. E., ed.), Academic, London, England, pp. 109–111.
56. Wroblewski, J. and Wroblewski, R. (1993) X-ray microanalysis of cultured mammalian cells, in *X-Ray Microanalysis in Biology: Experimental Techniques and Applications* (Sigee, D. C., Morgan, A. J., Sumner, A. T., and Warley, A., eds.), Cambridge University Press, Cambridge, England, pp. 317–329.
57. Lechene, C. (1989) Electron probe analysis of transport properties of cultured cells, in *Electron Probe Microanalysis. Applications in Biology and Medicine* (Zierold, K. and Hagler, H. K., eds.), Springer-Verlag, Berlin, Germany, pp. 237–249.
58. Hajibagheri, M. A. and Flowers, T. J. (1993) Ion localisation in plant cells using the combined techniques of freeze-substitution and X-ray microanalysis, in *X-Ray Microanalysis in Biology: Experimental Techniques and Applications* (Sigee, D. C., Morgan, A. J., Sumner, A. T., and Warley, A., eds.), Cambridge University Press, Cambridge, England, pp. 217–230.
59. Condron, R. J. and Marshall, A. T. (1990) A comparison of three low temperature techniques of specimen preparation for X-ray microanalysis. *Scanning Microsc.* **4**, 439–447.
60. Elder, H. Y. and Wilson, S. M. (1994) Preparation methods for quantitative X-ray microanalysis of intracellular elements in ultrathin sections for transmission electron microscopy: The freeze-dry, resin-embedded route, in *Cell Biology: A Laboratory Handbook* (Celis, J. E., ed.), Academic, San Diego, CA, pp. 186–192.
61. Pålsgård, E., Lindh, U., and Roomans, G. M. (1994) Comparative study of freeze-substitution techniques for X-ray microanalysis of biological tissue. *Microsc. Res. Tech.* **28**, 254–258.

62. Gupta, B. L. (1991) Ted Hall and the science of biological microprobe X-ray analysis: a historial perspective of methodology and biological dividends. *Scanning Microsc.* **5,** 379–426.

63. LeFurgey, A., Davilla, S. D., Kopf, D. A., Sommer, J. R., and Ingram, P. (1992) Real-time quantitative elemental analysis and mapping: microchemical imaging in cell physiology. *J. Microsc.* **165,** 191–223.

Index